FERTILIZING FOR MAXIMUM YIELD

Fertilizing for Maximum Yield

G. W. COOKE, C.B.E., Ph.D., F.R.I.C., F.R.S.

Chief Scientific Officer
Agricultural Research Council

Formerly Head of the Chemistry Department
Rothamsted Experimental Station

CROSBY LOCKWOOD STAPLES · LONDON

Granada Publishing Limited
First published in Great Britain 1972 by Crosby Lockwood Staples
Frogmore St Albans Hertfordshire AL2 2NF and 3 Upper James Street
London WIR 4BP
Second edition 1975
Reprinted 1976

ISBN 0 258 97042 1

Printed in Great Britain by
Fletcher & Son Ltd, Norwich

Contents

Notes on units

YIELDS, WEIGHTS AND MEASURES

The metric system of weights and measures used throughout the book is now usual in agricultural education in Britain and is used in most of the world's scientific journals. Chapters 19 and 20 are written to help British farmers and their advisers. Because metric units are still unfamiliar to many practical people, in these Chapters I have given fertilizer recommendations in Imperial as well as metric units and have retained the UNIT (of 1.12 lb) which has become so popular in giving advice to farmers in U.K.

Yields of cereals, except where otherwise stated, are of grain containing 15 per cent of moisture. Other yields are as stated, usually as fresh material or as dry matter.

Conversions

Factors for converting metric to British units are given below. For quick comparisons the following factors are accurate to two parts in 100:

1 lb/acre	=	1.1 kg/ha
1 UNIT/acre	=	1.25 kg/ha
1 cwt/acre	=	125 kg/ha
1 ton/acre	=	2.5 tonnes/hectare
1 gallon/acre	=	11 litres/hectare
1 kg/ha	=	0.9 lb/acre
1 kg/ha	=	0.8 UNITS/acre
100 kg/ha	=	0.8 cwt/acre
1 tonne/hectare	=	0.4 tons/acre
1 litre/hectare	=	0.09 gallon/acre

Notes on Units

Conversion of Metric to Imperial units

Metric		Imperial
1 centimetre (cm)	=	0.3937 inch (in.)
1 metre (m)	=	1.094 yard (yd)
1 hectare (ha)	=	2.471 acres
1 kilogramme (kg)	=	2.205 pounds (lb)
100 kg	=	1.968 hundredweight (cwt)
1 tonne (t)	=	0.9842 ton
1 litre (l)	=	1.760 pints = 0.2200 gallon (gal)

To convert	Multiply by
kg/ha to lb/acre	0.8921
kg/ha to cwt/acre	0.007966
tonnes/hectare to tons/acre	0.3983
litres/hectare to gallons/acre	0.08902

PLANT NUTRIENTS

To avoid ambiguity and the use of the obsolete 'oxide' terminology, plant nutrients should be stated in terms of the elements N, P, K, Ca, Mg, S and Na and these are used in all soil and plant analyses stated here. Statements required by Fertilizer Regulations in Britain and most other countries are still in terms of P_2O_5 and K_2O, and advice is given in these units too. Therefore in discussing practical recommendations for using fertilizers, I have generally used P_2O_5 and K_2O. A few of the experimental results involving nutrients removed by crops have been published by others as oxides of P and K; where these results have been quoted the units of the original publication are used. All tables of data indicate plainly whether nutrients are measured in elements or oxides.

Conversion factors

For quick conversions

(accurate to within 2%) the following factors may be used:

$2\frac{1}{3} \times P = P_2O_5$ $\frac{3}{7} \times P_2O_5 = P$

$1\frac{1}{5} \times K = K_2O$ $\frac{5}{6} \times K_2O = K$

$1\frac{2}{5} \times Ca = CaO$ $\frac{7}{10} \times CaO = Ca$

$1\frac{2}{3} \times Mg = MgO$ $\frac{3}{5} \times MgO = Mg$

For accurate conversions

To convert		multiply by	To convert		multiply by
P_2O_5	to P	0.4364	P	to P_2O_5	2.2915
K_2O	to K	0.8301	K	to K_2O	1.2047
CaO	to Ca	0.7146	Ca	to CaO	1.3994
MgO	to Mg	0.6031	Mg	to MgO	1.6581

Abbreviations

A.D.A.S.	Agricultural Development and Advisory Service (responsible for work in England and Wales formerly done by N.A.A.S.)
F.A.O.	Food and Agriculture Organisation of the United Nations
F.F.H.C.	Freedom from Hunger Campaign (of F.A.O.)
I.R.R.I.	International Rice Research Institute (Philippines)
N.A.A.S.	National Agricultural Advisory Service (for England and Wales; A.D.A.S. replaced N.A.A.S. in spring 1971)
W.H.O.	World Health Organisation
FYM	Farmyard manure: A rotted or partially rotted mixture of excreta and animals bedding (usually straw)
ppm	parts per million
T.V.A.	Tennessee Valley Authority

Acknowledgements

This book has developed from many discussions with colleagues at Rothamsted, members of staffs of Universities and the agricultural advisory services in Britain, and farmers. I am grateful to all these for help and advice. I thank particularly these colleagues who helped me greatly by reading and criticising chapters of the manuscript: T. M. Addiscott, Blanche Benzian, J. Bolton, Margaret Chater, S. C. R. Freeman, J. K. R. Gasser, A. E. Johnston, G. E. G. Mattingly, P. H. Le Mare, Brenda Messer, A. Penny, F. V. Widdowson and R. J. B. Williams. I also thank F. D. Cowland for arranging the figures and Lynda Woods for making the drawings. Several friends in the fertilizer industry and trade guided me on current prices.

I acknowledge the great help given by Maureen Maginnis in typing manuscripts and checking data and proofs.

G. W. COOKE

Rothamsted Experimental Station
Harpenden, Herts.
July 1971

I acknowledge the help of D. A. Boyd and B. M. Church of Rothamsted's Statistics Department who provided experimental results and data from the 1974 Survey of Fertilizer Practice for this Second Edition. Friends in the fertilizer industry have again helped me by discussing price trends. I also thank Miss I. P. Bond for preparing the new typescripts.

January 1975   G. W. C.

Preface to the Second Edition

In the four years since this book was written the scientific basis for advising on using fertilizers has changed little but the economic conditions which govern their profitable use have changed considerably. All fertilizer prices increased greatly between 1972 and 1974; in the same period cereal prices in the U.K. were roughly doubled but the prices of other crops did not increase in the same proportion and the values of some livestock products have tended to diminish. This new edition has given an opportunity to quote British fertilizer prices current in winter 1974/5. Important new information on the effect of fertilizers on crop yields is referred to in the text and recent publications, important to students and advisers, are listed in references 103 to 123. These range from accounts of a large series of experiments on the use of nitrogen fertilizer on grass (ref. 118), to F.A.O.'s review of fertilizer legislation in many countries (ref. 120) and to a book on field experimentation (ref. 110); 1973 was marked by the appearance of the Tenth Edition of Russell's famous text *Soil Conditions and Plant Growth* (ref. 122).

Fertilizer prices

Prices which British farmers paid in winter 1974/5 are quoted in the text and the tables wherever up-to-date figures are available. Not all the illustrations have been changed, however: for example, costs of fertilizers and crop prices in Fig. 5 remain at 1971 figures because they illustrate a principle and show how to allow for two levels of fertilizer price. The new prices seem high to our farmers but prices in other European countries are larger by 40 per cent or more.

Nitrogen fertilizer prices increased rapidly in 1973/4 due to the world-wide increases in prices of fuel oils and gases which are used to make hydrogen to combine with nitrogen in the fixation process and to provide power in the plants used. Coates (ref. 103) has described the background to these changes. The extreme example is ammonia at $25 to $40 per ton at U.S. ports in 1973 and $300–400 per ton in 1974. In Britain the cheapest sources of nitrogen cost 50 per cent more in 1974 than in 1971 (and are nearly twice as expensive to the farmer if account is taken of the subsidies previously paid, which ended in 1974). These British prices are for products made on existing plant; new plant is more costly and the fertilizers to be made in the new factories we need to meet increased demand for nitrogen will cost more too. It seems likely that during the next few years the prices of nitrogen in Britain will become double those charged in 1972.

Phosphate price increases are mainly due to changes in costs of the phosphate rock needed to manufacture these fertilizers. Costs of the rock have risen from about £7 per ton at U.K. ports in 1972 to about £30 per ton in 1974. The price of phosphorus in fertilizers offered to British farmers has doubled in this period; the real increase is two and a half times if the subsidy previously paid is taken into account.

Potassium fertilizers have also been affected by general increases in costs and the price paid at U.K. ports for potassium chloride (muriate of potash) has increased from £15 per ton in 1970 to £35 per ton in 1974. Prices to the British farmer have increased by about two and a half times in this period. A new development in Britain is the opening in 1974/5 of the potash mine in the Cleveland district of north-east England. This will provide more potasium fertilizer than is used on U.K. farms. There will be a surplus for export and home production should protect British farmers from having to pay high prices for imported potassium salts.

Compound fertilizer prices have reflected these increases in the prices of the ingredients used to make them; for example a 'High-nitrogen' NPK compound which cost about £40 per ton in 1971 is now £80 per ton.

Profitability

In spite of these large price increases fertilizers remain very profitable to British farmers; they are also essential for producing the large yields needed to repay the increasing costs of all the farmer buys. Prices of some crops have kept pace with increasing fertilizer prices and nitrogen used on cereals in Britain was never more profitable than at the 1974 harvest; some figures for the last sixty years are given below:

| | Price per unit of N | Wheat per cwt | Ratio of prices cwt of wheat |
	p	£	unit of N
1911–13	3.3	0.37	11
1936–9	1.8	0.35	19
1955–7	5.0	1.47	29
1971	5.4	1.56	29
1974	7.0[1]	3.00[2]	43

[1] Price of N in Spring 1974;   [2] Market price in Autumn 1974.

Dr D. A. Boyd of Rothamsted's Statistics Department recently calculated ratios of (value of extra yield produced by unit fertilizer) + (cost of nitrogen fertilizer); his figures showed the ratio was between 5 and 7 for cereals, 8 for potatoes and 5 for sugar beet, but only $2\frac{1}{2}$–3 for nitrogen used on grass made into silage. These figures illustrate the good profitability of nitrogen fertilizer used on arable crops; they also show the greater difficulty of securing adequate return from nitrogen used on grass which has to be eaten by animals before the value of the extra yield is realized.

The rational (and therefore profitable) use of fertilizers involves making full allowance for local conditions of soil fertility; these make far greater changes in the amounts that should be applied than do considerable changes in crop or fertilizer prices. Dr Boyd calculated that even a doubling of the ratio of the prices of (1 kg of N)/(1 kg of barley) only diminished the optimum dressing by 11 kg of N/ha (10 units of N/acre). Adjusting dressings to suit local soils and cropping systems, and the weather, has been made easier by the publication by A.D.A.S. of their 'Fertilizer Recommendations' (ref. 43). Farmers often ask how they can diminish fertilizer dressings to allow for increasing prices; the only worthwhile savings that are possible come from a careful assessment of local conditions of soil and farming system, aided by soil analyses, which leads to more rational manuring. These principles are well explained in Chapters 11 to 16 of this book.

There is no justification for reducing fertilizer or lime dressings on land where to lessen the amounts normally given will reduce yields. The greatest savings, and these can be made by many farmers, will come from fitting dressings to previous cropping and manuring and by allowing for the effects of weather on the efficiency of nitrogen fertilizers, and particularly the effects of winter rainfall on leaching of nitrate (Chapter 13). Considerable savings in phosphorus and potassium fertilizers can be made by regularly using soil analysis results to guide the amounts applied (Chapter 14). Further savings are possible on farms where livestock are kept by making the most economic use of farmyard and other organic manures, slurries, liquid manures and crop wastes. All these contain plant nutrients which should be *used* to produce another crop, so diminishing the need for fertilizers; wastes containing nutrients for plants *or* animals should be regarded as assets, not as disposal problems. The excreta of all livestock on U.K. farms in 1974 contained nearly as much nitrogen, as much phosphorus, and twice as much potassium, as was in the fertilizers farmers bought.

New cultivations and cropping systems

In 1974 several articles drew the attention of the public to the amounts of energy used in farming. Published estimates of the fossil fuels used in British agriculture varied from 3 to 6 per cent of the total energy used in the country. One and a half per cent of all the gas and oil used in the U.K. was needed to make nitrogen fertilizers. Although this is a small need to maintain production from our agriculture, which produces about two-thirds of the nation's food, nevertheless we must attempt to lessen the demand by farming on fossil fuel reserves. One important way is by using less power in cultivating the soil. Systems which avoid ploughing save much tractor fuel and often speed the work of sowing a new crop. Minimum cultivation systems where the surface of unploughed soil is only stirred to produce a seedbed have been used for years in some areas. New herbicides, which are applied just before sowing, kill all surface vegetation but do not damage the seed or seedlings; these chemicals have made it possible to establish crops by sowing directly into uncultivated soil.

Methods of sowing involving 'direct drilling' or reduced cultivation systems make little difference to the amounts of fertilizers needed by crops, a point discussed on page 198.

New arable crops are being grown to provide 'breaks' against the transmission of soil-borne disease of cereals and to save imports of human and animal food. Two of these, maize and oil-seed rape, are discussed on page 227. More experimental work is needed on both crops to determine how much fertilizer is required for maximum yields.

Recent experimental work on arable crops in Britain has concentrated on cereals. Several accounts have appeared of recent work on fertilizing wheat and barley (refs. 105, 116, 119). One important question is whether extra nitrogen fertilizer is needed when long runs of cereals are grown; further evidence on this point has been provided recently (ref. 119). Interest continues in ley-arable systems and an account of the results of a large series of long-term experiments made by A.D.A.S. has been published (ref. 115).

Farmers are always interested in soil organic matter and particularly in the value of straw when it is ploughed into the land after harvest. Older experiments showed little gain in soil organic matter or crop yields from ploughed-in straw. These results have now been confirmed by an extensive investigation on A.D.A.S. farms (ref. 123). But perhaps this is now an academic question! The increased prices of all animal feeding stuffs made straw a much more valuable commodity in 1974. In addition other uses are proposed, ranging from using straw to make paper, or as fuel, to fermenting it on the farm to produce methane!

Grassland. The returns from using nitrogen fertilizer on grassland used for animal feeding are likely to be less than the returns from fertilizing arable crops grown for sale. Nevertheless farmers have continued to use more as they have realized that the extra growth which nitrogen gives allows them to produce more home-grown feed and to increase their stocking density. In the eight years 1966–1974 the amounts of nitrogen used on both leys and permanent grass doubled, while the phosphorus and potassium used diminished slightly. The average amount of nitrogen used on permanent grass is still only half that used on leys; this is unfortunate as the extra yields from nitrogen given to old grass are as large as the increases obtained from leys. The value of nitrogen on grass, and the factors which affect the returns from using it, have been investigated in A.D.A.S. experiments recently described (ref. 118). When pastures are grazed most of the P and K in the herbage is returned to the soil in excreta; increasing production by fertilizers depends on using much nitrogen with the small quantities of P and K needed to maintain reserves in the soil. New compound fertilizers are now available to meet these needs; examples are mixtures with 28 or 29 per cent of N and only 5 per cent of P_2O_5 and K_2O. When grass is cut for silage or hay some P and much K is removed in the herbage and these amounts must be replaced to maintain soil fertility; this need is also catered for by new fertilizers such as 24 per cent N, 4 per cent P_2O_5, 15 per cent K_2O; or 20 per cent N, 8 per cent P_2O_5, 14 per cent K_2O. These newer compound fertilizers help in applying the recommendations made in Chapter 20.

Forms of fertilizers

The forms of fertilizers used in Britain have changed little. Some, such as sodium nitrate, have become much more expensive (Table 12). Changes in steel-making processes have diminished the amounts and altered the grades of basic slag available. The use of liquid fertilizers has continued to increase and the amount of N applied in liquid forms has doubled in the last four years. Fluids containing fertilizers in suspension have now been introduced into the U.K. but their use is still on a small scale; they contain polyphosphates which have not been previously used in British agriculture although they are well-known in the U.S.A.

Increasing yields

In the title of this book, and throughout its text, the importance of striving to grow maximum yields is stressed. This is even more important than previously. Values of agricultural land and rents have increased greatly in the U.K. in the last few years; all the farmer's supplies, such as machinery, fuels and feeding stuffs have increased in price too. These extra costs *may* be offset by increased values of products but the farmer has little control over selling prices. He can do much to maintain his income by increasing yields as these spread fixed costs over the larger crop that is sold. Recent improvements in average yields have resulted from improved varieties, new chemicals for disease and weed control, and better cultivations, as well as increases in the fertilizers used. One of the significant events of 1974 in England was the certified record yield of 11 t/ha (88 cwt/acre) of wheat from one field in Suffolk. The level of yield is a result of the interactions of many factors of crop and soil management; record yields are achieved in favourable seasons when farmers assess these factors correctly and manage the crop accordingly. The potential for high yields, which many of our soils possess, will only be achieved, even in good years, by crops which receive adequate nutrition. The 10 t/ha crop of wheat, which will be grown on more of our farms in future, needs twice as much plant nutrients as a crop yielding 5 t/ha—which is still more than our national average yield!

Rothamsted   G. W. COOKE
January 1975

Introduction

Fertilizers have been used for a century or more in most countries where agriculture is now well developed, but their *essential* role in modern farming has become clear only in the last 30 years. In 1939 the world's farmers used 9 million tonnes of plant nutrients (measured as $N + P_2O_5 + K_2O$); in 1970 about seven times as much was used. There have been similar changes in amounts of fertilizers used in Britain and most countries of Europe. These large increases in the nutrients supplied to crops have been essential to support the agricultural revolution which began in many temperate countries in the 1930s and which has greatly increased production. New varieties of crops, chemicals for better control of weeds, pests and diseases, adequate power for mechanisation, and the development of new farming systems, have all raised the *potential* of agriculture. But it would have been impossible in most places to realise the increased yields if fertilizers had not been available to supply the extra nutrients needed by larger crops.

To help British farmers make best use of the £90 million spent annually on fertilizers in the late 1950s, I produced a small book (*Fertilizers and Profitable Farming* (London: Crosby Lockwood Ltd.)) which was also found useful in other countries. This book is now out of date because our information about the effects of fertilizers has increased greatly in the last 10 years, and also because other advances in agricultural science have raised the potential yields of many important crops. Advances as diverse as the large gains in yield made possible by removing the shade trees over cacao in West Africa, new chemicals for controlling pests and diseases of crops in many countries, the development of hybrid wheat in Mexico, and the breeding of stiff-strawed varieties of rice suitable for tropical Asia, have made it possible to produce much more food from the land. All advances of these kinds have been consolidated by exploiting the *interactions* between extra nutrients supplied by fertilizers, and the control of other factors that alter plant growth. The use of interaction effects in building up agricultural production made me use the title 'Fertilizing for *Maximum Yield*'. Only by applying all new advances in scientific work on crop production, and backing them with extra nutrients, can we hope to increase yields quickly enough to feed increasing populations, and to lessen the price of food.

I hope this book will be as useful to British farmers, advisers and students as the previous book was. Chapters 19 and 20 are written specially for British conditions and, to help practical men apply advice, English weights and measures are given as well as metric units. Advice on fertilizing British crops can also be applied in most temperate countries where the same crops are grown. The *principles* of plant nutrition discussed apply to all crops and so I hope the book will be generally useful to students in many parts of the world. The last chapter does not attempt detailed recommendations for manuring sub-tropical and tropical crops; but it discusses recent advances that affect the fertilizing of many of the crops important in warm countries. Many students trained in temperate agriculture afterwards work in the tropics; I hope that Chapter 22 will be an introduction for them. I hope, too, that they will discern from this and other chapters how far the knowledge gained in one part of the world can be used advantageously in other parts.

The remainder of this introduction discusses progress in fertilizer use in Britain and some of the problems that occur when farming is intensified.

FERTILIZER PRACTICE IN BRITAIN

Twenty-five years ago British farmers used about 200,000 tonnes of N and K_2O and over 400,000 tones of P_2O_5. Ten years after, in 1960, both the nitrogen and potassium used had doubled but phosphate had changed little. In 1970 we used nearly twice as much nitrogen as in 1960, but only a little more phosphorus and potassium. The background of these changes has been recorded by surveys which show how fertilizers are used on British farms. Some results with arable crops are given below:

AVERAGE AMOUNTS OF FERTILIZERS USED ON CROPS
IN ARABLE DISTRICTS OF ENGLAND AND WALES

kg/ha of N, P_2O_5 and K_2O

	N	P_2O_5	K_2O	N	P_2O_5	K_2O
		Winter wheat			Spring barley	
1943/5	19	30	2	21	36	4
1950/2	33	28	15	25	30	20
1957	51	30	33	35	34	45
1962	74	36	43	56	36	50
1966	90	44	44	78	40	44
1969/70	90	41	35	82	40	44
1974	92	45	38	74	39	39

	N	P$_2$O$_5$	K$_2$O	N	P$_2$O$_5$	K$_2$O
		Potatoes			Sugar beet	
1943/5	79	92	100	92	88	72
1950/2	117	124	166	113	115	138
1957	124	126	200	134	119	203
1962	157	141	231	154	118	202
1969/70	166	181	250	161	117	191
1974	177	184	242	148	92	182

Cereals

The nitrogen (N) used has increased four or five times in the last 30 years. Farmers have realised that adequate N is essential for maximum yields of cereals grown in arable farming systems. Plant breeders have made this change in manuring possible by breeding new varieties with shorter straw on which the extra nitrogen can be used without the risk of lodging the crops. Predominantly arable farming systems have developed in many areas since the 1939-45 War and nutrient reserves in the soils are exhausted when all produce is sold, as often happens. Previously potassium (K) was replaced by farmyard manure, often the slow release of potassium that takes place from most clay soils was sufficient to replace the K removed in the small crops grown. As nitrogen dressings and yields increased in the 1950s, the amount of K released from soil became insufficient, particularly where cereal growing developed on farms where no animal manures were made; more potassium fertilizer had to be used to secure the extra yield that nitrogen could give and to maintain soil productivity. The nitrogen used in 1974 was, on *average*, enough for maximum yields and the phosphorus and potassium used were enough to replace the amounts removed in the grain; these average amounts used are close to the *average* recommendations of the advisory services.

Root crops

Changes in manuring of root crops have been proportionately less than changes in fertilizers used on cereals. The amounts given to sugar beet and potatoes have roughly doubled in the last 30 years. Because they were important sources of food in the 1939–45 War these crops received good allocations of the restricted supplies of fertilizers and were moderately-well manured. Subsequent changes in fertilizing have secured the larger potential yields made possible by earlier planting, better husbandry, non-bolting varieties of beet, new herbicides and the control of diseases such as virus yellows in sugar beet, and of virus and fungal diseases in potatoes. The average amounts of nitrogen now applied are enough to secure maximum yields; the phosphorus and potassium fertilizers used supply more P and K than the roots remove.

The need for discrimination in fertilizer recommendations

British farmers use roughly as much fertilizer as is recommended by their advisers for arable crops but field experiments show the amounts *actually needed* on particular fields range from none, to twice as much as the averages used. Surveys of fertilizer practice show that farmers do not make sufficiently large changes in manuring to allow for the fertility of individual fields. Some crops receive too little and yield is lost; many other fields receive too much fertilizer, at best the extra fertilizer is wasted, but often it also diminishes yields.

The main need in rationalising the fertilizing of arable crops in Britain (and in most other countries with well-developed agriculture) is not a wholesale increase in fertilizer dressings, but for more discrimination in the amounts used. Much more fertilizer than average should be used on poor land and less where reserves of nutrients have accumulated. It is difficult for farmers to make these adjustments rationally because many factors—soil, weather, farming system and crop—have to be considered. Several chapters of this book deal with the background information needed in making manuring more efficient. Previous cropping and manuring, and the weather, all affect the N that soil can supply (Chapters 11, 12 and 13). The amounts of P and K already present in soils determine how much P and K must be applied as fertilizer. The effects of farming systems on reserves of P and K can be assessed by using soil analyses (Chapter 14) and by drawing up nutrient balance sheets (Chapter 16). Field experiments on some perennial crops are difficult and take too long, the composition of the crop itself must be used to show its nutritional status. Leaf analyses (Chapter 15) are used in many countries to advise on fertilizing trees and other perennials. They may also help to make nitrogen fertilizer more efficient in arable farming when the crop's composition is used to indicate whether it has enough nitrogen, or whether extra top-dressings are needed.

Fertilizers used on grassland in Britain

Changes in the average amounts of fertilizers used on grass (in districts where this is the dominant crop) which have occurred during the last 30 years are shown below:

| | Temporary grass | | | Permanent grass | | |
| | N | P_2O_5 | K_2O | N | P_2O_5 | K_2O |
		kg/ha			kg/ha	
1943/5	4	11	0	4	11	0
1950/2	16	35	15	6	24	4
1957	26	34	21	11	20	9
1962	54	43	33	23	28	15
1966	66	50	30	29	29	16
1969/70	95	44	36	51	28	20
1974	137	36	29	66	21	15

The nitrogen used on grass has more than doubled in the last 10 years while the P and K used have changed relatively little. The traditional sources of nitrogen for our grassland are that fixed by legumes and by other biological processes, and the N returned by animal excreta and manures. Most farmers recognise that these sources supply too little for maximum growth. Yield and stock carrying capacity of our grassland can, on average, be nearly doubled by using about three times as much N as the *average* now used on temporary leys, and four or five times as much as the average used on permanent grassland. Less N is used on permanent grass than on leys; partly this is because much permanent grass occupies poor and inaccessible land. Many farmers, however, do not realise that properly fertilized permanent grass can yield as well as most leys; the old grass also has the advantage that it carries stock in wet weather better than new grassland which is liable to be damaged by 'poaching'.

Problems in deciding the correct nitrogen dressings on grassland are less difficult than are the problems of fertilizing arable crops correctly. Progress with grass has however been slower. This is inevitable even when farmers are convinced of the value of N on grass; they must reshape their farming policy and practice, and acquire capital to buy, house and handle the extra livestock that nitrogen enables them to feed. Few now question that nitrogen is the key to increased yields from grassland. Put crudely—with its aid we can move from keeping one cow on 2 acres (0.8 ha) (the limit if we rely on clover to fix the nitrogen) to 1 cow living on 1 acre (0.4 ha) of grass.

The amounts of P and K used on grassland have changed little for several years; few farmers realise that it is essential to vary the P and K dressings with changes in the amounts of nitrogen applied and the way the grass is used. Surveys show that no more P and K is used by British farmers who cut or graze six times in a year than is given on other farms to grass used only twice. Mown grass, whether for hay or silage, removes much K and some P. It seems that farmers do not appreciate that a good hay crop can take up 160-200 kg K_2O/ha; this potassium is removed from the field. The surveys show that mown leys or permanent grass get less K than grass which is strip grazed, in spite of the fact that most of the K in grass that is grazed is returned in animal excreta.

Surveys of fertilizer practice in England and Wales show that some farmers do not fully appreciate the value of lime for grassland (more should be used in some areas), and that many can have no sound policy for using phosphorus and potassium. On much grazed grassland, excreta and farm manures that are returned will maintain potassium supplies in soil, but some phosphate will have to be used. The returns from N used on grass that is cut, and also on some light soils where grass is grazed, will never be achieved without adequate potassium.

FERTILIZERS SUPPLY OTHER NUTRIENTS BESIDES N, P AND K

Most discussions of fertilizing centre on nitrogen, phosphorus and potassium, the three nutrients traditionally sold in fertilizers. The other nutrients needed by plants are often ignored in Britain because crops receive enough from soil, the rain and from animal manures. When

planning to fertilize for maximum yields, however, the other major
nutrients (calcium, magnesium and sulphur), and the micro-nutrients
(manganese, iron, boron, zinc, copper and molybdenum) must be considered; these elements are deficient in many of the world's soils. Sometimes they are supplied as secondary components of NPK fertilizers,
but usually a dressing of a suitable salt containing the element must be
applied as a fertilizer to correct a deficiency.

Many fertilizers supply nutrients besides the N, P and K on which
their price is based. Ordinary superphosphate and triple superphosphate
supply calcium as well as phosphorus; ammonium nitrate-limestone
fertilizers supply lime as well as N. Ordinary super supplies sulphur, as
does ammonium sulphate and potassium sulphate. Some potassium
fertilizers supply other nutrients that are useful; many potash salts
supply some sodium, important for sugar beet and also for the health of
stock.

Many grades of kainit supply magnesium and sodium as well as potassium. Basic slag supplies considerable amounts of lime and a little magnesium, some batches also contain small amounts of micro-nutrients.
All these components of fertilizers alter their value to farmers who need
the extra nutrients. The other elements present in a fertilizer should
always be taken into account when choosing the best material to buy.

Calcium and magnesium: Wherever rainfall is greater than transpira-
tion and some water drains away, *calcium* and *magnesium* are removed
so that soils become acid. Much yield is lost in most temperate countries because soils do not receive enough lime, soil acidity then prevents
the potential gains from fertilizers from being fully realised. Although
the British Government assists farmers to buy and spread lime the
amounts used are barely enough to maintain the existing lime status and
acid soils are still common. Magnesium deficiency is often found where
most crops are sold from light soils and no animal manures or magnesium fertilizers are applied.

The growing importance of fertilizing with *magnesium* and *micronutrients* in Britain is shown by these figures from Surveys of fertilizer
practice made in 1970:

| | Percentage of the area surveyed in England and Wales which received ||||
	Magnesium	Boron	Copper	Manganese
Spring cereals	0.2	0.1	0.1	0.2
Potatoes	7.5	0.0	0.0	0.3
Sugar beet	26	3.0	0.0	0.6

Sulphur (S) is an essential nutrient; crops need amounts of S roughly
equal to the amounts of magnesium and phosphorus taken up. In most
industrialised countries rain deposits more than enough, it is derived
from the sulphur-containing fuels burnt. Large areas of the world are
now known to have too little sulphur in circulation between air and soil

to grow the yields that the climate makes possible. For example, large areas of tropical Africa need sulphur fertilizers for maximum yield. If nuclear power replaces fossil fuels for making electricity, or if we recover more of the sulphur from furnace gases, sulphur deficiency may become common in temperate countries: it already occurs in parts of north-western Europe remote from industry.

Micro-nutrients: In some countries many soils contain too little of one or more micro-nutrients to provide for maximum yields. These problems are likely to be most serious where light soils are derived from very old parent materials which have been leached for very long, as in parts of Australia. Sometimes they occur where soils are formed from one special parent material; for example, copper deficiency occurs in England on soils derived from one of the Cretaceous limestone formations, and on peats and some glacial sands. Micro-nutrient deficiencies are less common on soils derived from younger sediments that have been little leached. They are rare in Britain in areas where the parent materials of soils were mixed and renewed during the last glacial period. Nevertheless the possibility of a micro-nutrient being deficient should never be ignored in attempting to identify causes of poor crops.

GROWING MAXIMUM YIELDS

When good prices are received for crops which are expensive to grow, fertilizing should be planned so that nutrition is not a limiting factor. I have emphasised the building of maximum yields, rather than the profitable use of fertilizers, because profitability in agriculture as a whole must depend increasingly on raising yields proportionately more than costs are increased. In highly developed agriculture large increases in yield potential will mostly come from interaction effects. Examples are interactions of fertilizers with irrigation, with new varieties of greater potential yield, or with new methods of growing crops. Often new maximum yields are achieved by controlling a pest or disease previously accepted as a 'natural' limitation to the crop. Farmers must be ready to test all new advances that may raise the yield potentials of their crops, and be prepared to try combinations of two or more practices. For example, some experiments in Eastern U.S.A. showed that neither potassium nor magnesium had any effect on yield of peaches when each nutrient was applied alone; but when *both* nutrients were applied yield was raised by 30 per cent. In West Africa fertilizers had only small effects on cacao yields when crops were grown traditionally under shade and were infested with insects. Freeing them from pests, and growing the crop in full light, has raised yield many times.

Modern farming systems that produce large yields have developed in most countries against an economic background which has made it necessary for all practical operations to be made more simple, and to be done more quickly and cheaply. In these conditions it is difficult to use fertilizers rationally if decisions are left until planting begins. Dressings should be planned well before planting by considering field experiment results that are relevant to the farm, soil analyses, and previous experience on the land. Decisions about amounts of phosphorus, potassium, magnesium and micro-nutrient fertilizers should always be made at leisure in winter. The effects of nitrogen fertilizer depend greatly on weather and decisions about amounts to use and times for applying the dressings may have to be modified at planting or afterwards. In most developed agriculture the growth of all non-leguminous crops depends on using the correct amount of nitrogen fertilizer, and applying it at times and in ways that avoid both damage to the yield and waste by leaching or denitrification. Decisions about nitrogen fertilizers must have priority.

Developing countries

All crops need the same six major nutrients and six minor nutrients. The amounts of each taken up vary with kind of crop; crops also differ in their abilities to extract nutrients from soil. Workers in developing countries where few investigations have been done are helped by results of research done elsewhere where fertilizers have been used for longer. Information in this book should therefore be useful in the early phases of agricultural development. It is, however, no substitute for local field work. In attempting to grow larger yields experiments on crops and soils must always be done to find the limitations caused by local soil and climate, by pests and diseases, and the effects of the farming systems used.

THE EFFICIENCY OF FERTILIZERS

Phosphorus and potassium are held in most soils which contain more than a few per cent of clay; the residues left by ordinary dressings are not lost by leaching. Therefore it is practical in developed agricultural systems to use fertilizers to maintain soluble P and K at concentrations that provide enough for the crops grown. Such methods are efficient when the returns from fertilizers are measured over a period of years. Calcium and magnesium fertilizers can be managed similarly, although some loss by leaching is inevitable in humid climates because these nutrients are more mobile than P and K.

These conditions apply to most modern farming where much money has already been spent on crop nutrition and impoverished soils have been improved. They do not apply in developing agriculture on poor soils; these farmers can rarely afford the large sums needed to build up reserves of soluble nutrients, they need the largest immediate return from the least expenditure. Methods such as placement near to seed or plant, pelleting nutrients with seed, special forms of fertilizers, and spraying nutrients on leaves, may all increase the proportion of the purchased nutrients that are taken up to produce extra crop immediately.

Nitrogen: All agricultural systems in humid areas, whether they are being developed on poor land or are producing crops intensively on good soils, suffer from the small efficiency of nitrogen fertilizers. Wherever leaching occurs nitrate is lost, and the largest losses are from the richest soils and where the most fertilizer nitrogen is used. Other losses occur when soils are temporarily waterlogged, nitrate being reduced and nitrogen is lost as gas. In temperate agriculture probably no more than half the N-fertilizer bought is useful in producing extra crops from arable land; the average proportion used by grassland is larger, but probably no more than two-thirds. Wherever nitrogen fertilizers are used greater efforts are needed to make them more efficient. Improvements will usually come from fitting N-fertilizing more closely to soil, farming system and weather, and by using plant analyses to determine when dressings are needed.

POLLUTION OF NATURAL WATERS

Because nitrate is lost in drainage from farmland, fertilizers have been considered responsible for the 'eutrophication' of natural waters which encourages the growth of micro-organisms, and particularly algae. This extra growth has been said to deoxygenate water and so kill fish, and to make the water tainted for drinking and difficult to filter and purify.

The name (eutrophication) is given to processes which add to natural water substances that are nutrients for micro-organisms and plants, or which stimulate their growth. We are only concerned with nitrogen and phosphorus (no serious suggestions have been made recently that potassium or other nutrients from fertilizers directly or indirectly impairs the quality of natural water); nevertheless, eutrophication includes additions of other major and minor plant nutrients, sources of carbon (probably the most important in increasing the growth of micro-organisms) and possibly growth substances too. Eutrophication is a natural process. Without plant nutrients a body of water is 'dead', water plants cannot grow, nor can the fish which graze them. Without eutrophication the water and swamp plants could not have grown in past ages to produce the deposits of gas, oil and coal we now enjoy.

One important aspect of eutrophication is the nitrate content of *drinking water*. Large crop yields can only be produced by much nitrogen (whether from fertilizer or from legumes); some of this nitrogen is inevitably leached and appears in land drainage that enters rivers, and in well and spring water. Babies cannot tolerate large nitrate concentrations in drinking water (nitrate poisoning of infants has occurred in European countries and in U.S.A.); the health of adults and livestock may be impaired by nitrate in water in hot climates where much is drunk. Natural water in regions where there is no agriculture is rarely free from nitrate, and where people or stock have suffered from nitrate in drinking water, the cause has often been a well or stream polluted by sewage. There is doubt about the acceptable concentration of nitrate in drinking water; the W.H.O. standard used to be 10 ppm of nitrate-N (NO_3-N) but this has now been raised in Europe to 23 ppm—much larger than the nitrate concentration in any drinking water we have examined at Rothamsted. (The U.S. Public Health Service limit is 10 ppm of NO_3-N.) None of the phosphate bought as fertilizer enters drainage water by being leached through normal agricultural soils. Some phosphate does enter streams and rivers in soil which has been eroded by water or wind. Much larger quantities enter rivers in effluent from sewage (and also in animal excreta where this is allowed to pollute watercourses).

Land drainage: On average, drainage from agricultural land in Britain, used in part or whole for arable crops, will contain about 10 ppm of nitrate-N, with larger amounts in spring and for a time in autumn. This concentration seems inevitable with modern farming systems using fertilizers. Phosphate in land drainage is *not* affected by fertilizer dressings but only by the solubilities of naturally occurring phosphates in subsoil, these are not well understood. Land drainage is often devoid of P, the amounts we have measured at Rothamsted, Woburn and Saxmundham, have fluctuated through the year but never exceeded 0.8 ppm.

Micro-organisms thrive in water that contains very small concentrations of nutrients, much less than often occurs in drainage from unfertilized land that is not used for farming. (At a symposium of the Society of Water Treatment and Examination on eutrophication in 1970 (*Water Treatment and Examination*, Vol. 19, pp. 223-238) Dr. A. L. Downing gave these threshold values for algal growth—0.3 ppm of N and 0.01 ppm of P.) Rain often contains 0.7 ppm of N, sufficient for algae if only a trace of P is present in water. The minimum concentration of P needed is less than is provided by the phosphate compounds naturally present in many unfertilized soils. It must be accepted that some N and P is likely to be present in natural water: sewage effluents supply much N and much more P than land drainage does. Well-treated sewage effluent has 30-40 ppm of inorganic-N and 2-5 ppm of P; much of the P comes from detergents; most rivers now contain more than 0.1 ppm of P, much more than the minimum needed by algae. While it is technically feasible to remove much of the N and P from sewage effluent, it is estimated that doing this for all effluent in Britain would cost £15-30m annually. Furthermore it is doubtful whether removing *some* of the N and P would lessen algal growth. By contrast the extra cost of filtering and treating water affected by eutrophication is likely to be much less than this.

The Royal Commission On Environmental Pollution has reported (Command No. 4585, London: H.M.S.O. February 1971): 'The social benefits of cleaner air and water, less noise and a more pleasant landscape have to be put into perspective . . . so long as resources are limited, choices have to be made between alternative ways of using them.' Dealing with fertilizers, the Report states that losses by leaching are 'contained in this country by good farming practice'. It draws attention to the need to use the wastes from intensive farming systems—'If this material is not returned to the fields, valuable fertilizer is wasted and there is a risk of soil deterioration; moreover special provision may have to be made for its treatment either on the farm itself or at the local sewage works. What is needed therefore is some economic inducement to farmers to use manure from intensive farming.'

It must be realised that people in densely populated countries are committed to using fertilizers to secure the large yields without which they cannot survive. A portion of the large amounts of nutrients involved in modern agriculture will 'leak' into water; it is a job for scientists to minimise these leaks. The scientific discussions held in many countries recently have shown that fertilizers are not *directly* responsible for any deterioration in the quality of water supplies. Problems associated with eutrophication which would not have arisen naturally are a result of large populations living on relatively small areas, needing regular supplies of food and being unable or unwilling to spend enough to dispose of their wastes so they make no difficulties for other people. Two important points have been emphasised in the many discussions that have been held: (i) The present small efficiency of nitrogen fertilizers is a challenge to agricultural scientists. (ii) The methods used to dispose of animal excreta affect both water pollution problems and also the amounts of fertilizers needed to produce our crops.

USING ANIMAL WASTES ON FARMLAND

One acre (0.4 ha) of well-fertilized land will support a milk cow for one year during which she might excrete about 20 lb (9 kg) of P, 100 lb (45 kg) of N, and 200 lb (90 kg) of K. Perhaps half of this N, P and K is returned to pasture while grazing; the other half is excreted indoors or near buildings and it must be handled as farmyard manure or slurry before it can be returned to land. In modern systems of keeping cattle where most of the manure *is* returned, some losses in handling are inevitable because a proportion is lost in wash water, by rain storms, and by seeping from gathering areas and gateways. Much larger losses of nutrients are inevitable when too little farmland is available for wastes to be used at correct rates, or when they are treated by sewage processes so that a clear effluent is discharged into a stream. These problems are discussed in Chapter 2.

Besides the major pollution caused when the excreta of cattle are removed in surface runoff in wet periods, or from frozen land, N and P reach streams from many less spectacular, but nevertheless, important sources. Drainage from manure heaps, from aprons where cattle are gathered and manure is handled, from yards where they stay overnight, and from gateways, often finds its way into nearby streams. Such local pollution by animal wastes can and should be prevented. It is much more difficult to avoid the simple overloading of the soil's capacity to hold nitrate which occurs when stocking density is increased. The most spectacular problems in dealing with animal wastes occur in U.S.A. where large 'feedlots' often have 1000-6000 cows on 100 acres (40 ha). In California, with a hot arid climate, manure is easily dried for sale, but disposal in wet areas is much more difficult. A large unit holding 10,000 cattle has a sewerage problem equal to that of a city of 45,000 people, even treatment lagoons cost $1 to $5 per head of feedlot capacity. These problems should be solved in countries such as Britain before they reach the size and urgency of problems in parts of U.S.A. Plant nutrients are expensive to buy and should be used more than once to produce crops. Farmers must also arrange manure handling to avoid polluting water supplies. Complying with these requests will be costly. Farmers may need help to buy capital equipment for handling and spreading manure and slurry; new contract services may have to be established for spreading manure in areas where farm workers are too few.

MODERN FARMING AND THE SOIL

Public concern about pollution has been related to fears that the inherent fertility of the soil is being eroded and its 'fundamental structure' damaged beyond repair. Difficulties that British farmers have had in arable cropping during recent wet harvests, and cold late springs, caused the Agricultural Advisory Council to initiate an inquiry. The Report (Modern Farming and the Soil, London: H.M.S.O. 1970) does not substantiate that many farmers are ruining their soil. In a section on crop nutrition the Report states:

'As far as the *nutrient fertility* of our soil is concerned, we have few misgivings. There is no evidence to show that the disappearance of livestock from certain areas and the replacement of ley-farming and farmyard manure by chemical fertilizers has led to any loss of inherent fertility. Nor is there any evidence that organic matter is intrinsically a better source of nutrients. However, we are somewhat concerned about the decline in the use of lime in some parts of the country and the lack of knowledge of the distribution of trace elements where intensive systems are practised. Thus, there is no fertility problem resulting from modern farming methods.'

Diminishing organic matter in certain unstable soils had led to difficulties when this land was used continuously for arable farming. Problems were often accentuated by bad drainage, and by weeds, pests and diseases. Where many stock were kept on grassland on these unstable soils, poaching damaged both soil and sward. *These problems are not new.* Some soils have always been easier to cultivate than others; drainage has always been essential for good crops on some land, elsewhere natural drainage has been sufficient. No one questions that large organic matter contents in soils make for easier cultivations and better seedbeds. I suspect that the areas of Britain where arable cultivation is now difficult, have always been 'difficult'. A century ago they were the first to fall to grass and much of the land remained in grass until the 1939-45 War, being slowly improved by the organic matter accumulated under grass or scrub. Problems of compaction did not begin when heavy machinery was introduced; they occurred when horses provided traction. Nevertheless modern power does allow soils to be damaged more severely because heavy tractors can plough and cultivate and haul heavy loads where horses could not. The tractors *should not* be used in such conditions, but this advice is not easy to follow when crops have to be harvested in a wet autumn.

FUTURE POSSIBILITIES FOR USING FERTILIZERS IN BRITAIN

An interesting task for agricultural scientists in Britain is to see if it is possible to feed our present population with all the foods now eaten which can be grown here. Four years ago ('The carrying capacity of the land in the year 2000.' In: *The optimum population for Britain*, (Institute of Biology Symposium No. 19) Ed. L. R. Taylor. London: Academic Press. 1970) I concluded that we could produce all the cereal grain, potatoes, vegetables and livestock products we need—

> if 30 millions tons of cereals could be harvested annually (—doubling present production)
>
> and if grassland could support twice as much stock as at present, since both the dairy herd and the ewe flock must be doubled.

There is little room to manoeuvre; because both grass and cereal yields must be doubled, we cannot take more grassland for wheat growing. The results of more research on control of cereal pests and diseases, new methods of cultivations, and new varieties, will all be needed if we are to approach the cereal target by the end of this century. We could go far to achieving the grassland target by using much more N, backed up with adequate P and K. It will however be essential to improve the efficiency of nitrogen fertilizers used on grassland if the extra crops are to be grown economically. The proportion of the N lost by leaching and denitrification must be diminished, both to raise efficiency, and also to prevent massive eutrophication of streams, rivers and lakes.

The table below estimates the plant nutrients in crops and grass grown now and those in crops we may produce in 30 years time:

Year (A.D.)	N	P_2O_5	K_2O
	thousands of tonnes taken up by crops and grass		
	In arable crops		
1970	380	150	350
2000	700	250	500
	In grass		
1970	1,200	370	1,400
2000	2,400	750	2,800

It is difficult to convert these amounts of nutrients in crops into the N, P_2O_5 and K_2O needed as fertilizers because most of the produce would go through animals, and much of the N and P and nearly all the K would appear in their excreta. I believe that N, P and K in excreta should go back to the land. If it does the fertilizers needed in Britain in A.D. 2000 might be roughly double as much N as we now use, but with no more P and K needed than at present. On the other hand future farming policies, labour supplies, prices of fertilizers and crops, and costs of sewage treatment, may all suit the development of intensive animal farming units housed on small areas. If most of the N, P and K in the excreta of these animals is discharged to rivers that flow to the sea, much more fertilizer will be needed, perhaps treble the N and K that we now use, and twice as much P.

In addition to uncertainties about the amounts of N, P and K fertilizers needed in future, caused by the uncertain use of animal excreta, the amount of N needed will depend on how much is recovered by crops. If we devise ways of using nitrogen fertilizer more efficiently less will have to be bought. For example: if the present small efficiency of the nitrogen used on arable land is not improved (only about one-third of the N applied appears in the yield), British farmers may need 1,200,000 tonnes of N annually by A.D. 2000 for arable crops alone. If we could improve fertilizer efficiency so that crops took up half of the nitrogen applied, 900,000 tonnes/year would be enough; this would save £45 m a year at present prices!

CHAPTER ONE

Crop Nutrition and Fertilizers

PLANT NUTRIENTS

The dry material of a crop contains much carbon, hydrogen and oxygen which come from carbon dioxide in the air and from water. It also contains *nutrients* taken from soil which were essential for the growth of the crop. These elements are in two groups: *Major* plant nutrients are needed in amounts ranging from a few kilogrammes to two or three hundred kilogrammes per hectare. *Minor* plant nutrients (called 'micro-nutrients' or 'trace elements') are just as essential, but the amounts needed are from only a few grammes to several hundred grammes per hectare. Table 1 lists amounts of the two groups of nutrients in average yields of typical arable crops (data for micronutrients are based on those published by R. L. Mitchell (ref. 1).

TABLE 1

APPROXIMATE AVERAGE AMOUNTS OF MAJOR AND MINOR NUTRIENTS IN BRITISH CROP YIELDS

Major plant nutrients	kilogrammes/ hectare	Minor plant nutrients	grammes/ hectare
Nitrogen (N)	100	Iron (Fe)	600
Potassium (K)	100	Manganese (Mn)	600
Calcium (Ca)	50	Zinc (Zn)	200
Phosphorus (P)	15	Boron (B)	200
Magnesium (Mg)	15	Copper (Cu)	100
Sulphur (S)	30	Molybdenum (Mo)	10
		Cobalt (Co)	1

Major plant nutrients

Nitrogen, phosphorus and *potassium* are the three elements usually sold in fertilizers. The other three, *calcium, magnesium* and *sulphur,* are just as essential and when they have to be supplied because soils contain too little, they also are fertilizers.

Shortages of calcium lead to acidity of soil and to crop failures. They are easily and cheaply remedied by dressings of lime. Most medium and heavy soils contain enough *magnesium* for many years of cropping but sandy soils often contain too little and magnesium fertilizers are commonly needed, particularly for horticultural crops, sugar beet and potatoes grown on light soils. *Sulphur* is provided by some fertilizers, such as ammonium sulphate (24% S) and ordinary superphosphate (12% S). These are dilute fertilizers and are less commonly used than formerly; crops depend more on the amounts of S in soil which are replenished by rain. The only examples of responses to sulphur fertilizers in Britain have occurred with quick-growing crops on very light soils near the sea in Dorset.

Sodium is not an essential nutrient for most plants, but it seems essential for satisfactory yields of sugar beet and related crops. A crop of sugar beet (45 t/ha of roots) may contain 100 kg Na/ha (obtained from the soil, where supplies are replenished by rain); if sodium fertilizers are given beet may contain much more than this and on most European soils giving sodium fertilizer increases yields of sugar.

Minor plant nutrients

The first six elements listed as minor nutrients in Table 1 are used as fertilizers in various countries. The needs of crops, and their abilities to secure supplies of these elements from soils vary greatly. *Iron* deficiency often occurs on limestone soils and in fruit trees; *zinc* deficiency affects fruit trees too. *Manganese* and *copper* deficiencies occur on some peat soils, copper deficiency also on coarse sands and on Chalk soils in Britain. *Boron* deficiency occurs on light soils and usually shows in roots like sugar beet and in brassicae crops. *Molybdenum* deficiency is found when legumes are grown on acid soils. Very few examples of molybdenum deficiency have been confirmed in Europe but it is common in New Zealand. *Cobalt* deficiency has been demonstrated in plants grown in glasshouse pot experiments, but not for field crops. *Chlorine* is essential for plants but they take up very much more than they need and this is supplied by rainfall. Other elements have been suggested as essential for plants and it is likely that some of these will be confirmed in future.

Some trace constituents of plants are essential for animals; they include cobalt, iron, manganese, copper, zinc, molybdenum and also iodine and selenium. Supplies for animals come from the plant materials they eat and extra may be needed to improve the health of livestock. Trace elements for animals are sometimes supplied as fertilizers to the crops (cobalt for example) or they may be given separately in a mineral supplement to the normal feed.

Nutrients in crops

The amounts of nutrients taken up by crops depend on type of plant and yield; Table 2 lists amounts removed in average harvested yields of a selection of crops. Such figures are useful for constructing nutrient balance sheets for farming systems as they show how much is *removed* from a field. They should however be used cautiously in assessing total nutrients needed; roots and unharvested parts of crops contain further

TABLE 2

AMOUNTS OF MAJOR NUTRIENTS IN AVERAGE YIELDS

	Yield/hectare (tonnes)		Nutrients in yield, kg/ha					
			N	K	Ca	Mg	P	S
Wheat*	4	(grain)	80	40	10	5	12	20
Barley*	4	(grain)	70	30	10	5	12	15
Sugar beet*	40	(fresh roots)	230	220	70	25	20	30
Beans*	2.5	(grain)	110	50	20	5	15	25
Kale	50	(fresh crop)	200	180	200	20	25	100
Grass	10	(dry crop)	250	250	70	20	30	15
Clover	5	(dry crop)	150	100	100	10	10	10
Potatoes*	50	(tubers)	180	200	10	15	25	20
Lucerne	10	(dry crop)	280	180	250	20	30	30

*Amounts stated are in:
  Cereals, grain + straw; beans, grain only; sugar beet, roots + tops; potatoes, tubers only.

amounts and some plants lose nutrients before harvest. For example when crops mature they often lose some of the nitrogen used to make their maximum growth; cereals grown in Europe contain much more potassium when they are flowering than at harvest. A wheat crop yielding as in Table 2 will contain 100 kg K/ha in the green plant during summer but leaching and leaf fall remove half or more so that the harvested grain and straw removes only 30–40 kg K/ha.

SUPPLIES OF NUTRIENTS TO CROPS

The nutrients needed by crops are taken by roots from the soil; three sources are important in replenishing the stock of soluble nutrients that the roots draw on, (1) rain, (2) soil reserves, and (3) fertilizers.

Nutrients supplied by rain

Where crops rely largely on natural sources for nutrients, rain may supply an appreciable proportion. Recent figures from Saxmundham (Suffolk), Sweden, Norway and Australia are in Table 3. Scandinavian data show how much rainfall composition may vary from place to place in one country and that rain may supply nearly as much of some elements as moderate crops contain. On the other hand figures from Northern Australia show that in such areas rainfall makes virtually no contribution (the little sulphur is typical for areas where sulphur-deficiency is common).

Sulphur is the most important nutrient supplied by rain and where sulphur-free fertilizers are used on well-leached soils, rain is the only source. Commonly British soils receive each year in rain between 15 and 100 kg S/ha; most of this comes from burning coal and oil, the rest from the sea. Where 10 kg S/ha or more falls annually in rain, crops should receive enough without supplying more S in fertilizers. Most of Western Europe receives 14 kg S/ha or more. In parts of Scandinavia, North America, Central Africa and in much of the Southern Hemisphere (particularly Australia and New Zealand) much less sulphur is provided by rain and, in many areas, sulphur fertilizers are essential for large yields.

TABLE 3

PLANT NUTRIENTS IN ANNUAL RAINFALL

(kg/ha of elements)

	Saxmundham (average, 1966-70)	Sweden	Norway	Western Australia
		(Some ranges of published data)		
N as ammonium	9	0.7–4.0	0.8–6	0.7
N as nitrate	7	0.15–0.8	—	0.5–0.8
Phosphorus	0.2	—	3–19	—
Sulphur	18	—	0.8–8	(<2)
Potassium	3.0	1.1–3.5	3–14	0.3
Calcium	13	6–19	0.4–17	—
Magnesium	4.0	—	0–148	0.2
Sodium	27	4–30	1–257	1.1
Chlorine	50	2.5–40		1.6–1.8

Nutrients from soil reserves

Where no fertilizers or manures are used the soil is the immediate main source of all nutrients except sulphur. The total amounts of nutrients in soil are many times as much as crops require but they are poor guides to soil fertility as only a small fraction of the total amount becomes soluble and useful to crops in any season. The amounts of phosphorus soluble in dilute alkaline or acid solutions, and the amounts of potassium and magnesium that are displaced by neutral salt solutions are used to forecast how much the soil may supply. Nitrogen is continuously released from soil organic matter and the amounts that can be extracted by water in spring do not show how much may be available later in the season (or how much must be supplied as fertilizers). Chemical measurements of the sulphur reserves in soils are not useful guides.

Measurements of nutrients available to plants from soils are of two kinds: (1) The concentrations of nutrients in soil solutions that surround roots, when expressed in standard ways, are called 'intensity' measurements. (2) More vigorous solvents, usually involving ion exchange reactions, or solution by acids or alkalies, determine the much larger quantities of nutrients which may be useful to plants; these are usually called 'quantity' measurements. The two kinds of measurements help in understanding processes of nutrient uptake from soils. Practical methods of soil analysis are discussed in Chapter 14.

Nutrients from fertilizers

Fertilizers are substances applied to soil to increase crop yields by providing one or more of the elements that are essential plant nutrients. Wastes from crops (such as straw and natural vegetation, and animal excreta, have been used to fertilize cropped land for thousands of years; marl and other liming materials were used in Roman times to supply calcium and improve soil. These were the only fertilizers until natural deposits of sodium nitrate and bones were used at the beginning of the Nineteenth Century. When bones are dissolved by sulphuric acid the calcium phosphate they contain becomes water-soluble and forms superphosphate. This process was developed by J. B. Lawes at Rothamsted in the 1840s and, with it, he laid the foundation of the chemical fertilizer industry.

Organic manures supply nutrients that have already served to grow the crops that produced the wastes or excreta used to make the manure. Chemical fertilizers have a unique function: they supply *extra* nutrients and make farmers independent of supplies in their soils; if fertilizers can be bought yields need not be limited by the natural fertility of soils. In most developed countries manufactured fertilizers now supply much more plant nutrients than organic manures.

Fertilizers raise soil fertility by increasing the plant nutrients in the cycle of growth and decay. With good farming practices much of the extra plant foods that are bought can be maintained in circulation, so raising cropping potential or fertility of the land.

Fertilizers have another function that is not always realised: they lessen costs of production per tonne since they raise yields without a correspondingly large increase in total costs per hectare.

KINDS AND AMOUNTS OF FERTILIZERS USED

The revolution in crop nutrition that fertilizers have made possible is illustrated by Table 4. The phosphate used has increased steadily for over a century. Deposits of 'natural' sodium nitrate and guano were first worked in the first half of the Nineteenth Century, and much N was used in U.K. by 1850. The nitrogen used increased little during the second half of the century; during this time ammonium sulphate became available as a by-product from gas-making. Processes for fixing nitrogen from the air were developed in the early years of this century, they made large supplies of N-fertilizers available and lessened their cost; the N used has increased steadily in the last 60 years. Potassium fertilizers were first available when German and French deposits were mined in the latter part of last century, but little was used until the early years of this century.

In the 70 years from 1870 to 1940 developments in fertilizer use were dominated by the poor and uncertain prices that farmers in most countries received for their crops. Apart from the period of the 1914–18 War (when fertilizer supplies were restricted) crop prices were so low that fertilizing rarely seemed to be justified. In this time the N used in U.K. doubled and the K used increased to equal the N. In the following 10 years the N, P and K used by British farmers were all roughly doubled. The amount of phosphate used has not increased greatly since 1948. Nitrogen used has increased very quickly; the amounts applied were roughly doubled in the two ten-year periods 1950–60, and 1960–70. Increases in use of potassium kept pace with increases in nitrogen until about 1960, since then the K used in U.K. has changed little. The 'explosion' in use of nitrogen in U.K. was paralleled in the World as a whole; the amount used in 1939 had doubled by 1954 but subsequently, doubling of N used has taken only 7 years. World use of phosphate and

TABLE 4

FERTILIZERS USED IN UNITED KINGDOM AND IN THE WORLD

United Kingdom	N	P₂O₅	K₂O
	N	P_2O_5	K_2O
	thousands of (long) tons used		
1837	—	15	—
1845	33	46	—
1874	34	90	3
1896	33	122	5
1913	29	180	23
1939	60	170	75
1949	185	419	196
1958	315	386	348
1965	565	479	425
1969	790	476	458
1973	932	474	428

World	Millions of (metric) tonnes produced		
1913	(1.3)	(2.0)	0.9
1939	2.6	3.6	2.8
1954	5.6	6.4	5.7
1960	9.7	9.7	8.6
1965	16.8	13.8	12.1
1969	26.6	17.5	15.9
1972	35.1	22.5	19.2

potassium is still increasing, the amounts doubling roughly every 10 years. Wherever fertilizer use has greatly increased, crop yields have been increased, and where much fertilizer has been used for long periods, the soils are now more productive because they have accumulated reserves of nutrients.

FACTORS AFFECTING THE NEED FOR FERTILIZERS
Nitrogen

After land has been reclaimed from forest or grass, enough nitrogen may be mineralised from organic matter reserves for perhaps 5 to 20 years; but land used continuously for producing crops or grass can rarely provide more than one-fifth to one-third of the N needed. Inorganic N is ephemeral in soil, much of any excess being lost by leaching, before another crop is grown, especially in cool, humid climates. Therefore, in temperate climates nitrogen is usually deficient and, because it regulates yield, fertilizer is needed to make the amount up to the maximum needed.

Phosphorus

Most unfarmed soils contain too little phosphorus for good yields of cultivated crops, and phosphates have usually been the first fertilizers used in improving land for agriculture. As phosphate does not move easily in soil, being precipitated in forms with only slight solubilities, crop roots never reach more than one-quarter or one-third of a dressing of P fertilizer in a single year. The remainder of the P accumulates as residues which, after many years, may account for half of the total P present in soil. These residues are useful to following crops, and most manuring schemes increase the reserves of soluble P in soil.

Potassium

With much total K in soils it may seem surprising that potassium salts are important fertilizers. Many clay soils have 1% or more of total K; some clays steadily release K and supply enough for crops for many years, but other clay soils, and most sandy soils, can supply little. Crops take up much K (from 100–300 kg K/ha is common); unless this is returned in organic manures, or excreta of grazing animals, the reserves of most soils are depleted so that sooner or later maximum yields cannot be grown.

Calcium and magnesium

Calcareous soils, and some clays, contain large reserves of calcium carbonate. In temperate climates calcium is leached by percolating rain and soils become acid unless they have large reserves. Liming of acid soils is essential as many cultivated crops are depressed in growth or killed by acidity. Calcium used in this way is a fertilizer, but liming is so old a practice that it is often regarded as different from fertilizing.

Magnesium supplies come from soil, and most land in this country still supplies the 5–25 kg Mg/ha/year needed by crops. In the lighter soils with least reserves, magnesium deficiency is becoming increasingly common and more Mg is being used as a fertilizer. This trend will continue wherever large crops are sold and organic manures (which supply Mg) are not used.

Sulphur

Crops take up 10–25 kg/ha of sulphur, most needing about as much S as P (some crops, such as kale, take up much more S (Table 2)). Because so much sulphur-containing fuel is burnt, the rain in Britain provides more S than crops need. If, in future, power is obtained from other sources, sulphur could become an important element in fertilizers as it is in New Zealand and Australia, and other parts of the world. Most of the sulphur needed is taken from the soil solution but plants can also take it directly from the air.

Micronutrients

These elements are, at present, only needed as fertilizers in a few parts of Britain. They have been mentioned above and are discussed in detail in Chapter 9.

PROCESSES INVOLVING NUTRIENTS IN SOILS AND CROPS

To grow the maximum yields that fertilizers can give, and to use fertilizers efficiently, it is essential to know how nutrient ions reach plant roots and about the processes in soil that cause waste of nutrients.

Nutrient uptake

Crops take their nutrients (as inorganic ions) from the solution that surrounds soil particles and roots. Rain percolating slowly through soil displaces this solution and if drainage is collected and analysed, the results give an indication of the concentrations of nutrient ions from which crops may take their nutrients. (Cations, particularly K, may be retained by subsoil, so drainage water analyses may not indicate the composition of soil solutions accurately.) The figures in Table 5 are average composi-

TABLE 5

MEAN CONCENTRATIONS OF IONS IN DRAINAGE WATER AND APPROXIMATE

AMOUNTS AVAILABLE TO A BARLEY CROP

(after R. J. B. Williams (ref. 2))

	Concentrations of ions in drainage, mg/l		Approximate amounts of nutrients (kg/ha) available to crops in 250 mm of water transpired	
	Woburn	Saxmundham	Woburn	Saxmundam
Na	22	14	55	35
K	1.3	1.9	3.2	4.8
Ca	190	166	475	415
Mg	10	9	25	22
NH_4-N	0.1	0.1	0.25	0.25
PO_4-P	0.03	0.08	0.08	0.20
Cl	44	47	110	118
SO_4-S	67	57	168	142
$(NO_3$-N)*	(13)	(20)	—	—

*NO_3-N concentrations vary greatly during the year and average compositions are no guide to the nitrate available in the growing season.

tions of drainage water from two of our farms. They illustrate the great differences in the movement of nutrient ions in soil which affect crop nutrition and fertilizer efficiency. Nutrients reach roots by '*mass flow*' in the water that they absorb; the ions also *diffuse* to roots from solid particles through the soil solution. (Some authorities consider that a proportion of the nutrients taken up are *intercepted* by roots growing through the soil.)

Mobility of nutrient ions

Woburn and Saxmundham average 650 mm of rain a year; about 250 mm can be assumed to be used by spring barley. (If the land is uncropped before sowing the barley and also after it is harvested, during the autumn, winter and early spring about 150 mm may be lost by evaporation and 250 mm by drainage.) On this basis approximate amounts of nutrients in the water transpired were calculated and are shown in Table 5. The Na, Ca, Mg, Cl and S supplied are much more than crops need. The potassium is only about one-twentieth or less of the amount needed by barley; the phosphorus a fiftieth or less. (At flowering time, barley contains much more K than the harvested grain removes.) Concentrations of nitrate vary greatly and uptakes are not shown in Table 5. Common N-fertilizers are completely soluble and all nitrate will be in the soil solution; if not leached the nitrate needed will be transported to the roots in mass flow. American workers (ref. 3) have shown how mass flow may supply enough Ca and Mg for crops but only 10% of the K and 1% of the P needed. In their calculations they used these average concentrations of ions in saturation extracts of 135 U.S.A. soils, expressed in mg per litre: Ca, 30; Mg, 25; K, 4; and P, 0.05. Calcium concentration is much less and Mg and K more than in the British drainage waters in Table 5; P concentration is about the same.

Because soil solutions are so rich in nitrate, calcium, and magnesium (and sodium and sulphur) the whole amounts of these elements needed by crops will be obtained by roots in the water taken up. There are therefore no 'availability' problems in supplying Ca, Mg, S and Na caused by bad soil structure, or poor root range, or other seasonal characteristics. If the total supply is sufficient, the soil solution will contain enough, poor growth may however be caused if the soil cannot supply enough water.

These facts about ion mobilities are reasons for yields being linearly related to amount of mobile nutrient added (the 'curves' relating fertilizers applied and yields are discussed in Chapter 12).

Phosphate and potassium ions are not mobile in soils and Table 5 shows that only small proportions of the P and K needed by crops are transported to their roots in mass movement of water. It is believed that much of the supply of K and nearly all the P that reaches plants moves by diffusion. As rates of diffusion depend greatly on soil moisture content it is not surprising that the amounts of P and K that soils can supply alter with moisture content, and with changes in soil structure which greatly alter the paths and channels by which ions diffuse to roots;

both soil moisture and structure vary with season. This is why the *actual value* to *crops* of a given amount of soil phosphorus or soil potassium (measured by chemical methods) varies from year to year. When soluble phosphate or potassium fertilizers are added to soil, the soil solution near the granule contains large concentrations of both ions which alter the relative importance of mass flow and diffusion. If these concentrated solutions are taken up by roots, mass flow may supply much of the P and K; the actual amount taken up will, however, depend on where the fertilizer is placed and the amount of moisture in the soil.

LOSSES OF NUTRIENTS FROM SOIL

Nutrients are leached from soil by percolating water. Large amounts of several elements are lost in this way. Nitrogen is also lost as gas to the air by two processes that involve *volatilisation*. Other nutrient elements are removed from the cycle of circulation between plant and soil by *fixation* processes that make them very insoluble; although they are not *lost* from the soil, fixed nutrients can only become useful to crops when they are able to dissolve in soil moisture and this takes a long time.

Leaching

Examples of the concentrations of nutrient ions in drainage water are in Table 5. Losses per hectare depend on both the concentration of nutrients and the volume of drainage. Assuming the average concentrations of ions in drainage water shown in Table 5, and annual drainage (say 150 mm) the amounts of nutrients leached can be calculated.

Anions. At both stations much nitrate-nitrogen is lost in drainage; the figures given suggest a fifth to a quarter of annual dressings of 100 kg N/ha may be leached. This is probably an underestimate of the losses that may occur in years when large rainfall in spring leaches much of the recently-applied dressings of fertilizer-N before they can be taken up by crops. The drainage water removes much chloride and sulphate; both are supplied by rain and by some of the fertilizers used; neither ion is retained by the soil. Practically no phosphate is lost in drainage from either soil.

Cations. The losses of *calcium* are usually much larger than losses of other cations. *Sodium* is not 'fixed' by soils and much is supplied by rain (Table 3); as most crops take up little Na, particularly when potassium is supplied, concentrations of Na in drainage depend largely on amounts in rainfall.

Calcium and *magnesium* are leached in amounts depending on rainfall, soil texture, supplies of Ca and Mg in soil, and on other fertilizers used. There is no 'fixation' mechanism to hold non-exchangeable but potentially useful reserves of Ca and Mg (such as hold potassium reserves in soil). Heavy clay soils may have several thousand kg/ha of exchangeable Ca; in addition all calcareous soils contain further reserves as calcium carbonate. In southern and eastern England where drainage averages 150 mm/year losses range from about 50 kg Ca/ha from unfertilized light acid soils (which have little reserves) to 300 kg/ha of Ca from slightly-calcareous clay-loam soils receiving much fertilizer (such as many Rothamsted soils). An average annual loss of about 150 kg/ha of Ca seems inevitable in the kinds of intensive agriculture now common in north-western Europe. Annual losses of magnesium are much less, estimates for agricultural soils range from 5 to 30 kg Mg/ha. Losses of cations are largest from soils with the most reserves and many experiments on liming show losses of calcium were greatly increased by giving more than was needed. Losses of Ca and Mg are accelerated by fertilizers which supply nitrate, sulphate and chloride. These anions are not retained by soil and when dissolved in percolating water they are accompanied by equivalent amounts of cations. If the anions added which are not taken up by crops are increased, the losses of cations will increase proportionately.

Little *potassium* is lost from most agricultural soils in Britain that contain more than a few per cent of clay. The clay minerals retain surplus potassium in non-exchangeable forms, it is also held in exchangeable form by clay and organic matter; both processes protect it from leaching.

These comments on losses by leaching apply to most British soils used for agricultural crops. We have found, however that some coarse sandy soils with only 1% clay derived from Eocene Bagshot Beds in Dorset retain little nutrients. N is quickly leached by spring and summer rain and 70% of the K applied has been leached. Even water-soluble phosphates are leached from this soil; only 10% of the P applied over 4 years as superphosphate was retained in topsoil, but most of that given as basic slag or rock phosphate was held. These results for phosphate are unusual; such light soils are not commonly used in agriculture.

Volatilization

Nitrogen applied as fertilizer is lost from soils as gases evolved by micro-biological processes ('*denitrification*') in which nitrate is reduced to nitrogen and nitrogen oxides. The processes occur rapidly in water-logged soils containing much easily decomposed organic matter (straw, roots etc.) which provide carbohydrate for the denitrifying bacteria. Denitrification is a well-known process but the amounts of N lost are difficult to estimate; it is possible that when soils are often water-logged as much N may be lost by denitrification as is leached.

Ammonia gas is also lost from soils; this is most likely when ammonium salts applied to calcareous soils are not buried *in* the soil. The same reactions often cause a serious loss of ammonia when urea is applied as a top-dressing. Plant materials decomposing in the soil may also liberate ammonia; if this occurs near the soil surface the ammonia may be lost to the air.

Fixation

Most soils have mechanisms for retaining phosphate and potassium ions in insoluble forms. These mechanisms are beneficial because they make the ions so insoluble that little potassium and practically no phosphate is lost by leaching. Fixation processes make the nutrients very immobile so that more than crops need immediately must be added as fertilizers to very poor soils if the roots are to reach enough. This, however, is a short-term disadvantage. Nutrients conserved by fixation processes are slightly soluble and can feed later crops; large reserves of fixed P and K can supply sufficient for a sequence of crops and act as a buffer against shortages caused by unfavourable seasons (when plants do not use fresh fertilizers efficiently) or temporarily stopping the supply of fertilizer. Soils with good reserves of P and K are often better media for crop growth than poorer soils however much fresh fertilizer is applied. The following comparison is of amounts of fertilizer nutrients applied in the U.K. from 1837 to 1973 with rough estimates of the total amounts now in the soils:

	N	P	K
	millions of tonnes		
Added as fertilizer 1837–1973	16.9	12.0	9.8
Present in soil	60	20	260

The supplies of N and K from fertilizers are only small proportions of the total amounts in soil, but the P applied, much of which still remains as residues, is a sizeable part of the total.

CHAPTER TWO

Organic Manures and Fertilizers

Organic manures are composed mainly of wastes and residues from plant and animal life. They contain much carbon and relatively small percentages of plant foods, usually these come from the plants that fixed the carbon.

Organic fertilizers are usually wastes from industrial processing of parts of plants or animals. They contain more nitrogen and phosphorus than manures do and are classed as fertilizers rather than manures.

ORGANIC MANURES

Organic manures supply some nutrients for plants, and the carbon-containing compounds are food for small animals and micro-organisms. Manures often improve the structure of soils; they may do this directly through their action as bulky diluents in compacted soils, or indirectly when the waste products of animals or micro-organisms cement soil particles together. These structural improvements increase the amounts of water useful to crops that soils can hold; they also improve aeration and drainage, and encourage good root growth by providing enough pores of the right sizes and preventing the soil becoming too rigid when dry or completely waterlogged and devoid of air when wet.

The waste from mixed arable and livestock farming called farmyard manure (FYM) is used everywhere, it is partially rotted straw containing urine and faeces. But other rotting plant remains (composts) are manures too, and undecomposed materials such as straw should also be included. Organic wastes from industrial processes, town refuses and sewage sludges are used as manures. This chapter is concerned with nutrients supplied by manure; the role of organic manures in building up soil fertility has been discussed elsewhere (ref. 4).

FARMYARD MANURE

Farmyard manure supplies both major and minor plant nutrients, and its effects on crops are due to (1) its physical effects on soil condition, (2) the nutrients it supplies and (3) the way it supplies them; these three effects are, however, difficult to separate in field experiments. The nutrients supplied by single dressings of FYM are usually more important than physical effects for crops, but the improvements in soil condition caused by cumulative effects of many dressings may dominate its effect on some horticultural crops. In England we produce just under 5 tonnes for each hectare of crops and grass. (The total amount probably contains about 200,000 tonnes of N, 170,000 tonnes of K and about 40,000 tonnes of P.) About 4 t/ha is used in the arable districts, 5 tonnes in the upland districts, and 6 tonnes per hectare in the lowland grass districts. Where root crops are common most FYM is reserved for them. In England about 25 per cent of the sugar beet area had FYM (at 37 t/ha) in 1969, a third of the potato area had manure at an average of 42 t/ha. In 1939 FYM was the major source of the plant nutrients used on most farms. The *effective amounts* of N, P and K now supplied are equivalent to about a tenth of the N, a fifth of the P and two-fifths of the K applied as fertilizers.

Composition of FYM

On average dry FYM contains about 2 per cent N, 1.7 per cent K and 0.4 per cent P; but different batches may contain very different percentages of nutrients depending on origin and storage. Samples used in experiments at Rothamsted and Woburn have varied greatly; %N has shown two-fold, %K three-fold and %P four-fold variation. Many series of analyses of batches of FYM have been published. Sometimes the nutrients are expressed on a dry matter basis (and % dry matter must be known to interpret these in terms of field dressings of fresh manure). Very many FYM analyses made by N.A.A.S. (ref. 6) are summarised in Table 6, together with analyses of other British organic manures and figures for some samples from U.S.A.

Value of nutrients in FYM

Chemical analyses measure the total quantities of N, P and K in FYM, but not their availabilities to crops, these can only be measured in field experiments. Almost all of the nitrogen in FYM is combined with organic substances and is released only when they decay; in practice, about a third of the nitrogen is released quite quickly, but much is very resistant and persists long in the soil. Much of the phosphorus is also combined with the organic matter and little is known of its value; but roughly half of the total P present is quickly available to crops. Most of the potassium in FYM is soluble in water and nearly all can be rated as available to crops. In contrast, in some samples of FYM more than half of the magnesium is not even soluble in dilute acid. The field experiments done on FYM in Britain have given the estimates of the nutrients supplied that are in Table 7. For many circumstances 25 t/ha of FYM will supply to a first-year crop about 40 kg N, 20 kg P and 80 kg K/ha (the P and K are equivalent to about 45 kg P_2O_5 and 96 kg K_2O).

TABLE 6

ANALYSES OF ORGANIC MANURES USED IN BRITAIN AND U.S.A.

	Range				Mean values			
	Moisture	N	P	K	Moisture	N	P	K
			Percentages in materials as received					
British analyses								
Poultry manures								
Deep litter ⎱ N.A.A.S.	6–71	0.3–3.5	0.04–2.3	0.17–2.1	32	1.7	0.9	1.1
Broiler litters ⎰ data	9–75	0.4–3.6	0.09–1.7	0.25–2.0	32	2.3	0.9	1.1
Battery (ref. 6)	12–88	0.5–4.5	0.13–2.1	0.17–3.3	66	1.5	0.5	0.6
Poultry-droppings compost ⎱ Reading	—	—	—	—	75	1.2	0.4	0.4
Straw-droppings compost ⎰ University	—	—	—	—	65	1.1	0.9	0.8
Deep litter manure (ref. 8) ⎰ data (ref. 8)	10–81	0.4–5.7	0.22–1.9	0.08–1.4	35	1.9	1.6	1.2
Turkey manures	—	—	—	—	55	1.2	0.6	0.7
Cattle manures								
FYM	8–86	0.3–2.2	0.04–0.9	0.4 –1.2	76	0.6	0.1	0.5
Faeces (fresh)	—	—	—	—	85	0.4	0.1	0.1
Faeces + urine	86–93	0.2–1.7	0.04–1.0	0.08–1.9	89	1.3	0.1	0.7
Pig slurry	85–99	0.02–1.0	0.01–0.35	0.08–0.33	97	0.2	0.1	0.2
Sewage sludge	5–94	0.1–2.7	0.04–2.1	0.01–0.7	55	1.0	0.3	0.2
Town refuse								
Municipal	4–78	0.3–1.0	0.04–0.9	0.17–1.3	35	0.5	0.2	0.3
Gondard process	—	—	—	—	20	0.6	0.3	0.2
Wood ash	—	—	—	—	2	0.1	0.3	1.0
Analyses made in U.S.A. (ref. 7)								
Chicken manures								
From boards, no litter	—	—	—	—	54	1.6	0.4	0.4
With litter	—	—	—	—	61	1.7	0.6	0.6
Animal manures								
Dairy cattle	—	—	—	—	79	0.6	0.1	0.5
Fattening cattle	—	—	—	—	80	0.7	0.2	0.4
Pigs	—	—	—	—	75	0.5	0.1	0.4
Horses	—	—	—	—	60	0.7	0.1	0.6
Sheep	—	—	—	—	65	1.4	0.2	1.0

POULTRY MANURE

Average compositions of poultry manure are in Table 6; one set of data are from N.A.A.S. workers (ref. 6), others were obtained at Reading University (ref. 8). Fresh poultry droppings contain twice as much nitrogen as farmyard manure; they are much richer in phosphorus and contain about as much potassium as FYM. Composts of droppings and straw are richer than FYM in all three nutrients, and deep litter manure is richer still. In moist soil about half of the total nitrogen in droppings and in deep litter is equivalent to inorganic nitrogen fertilizer, but the proportion that is immediately useful in composts varies with their composition and maturity.

TABLE 7

NUTRIENTS PROVIDED FOR A FIRST YEAR CROP BY 25 TONNES/HECTARE OF FYM

Kilogrammes per hectare

	N	P	(P_2O_5)	K	(K_2O)
Swedes and turnips (Scotland)	15	7	(16)	63	(75)
Kale, potatoes, grass (Herts)	34	4	(9)	83	(100)
Sugar beet (Eastern England)	38	22	(50)	134	(160)
Potatoes (Eastern England)	38	22	(50)	78	(93)
Potatoes (Herts)	38	22	(50)	78	(93)
Sugar beet (Herts)	>38	>22	(>50)	>78	(>93)

The uric acid in fresh droppings is decomposed by micro-organisms to give ammonia which is easily lost if the manure is exposed. Drying fresh droppings quickly by heat stops these losses but special equipment is necessary; another way is to add 50 kg of superphosphate per tonne of droppings. Composting lessens the losses of nitrogen. Reading workers found (ref. 8) a ratio of 1 part (by weight) of straw to 1.5 parts of fresh droppings, with some extra water, is satisfactory. Nitrogen is also easily lost from poultry manures taken from houses where cereal straws have been used in the 'deep litter' system. Much loss *in* the house is inevitable, but when deep litter is stored after clearing it should be moistened to prevent ammonia escaping and be protected from leaching by rain. Where straw is limited, deep litter systems are better than composting; one tonne of straw can deal with the residues from 20 tonnes of droppings added gradually, but in composting 1 tonne of straw can deal with only $1\frac{1}{2}$–2 tonnes of droppings.

Experiments made by N.A.A.S. (ref. 8a) tested the value of poultry manure as a source of N for grass. The fresh manures used were very variable, but averaged 29 per cent dry matter, 1.4 per cent N, 1.2 per cent P_2O_5 and 0.6 per cent K_2O (in fresh manure). Spring dressings of poultry manure were about 50 per cent as efficient in terms of total nitrogen as fertilizer put on at the same time. Continuing the experiments suggested that poultry manure had appreciable residual effects. Pre-war experiments reported by H. V. Garner (ref. 8b) tested kiln-dried poultry manure on agricultural crops and vegetables. The manure contained 86 per cent dry matter, 3.6 per cent N, 3.6 per cent P_2O_5 and 1.8 per cent K_2O. The nitrogen it contained had the following percentage efficiencies in terms of N in ammonium sulphate as standard: 53 per cent for potatoes, 65 per cent for kale, swedes and Brussels sprouts, 75 per cent for mangolds, 100 per cent or more for sugar beet, red beet, French beans, broccoli and marrows. Residual effects in these experiments were small.

LIQUID MANURES

Many livestock farms were equipped in the last century with underground tanks in which urine and drainage from buildings were stored for manure. Later it became uncommon for these tanks to be installed or used in Britain. In contrast, in other European countries liquid manures are generally collected and used and are important sources of plant nutrients. British farmers know that in disposing of effluents by normal drainage they are wasting plant nutrients, but because returning liquid manures to the land often needs different management and extra labour, many consider it cheaper to replace the plant nutrients lost by bought fertilizers. Legislation we now have is intended to prevent farm effluents being discharged in ways that may pollute rivers or water supplies; but it can be very expensive to discharge effluent into public sewers. For these reasons, and also because systems of 'zero grazing' and other modern methods of keeping and handling livestock involve little or no bedding that can be the basis of FYM, many farmers are having to consider how best to dispose of liquid manures and slurries on their own land.

Use of liquids in Europe

The 'Gülle' system is often used; a mixture of faeces, urine and litter is matured and diluted with water. On all-grass farms in *Switzerland*, where all the herbage is used on the farm either for grazing or silage, over 80 per cent of the N and P, and over 90 per cent of the K contained in forage are returned to the land if solid and liquid manures are properly used. The system results in intensive farming, producing mainly meat and milk, being carried on with little need for fertilizers since the plant nutrients are recirculated. Stocking is heavy, usually about $2\frac{1}{2}$ cows/hectare. Two dressings of liquid manure, diluted 3–5 times with water, are given in a year to grass; this supplies about 130 kg N, 17 kg P, and 200 kg K with 3000 kg organic matter per hectare. In Swiss experiments the N in 'full liquid manure' was only half as effective as the same total amount of N supplied as urea; but urine tested alone was nearly as good as equivalent amounts of NPK fertilizers. Swiss work also shows that the P and K in liquid manures is about as effective as in ordinary fertilizers.

The Gülle system is also used in *Germany* where it is developed in mountain districts that are wholly in pasture and have little straw for the bedding needed to make good FYM. German estimates are that a cow excretes about 2 tonnes of dry matter each year, containing about 68 kg N, 10 kg P, 95 kg K and 33 kg calcium. Nutrient ratios in excreta are:

	N	P	K
Urine	100	0.4	250
Dung	100	26	44

Solid excreta is added to urine to increase the P content of liquid manures and this provides the slurries used in the Gülle system. In a survey of farms average composition of the diluted liquids applied was 0.13 per cent N, 0.1 per cent P, 0.2 per cent K, 0.06 per cent Ca. On the surveyed farms two-thirds of all the plant nutrients applied came from Gülle; the ordinary FYM made provided less than one-fifth of the total nutrients, and fertilizers one-seventh—mostly phosphate.

Much of the nitrogen in liquid manure is ammonia which, being volatile, is easily lost. Losses are greater the more concentrated the manure and the drier and windier the weather. Diluted liquid manure spreads more evenly than concentrated liquids, less nitrogen is lost, and clovers are less easily damaged. The loss of ammonia may be lessened by adding superphosphate.

British experiments with liquid manures

All British work emphasises the great variability of liquid manures and slurries. Some published analyses are in Table 8. Much of the variation is caused by season and few comparisons of slurries made in summer and winter have been published.

It will rarely be possible for farmers to have their slurries analysed but the figures in Table 8 help in making a rough estimate of the nutrients in the slurry they apply to land. Rates of application also vary greatly. The liquids listed as A, B, C were applied at 17,000 to 75,000 litres/hectare, mostly to grassland, but some was given to cabbage and kale. The richer winter slurry (F) produced while cows were housed was applied at very large rates, 300,000 litres/hectare for grassland that was cut, 800,000 litres/hectare on arable land that grew maize. The poorer summer slurry listed (G) was given to cow paddocks at 145,000 litres/hectare.

Pig slurries are much produced in *Northern Ireland*. Their value is said to depend on the age and type of animals and the food they receive, on methods of collecting, storing, and spreading, and on the amount of water used in washing or as diluent. On average, of the slurries surveyed, 1,000 litres contained 4 kg N, 0.9 kg P, and 1.8 kg K. The N and K figures were consistent, the whole ranges being 3–5 kg N and 1.4–2.2 kg K; but P varied widely from 0.2–2.8 kg P/1,000 litres.

Most liquid manures are produced and used on grassland farms.

Applying liquid manure may affect the response of stock to the grass grown. Herbage growing on urine patches is often neglected by grazing stock for a short time, but herbage growing on dung patches is neglected for a year or more. Such taint effects have been reported where slurries were used.

TABLE 8

PLANT NUTRIENTS IN SOME LIQUID MANURES AND SLURRIES
PRODUCED IN BRITAIN

	Type of stock and liquid	Sample	N	P	K
			kg per 1,000 litres		
Staffordshire	Cattle slurries	A	4	0.1	2.5
		B	1	0.05	1.7
		C	7	1.8	8.0
	Pigs (slurry)	D	9	3	6
Hampshire (ref 9)	Dairy Cows				
	Winter slurry	F	1.1	0.3	0.8
	Summer slurry	G	0.3	0.04	0.09
British survey	Poultry, fresh droppings	H	14	3	6
	Cows, faeces & urine	J	4	0.4	5
	Pigs, faeces & urine	K	3	0.4	2
Northern Ireland survey	Pig slurries	L	4	0.9	1.8

Experiments made by N.A.A.S. in England (ref. 10) tested cow slurry, poultry and mixed slurries on grassland and pig slurry on spring barley. The cow slurry contained 0.35% N, 0.15% P_2O_5 and 0.51% K_2O and the experiments on grassland tested its value as a source of nitrogen usually when applied at between 33,000 and 66,000 litres/hectare. The efficiency of the N in slurry was calculated from yields and amounts of N in the grass by determining how much fertilizer-N would be needed to give the yield given by slurry. This amount was expressed as a percentage of the N actually applied in the slurry. Efficiencies varied with the time when the slurry was applied; they are summarised below. The best time of application was late winter (March) when half the N was recovered, later and earlier dressings were less efficient:

Time of application	Percentage efficiency of N in slurry	
	for yield	for % N in herbage
Early winter (December)	18	14
Mid-winter (January-February)	27	26
Late-winter (March)	56	49
Spring (April)	37	39

In experiments with barley the value of pig slurry also depended on time of application. When applied in mid-winter about a quarter of the nitrogen became available to the barley crop; when applied in spring immediately before sowing between a quarter and three-quarters was available. Slurry was more efficient when applied at small rather than large rates.

Practical use should be made of liquid manures and slurries wherever possible so that the nutrients they contain are returned to the land. They should be regarded as an asset, not as a disposal problem. Farmers will often be uncertain of the value of the slurries they make. Rough guidance can be obtained by using the average analyses given by H. T. Davies (ref. 10) for a very large number of British and Swiss slurries:

$$0.7\% \text{ N}, \qquad 0.2\% \text{ P}_2\text{O}_5, \qquad 0.7\% \text{ K}_2\text{O}$$

Such a material would supply 7 kg N, 2 kg P_2O_5 and 7 kg K_2O in 1000 litres (or 70 lb N, 20 lb P_2O_5 and 70 lb K_2O in 1000 gallons). Applied late in winter (say March) the N applied should be equal in effect to N-fertilizer supplying half as much N, perhaps more. The P and K should be fully useful in maintaining soil P and K reserves.

MANURES FROM TOWN WASTES

Modern sewage systems produce sludge that contains enough N and P to be useful as manure. Solid wastes collected from households also contain organic matter and small proportions of plant nutrients. Town refuse is sold either as screened dust or as pulverised refuse (which contains slightly more organic matter). Municipal composts are also made by fermenting separated refuse with sewage sludge or ammonia liquors from gas works. The commonly-used Dano process involves continuous mechanical composting lasting 5 days. For example a compost made in Scotland had 1.5% N, 0.12% P and 0.16% K in its dry matter; its N content was similar to that of FYM, its P about half, but it supplied negligible K.

Nutrient contents

Manures from town wastes vary greatly in the plant nutrients they supply. Some analyses have already been given in Table 6. Many of the wastes used in British experiments (ref. 13) were poor in all nutrients; only sewage sludges contained as much nitrogen and phosphorus as FYM does. The outstanding difference between all town waste manures and FYM is that the wastes supply much less potassium.

Possible toxic effects

A serious disadvantage of wastes from towns and from industry is that they may contain certain elements in quantities toxic to crops. These elements may occur naturally in soils but most pollution of this kind is from factory chimneys, some waste limes, sewage sludges (from industrial towns), and some town refuses. Where contaminated wastes are used regularly, toxic metals accumulate and may damage crops; as these metals do not move easily in soil it is impossible to remove them. Crops grown on such soils contain more toxic elements than normal, but the possible effects when they are used as food for man or animals are not known. Because the availability to plants of these metals varies from one manure to another, and depends further on the acidity of the soil on which they are used, safe limits of metals in waste limes, sewage sludges or town wastes, cannot be stated. Farmers should take advice before using wastes proposed as manures.

Comparisons of FYM with sewage sludge and refuse manures for agricultural crops

In British field experiments on organic manures (ref. 14) FYM was outstanding in containing about 3 to 6 times as much K (most was available to crops) as the other manures tested; FYM also produced largest yields. Strawy manures all supplied some K, but K in sewage sludges appeared to be unavailable and these were mainly useful as sources of N and P. All the manures tested supplied N, and all except the straw composts made only with inorganic N, supplied P. The main value of all the organic manures tested was in the plant nutrients they contained, and content of organic matter *as such* was not important; sewage sludges supplied about twice as much *organic matter* but were far less effective than FYM.

In other British experiments comparing bulky organic manures, FYM was better than raw and digested sewage sludges, treated town refuses, and screened refuse dust. Sewage sludges supplying 25 tonnes dry matter/hectare gave one-third or one-half of the increase given by 40 tonnes of fresh FYM. Fermented town refuse, and composts of sewage sludge and town refuse (also applied at 40 t/ha) had from one-third to two-thirds of the effects of a similar weight of FYM. Screened dust from town wastes was variable and generally inferior to other wastes. The main value of sludges for potatoes was in supplying N and P; screened dusts and fermented refuses supplied little N or K. For most vegetables FYM was much superior to all other organic manures, but sludge was effective for savoy cabbages and kale. The average percentage increases in yield from refuses in relation to the increase from an equal weight of FYM were:

	Increase from	
	pulverised refuse	screened dust
	as per cent of effect of FYM	
Sugar beet	22	40
Mangolds	83	28
Carrots	85	41
Potatoes	51	— (the dust depressed yields)
Average of 16 experiments	64	25

Horticultural crops

Organic manures are traditionally prized for vegetable growing and many experiments have tested them on outdoor and glasshouse crops. As in experiments on agricultural crops both the materials tested and results have been very variable. Farmyard manure usually gave the largest yields (although often other materials were not far behind), perhaps because of the potassium it supplied. Sewage sludge, as in other experiments, acted mainly as a source of nitrogen. There was no advantage in composting straw before applying it, equally good results were obtained when it was ploughed in with additional nitrogen fertilizer. Any effects that seemed due to physical improvement of soil were not easy to distinguish from those due to the supply of nutrients.

STRAW

Straw is a dominant constituent of farmyard manure, therefore it is often suggested as a substitute for manure. Many farmers ask about the possible benefits from ploughing in, removing and selling, or burning the large quantities of cereal straw produced in arable farming systems when it is not required for making into FYM. Straw itself is poor in plant nutrients and the richness of FYM comes from nutrients in the foods eaten by stock that make the manure. In the nineteenth century, nitrogen and potassium were brought on to British farms mainly in feeding stuffs for cattle which were fed in yards. Treading straw into manure provided the only feasible means of transferring the nitrogen and potassium in the feeding stuffs, roots and straw, back to the land. Rotted straw conveyed nutrient elements from one branch of the farm to another, but this function is not so essential under modern systems of farming when chemical fertilizers are available.

An average crop of straw (about 3,500 kg/ha) only contains about 17 kg of nitrogen, 3 kg of phosphorus and 30 kg of potassium. When straw is ploughed back into the land the bacteria and fungi which decompose it need nitrogen and, if extra nitrogen is not supplied, they will take it from the soil and rob the following crop. In such bulky manures about one part of nitrogen for every 20 parts of carbon is useless, only when proportionately more nitrogen is present can it serve the immediate needs of crops. When straw is rotted to farmyard manure or compost, 0.75 parts of N are locked up for each 100 parts of straw and, when straw is ploughed in, this amount must be added if the following crops are not to be robbed. Many experiments have examined the benefits on crops from ploughing in straw; some have also tested burning or composting.

Burning straw seems wasteful. All the organic matter and the nitrogen in it are lost, but the phosphorus and potassium return to the soil. Experiments so far done show neither increases nor losses in yield from burning straw as compared with ploughing it in. When ploughing in straw is difficult mechanically, or when it is infested with weed seeds (for example wild oats), it *should* be burnt. On farms without stock, ploughing in straw *may* be justified on soils where structure appears to deteriorate with continued arable cropping, *but* there is no certain evidence from experiments of such improvements. It may be that future experiments will show that straw should be ploughed in where bad soil structure causes difficulties; but until such evidence is obtained there seems no reason why straw that cannot be used this way should not be sold or burnt. The results of recent experiments made by A.D.A.S. have been published (ref. 123).

OTHER ORGANIC WASTES

Organic wastes from processing natural products, and from industry, are often offered to farmers. Their value as manure depends much on the per cent of nitrogen they contain. If large quantities are to be used on farms their value must be assessed by analysis. Usually mineral nitrogen will be formed in the first few weeks after adding organic materials to soil if the dry weight has more than 2% N; mineral N is not generally released initially if the waste has less than 1.5% N or if the carbon-nitrogen ratio is greater than about 25:1.

Sawdust is sometimes used as an organic manure; it is also added to soil in litter from poultry and animal houses. Its rate of decomposition depends on species of tree and the per cent N it contains (from 0.05–0.25 %N are recorded). For example a pine sawdust containing 0.13% N and 45% C needed 8–10 kg of N per tonne of air-dry sawdust to produce maximum decomposition. Most conifer sawdusts decompose slowly and, because of this, the soil supplies enough nitrogen and supplements are not needed.

ORGANIC MANURES IN OTHER COUNTRIES

The analyses of manures and results of experiments described in this chapter are relevant to most humid temperate countries where farming systems are similar to those in North-West Europe. Examples of analyses of manures in U.S.A. are in Table 6, they are similar to British figures. But farm manures made by animals in other countries are sometimes very poor because the animals' diets are poor; poor feed means poor excreta and poor manure. For example faeces of sheep and goats grazing annual weeds and the sorghum straw and scrub in the Sudan had only 1.3% N in dry matter. In dry parts of the unirrigated Gezira only 1 kg of N/ha was deposited in faeces; 7 kg of N/ha was dropped by cattle grazing in irrigated areas. Such manures release nitrogen very slowly; often applying them to soil depresses the nitrate available to plants for a time. Urine contains more nitrogen, but much of this is easily lost to the air or by leaching. Animal excreta make only a small contribution to soil fertility in the irrigated Sudan Gezira even where stocking seems moderately dense.

Some experiments in tropical countries have tested local animal manures and composts which were as effective as manures made in temperate farming systems. For example an investigation under continuous cropping in Ghana (ref. 16) showed that a mulch of 12½ t/ha of dry grass, or 'Kraal manure' at 5–10 t/ha, nearly always gave better yields than inorganic fertilizers (these supplied small amounts of nutrients, *about* 25 kg/ha each of N and P_2O_5 and 35 kg/ha of K_2O). The manures acted mainly by supplying nutrients but had some additional effects caused by a slow release of nutrients or by physical improvement of soil.

Relatively few analyses have been published for manures made in the tropics. One investigation in *Malawi* examined FYM made in open and covered kholas; average analyses of many samples were (in dry matter):

	%N	%P	%K
Open khola	0.95	0.30	1.26
Covered khola	1.28	0.48	2.29

Manure made in open yards was similar in composition to average British samples. FYM made in covered yards was much richer. Some compositions of manures made in *Tanzania* have been published, together with analyses of some composts made from local plant materials. These data are a summary:

	N	P
	per cent in air-dry material	
FYM	1.50	0.30
Compost	0.93	0.23

The manure was as rich as British samples, compost was poorer. Another report from *Tanzania* compared compositions of two samples of compost made from crop wastes with samples of cattle manures collected from African kraals; these ranges of data are expressed on the air-dry materials:

	%N	%P	%K
Composts	0.58–0.74	0.16–0.23	0.83–1.03
Cattle manures	0.53–0.85	0.25–0.42	1.31–2.22

The cattle manures were richer in all nutrients, but particularly in P and K. The manures from the kraals contained no bedding; they had much less organic matter and much more inorganic material (due to soil contamination) than manures of the kind made with bedding. These few analyses show that local manures made in the tropics are very variable in composition, depending on how they are made. Nevertheless some that have been used in experiments are as rich or richer than British FYM and the NPK provided must be largely responsible for the gains in yield which have resulted from their use.

ORGANIC FERTILIZERS

Some wastes from processing animal and plant products contain several per cent of nitrogen and phosphorus and are sold as fertilizers. Generally they are used in intensive horticulture, but some mixed fertilizers used in

TABLE 9

SOME CONCENTRATED ORGANIC FERTILIZERS

	Origin	Approximate amounts of the principal constituents		Particulars to be given in Statutory Statement under the British Fertilizer Regulations (ref. 17)
		%N (total)	%P_2O_5 (total)	
Hoof and horn meals	} Slaughterhouses	12–14	–	%N
Dried blood		12–14	–	%N
Shoddy	Wool waste	3–12	–	None
Meat and bone meal, meat meal, carcase meal, meat and bone tankage	Slaughterhouses	6–10	18	%N, %P_2O_5
Fish meals, fish manures, fish guano	Fish processing	7–14	9–16	%N, %P_2O_5
Leather wastes	Leather making	7	–	(Not listed)
Castor meal	} Residues from oil-seed processing	5–6	1–2	} %N
Rape cake		5	small	
Bone meals	Grinding or crushing bones	3	20	%N, %P_2O_5
Steamed bone flour and steamed bone meal	Steaming bones	0.5–0.8	26–29	%N, %P_2O_5
Guano	Excrement and remains of birds (except poultry)	10	12	%N, %P_2O_5, %K_2O

agriculture contain a proportion of organic materials. Potassium in such fertilizers is not in organic combination and usually it has been added as an inorganic salt. A selection of common organic fertilizers is listed in Table 9.

Some merits claimed for organic fertilizers are:

(1) The nitrogen and phosphate present are not water-soluble. As the fertilizer decays in the soil these nutrients may be released slowly at a rate that matches uptake by the crop. The process also protects plant nutrients from leaching.

(2) Organic fertilizers contain little or no soluble salts (unless inorganic fertilizers have been added) and they can be applied at large rates without the risk of damage to crop roots that may occur if inorganic fertilizers are used to supply corresponding quantities of plant nutrients.

Comparisons of organic fertilizers

In experiments on agricultural and horticultural crops grown out-of-doors concentrated organic nitrogen fertilizers, like hoof and horn and dried blood, gave similar yields to those obtained with ammonium sulphate. The advantage was with the inorganic materials for crops such as tomatoes, potatoes or cabbage which required a lot of nitrogen fertilizer to give full yields. For crops such as leeks and lettuce which need less nitrogen the organic fertilizers tended to be a little superior. Leather wastes that had been correctly processed, formalised casein (a waste from making plastics) and dried blood, all behaved in much the same way as crushed hoof in these experiments. In other experiments bone fertilizers

TABLE 10

PRICES, ANALYSES AND UNIT COSTS OF ORGANIC FERTILIZERS

(Quotations from British merchants, December 1974)

		Price per (long) ton (£)	Cost of 1 kg of plant nutrients (new pence)
Hoof and horn meal	14% N	260	183 (for N)
Dried blood	12% N	350	287 (for N)
Steamed bone flour	0.6% N, 26% P_2O_5	120	45 (for P_2O_5 only, neglecting N)
Bone meal	3.6% N, 21% P_2O_5	100	–
Meat and bone meal	6% N, 12% P_2O_5	80	–
Fish meal	6% N, 4% P_2O_5	70	–

have been tested as sources of phosphate. They acted more slowly than superphosphate, but they were quite effective on acid soils, particularly for crops with a long growing season. There were no instances where bone phosphates were more efficient than superphosphate.

There is no experimental evidence that organic nitrogen and phosphate fertilizers are markedly superior to inorganic fertilizers supplying equivalent amounts of plant nutrients for crops grown out-of-doors. The right amounts of inorganic fertilizers used at the correct time and in the right way were at least as effective as organic materials.

Apart from the considerations discussed above, which may lead a farmer to choose an organic fertilizer, the materials should be bought on their declared analysis and on the price per kg of plant nutrient this represents. A few market quotations for prices and analyses in December 1974 and the corresponding unit costs are given in Table 10. Comparisons of Tables 10 and 12 show that organic fertilizers are much more expensive per unit of N and P_2O_5 than inorganic materials. There was no subsidy on 'organics' in Britain.

CHAPTER THREE

Inorganic Fertilizers

Any substance that is added to soil to supply one or more plant nutrients and intended to increase plant growth is a fertilizer. Inorganic fertilizers are usually simple chemical compounds, made in a factory, or obtained by mining, which supply plant nutrients and are not residues of plant or animal life. In most countries the term 'inorganic fertilizers' means materials supplying nitrogen, phosphorus and potassium; sometimes liming materials that supply calcium are included, magnesium is less often. However, *all* essential plant nutrients (major and minor) when added to soil to benefit crops are fertilizers.

REGULATIONS AND DEFINITIONS

The trade in fertilizers is governed in Britain by Regulations made under the Fertiliser and Feeding Stuffs Act (1926); the latest were issued in 1973 (ref. 17). They include materials usually sold to supply any one, two or three of the elements, N, P and K, and liming materials. When magnesium-containing liming materials are used, the Regulations define the amounts of Mg they must contain if they are to be described as magnesian limestone' or 'burnt magnesian lime', but straight magnesium fertilizers are not included. The First and Second Schedules in the Regulations list the articles to which the provisions of the Act apply and give the particulars (chemical analysis or fineness or both) that must be in the Statutory Statement describing the fertilizer when it is sold. The Fourth Schedule defines the materials that are named as fertilizers. Other Schedules of the Regulations prescribe ways of taking samples and making the analyses, and the limits of variation in declared analyses that are permitted.

Present Regulations make provision for declarations of other nutrients that may be added to fertilizers by requiring that 'the name of any pesticide or herbicide, or any of the substances, boron, cobalt, copper, iron, magnesium, manganese and molybdenum . . . which has been added as an ingredient in the course of manufacture or preparation for sale' should be stated. (If a material is added only to improve the handling of the fertilizer, amounts of these elements need not be stated even if they are

28

present in the conditioner.) The information that has to be stated for different materials is given in later chapters dealing with kinds of fertilizers.

F.A.O. has published a study of fertilizer legislation in many countries (ref. 120).

FERTILIZER ANALYSES

In valuing fertilizers for sale only nitrogen, phosphorus and potassium are considered useful and amounts of these are given as percentages of

> *nitrogen:* expressed as N (i.e. the element)
>
> *phosphorus:* expressed as P_2O_5 which is phosphorus pentoxide (sometimes called (incorrectly) *'phosphoric acid'*); in the British Regulations it is called 'phosphoric anhydride'
>
> *potassium:* expressed as K_2O (often called *'potash'*); in the Regulations described as potassium oxide.

These units of measurement for P and K (P_2O_5 and K_2O) are relics of very old methods and ideas in analytical chemistry. P_2O_5 is phosphorus pentoxide (not phosphoric acid) and fertilizers do not contain this substance. Nor do they contain potash (K_2O) which is potassium oxide; in fact the most common K-fertilizer is potassium chloride (KCl) which contains no oxygen. But most countries that regulate fertilizer sales use P_2O_5 and K_2O and international trade is in terms of these oxides, as are FAO statistics (ref. 18). Therefore they are used in this book when describing actual fertilizers and in discussing fertilizer dressings. Almost all scientific measurements on crops and soils are now reported in terms of %P and %K; this convention is used here in stating soil and crop analyses. A few countries already use %P and %K in stating the nutrients in fertilizers and it is certain that others will change in future; Table 11 lists some of the 'straight' fertilizers that are now available in Britain and gives their usual analyses in both units.

The term *'straight fertilizer'* has for long been used in Britain to describe simple materials which are usually intended to supply only one major nutrient. As Table 11 shows, many simple fertilizers do supply other nutrients besides N, P and K, these may be useful to crops and should not be ignored. It is probably best to describe manufactured materials that only contain one of the three elements N, P or K as 'simple fertilizers'.

When amounts of nutrients other than N, P and K are declared they should be as percentages of elements (or as parts per million (ppm) of micronutrients when fertilizers contain less than 0.1% of an element). Some countries use the old oxide terminology for reporting other elements (e.g. CaO, MgO, Na_2O); this has no advantage and has obvious disadvantages for understanding crop nutrition as the elements are taken up as ions, not as oxides.

TABLE II

SIMPLE FERTILIZERS, SUPPLYING NITROGEN, PHOSPHORUS AND POTASSIUM LISTED IN FERTILIZER REGULATIONS (1973) OR AVAILABLE IN BRITAIN IN 1975

Names	Main nutrient (percentages)		Other useful nutrients* (percentages)		
	N		Na	S	Ca
NITROGEN FERTILIZERS					
Nitrate fertilizers					
Calcium nitrate (nitrate of lime)	15.5		–	–	20
Sodium nitrate (nitrate of soda)	16		27	–	–
Ammonium fertilizer					
Ammonium sulphate	21		–	24	–
Ammonium nitrate fertilizers					
Ammonium nitrate	32–34.5		–	–	–
'Nitro-Chalk'	25‡		–	–	11
'Nitra-Shell'	34‡		–	–	–
Ammonium sulphate-nitrate	26		–	12	–
Amide fertilizers					
Urea	46		–	–	–
Calcium cyanamide	22		–	–	38
Liquid fertilizers					
Nitrogen solutions	26		–	–	–
Anhydrous ammonia	82		–	–	–
Aqueous ammonia	27		–	–	–
	P_2O_5	P			
PHOSPHORUS FERTILIZERS					
Water-soluble					
Superphosphate	18–20	8–9			
Triple (concentrated) superphosphate	47	20		12–14	20
Citric-soluble					
Dicalcium phosphate	40	17.5			22
Basic slags, various grades	9–16	4–7	–†	–†	–†
Insoluble					
Ground rock phosphate	29–36	12.5–15.5			31–35
	K_2O	K	%N	%Na	%S	%Ca
POTASSIUM FERTILIZERS					
Muriate of potash (potassium chloride)	60	50	–	–	–	–
Sulphate of potash (potassium sulphate)	50	42	–	–	17	–
Nitrate of potash (potassium nitrate)	44	37	13.4	–	–	–
Potassic nitrate of soda (Chilean potash nitrate) (a mixture of sodium and potassium nitrates)	10	8	15	20	–	–

*Basic slags contain much calcium and smaller amounts of other nutrients. They vary in their...

*Approximate.   †Basic slags contain much calcium and smaller amounts of other nutrients. They vary in...

UNITS OF PLANT NUTRIENTS

While advice to farmers may often be given in terms of weights of simple or compound fertilizers, these have to be based on the amounts of nutrients contained in the fertilizers. To aid both understanding of a recommendation, and choice of a fertilizer to carry the recommendation into practice, it is essential to use a system for expressing weights of nutrients per unit area of land so that they are easily and quickly converted into weights of fertilizers; it is an added advantage if the figures can refer easily to the packages in which fertilizers are bought.

Metric units

The percentage of N, P_2O_5 or K_2O stated on bag or invoice gives the number of kilogrammes of these plant foods in a package of 100 kg. Ten times this figure gives the kilogrammes of N, P_2O_5 or K_2O in 1 tonne of fertilizer. Bags containing 100 kg are too large for easy handling and fertilizers are most likely to be packed in 50 kg bags (if they are not sold in bulk); if the percentage of N, P_2O_5 or K_2O is divided by 2, this gives the number of kilogrammes of N etc. in one 50 kg bag. For example a recommendation might be to apply 80 kg of P_2O_5/ha to grassland, the fertilizer available being superphosphate with 20% P_2O_5, so 100 kg of superphosphate contains 20 kg P_2O_5; the recommendation is therefore put into practice by applying $80/20 \times 100$ kg = 400 kg of superphosphate (0.4 tonne) per hectare. If packed in 50 kg bags, 8 would be needed to supply the 80 kg of P_2O_5. The same system is applied to compound or mixed fertilizers. A fertilizer containing 20% N, 10% P_2O_5, 10% K_2O has 200 kg N, 100 kg P_2O_5 and 100 kg K_2O in 1 tonne.

Units based on the short ton

Some countries use the short ton of 2000 pounds which is standard in U.S.A. The units for selling fertilizers are either the ton (2000 pounds) for bulk consignments, or bags of 100 pounds. The actual percentages of plant foods show the numbers of *pounds* of N, P_2O_5 and K_2O in a 100 pound bag. If these percentages are multiplied by 20 they give the pounds of plant food in 1 short ton of fertilizer. Advisory recommendations made in pounds per acre are quickly converted to the practical units in which farmers buy their fertilizer. For example the recommendation for applying phosphate used in the last (metric) example is (roughly) equivalent to 70 lb P_2O_5/acre. This amount would be in $70/20 = 3\frac{1}{2}$ bags each holding 100 pounds of superphosphate. A ton of super will contain $2000 \times 20/100 = 400$ pounds of P_2O_5; so to apply 70 pounds/acre will need $70/400$ ton = 0.175 ton; a 10-acre field will need $1\frac{3}{4}$ tons.

Imperial units

The long ton of 2240 pounds makes calculations difficult if pounds are retained as the basic unit. Since small packages of fertilizers are usually

(in 1971) in 'hundredweights' (cwt) of 112 pounds, it has been foun(
convenient in advising farmers to use one-hundredth of 1 cwt = (1.12 lb
for giving weights of plant foods; this amount is called a UNIT. Th(
reason is that the percentage composition stated on bag or invoice give
the number of UNITS of N, P_2O_5 and K_2O in a one-hundredweight bag
This system is as simple to use as those based on the 2000 pound shor(
ton, or the metric tonne of 1000 kg. It is described here because the lon,
ton is likely to persist for some years in Britain for advising on fertilizer
and for trading in them.

Another older British system of measuring plant foods is still used:
fertilizers were valued, and compensation for manurial residues wa(
awarded, in terms of 'unit prices' taken as the price of 1% of plant food
in a (long) ton of fertilizer, i.e. 22.4 lb of N, P_2O_5 or K_2O. (These uni(
prices are used in the latest British compensation Regulations issued i(
1969 (ref. 19)).

VALUATION OF FERTILIZERS

The choice between alternative fertilizers on sale should be made by
picking that which offers best value for money and provides N, P_2O_5 o(
K_2O at lowest cost per kg of plant nutrient and in the form thought to be
best.

For simple fertilizers valuation is easy. By multiplying the figure fo(
price in £ per tonne by 10 we get the price in new pence per 100 kg. I(
this amount is divided by the percentage of plant food in the fertilize(
the result is the price of 1 kg of plant food. A recent quotation fo(
ammonium sulphate at £42 per (long) ton represents 19.7 np/kg of N.

Current prices calculated in this way for a few simple fertilizers ar(
given in Table 12. Comparisons of these prices per kg of plant food
show which fertilizer offers the best value for money so far as plant food
content is concerned. Allowances may be needed for the convenience
of granulated fertilizers, for the effects on soil acidity of fertilizers like
ammonium sulphate (which causes a loss of lime) or basic slag (which
supplies lime) and for the other constituents present such as the sodium
in Chilean nitrate of soda or the sodium and magnesium in lower-grade
potash salts. In other countries the value of the sulphur in superphosphate
or ammonium sulphate should be included where this nutrient increases
yields.

Comparisons of fertilizer values may be made in exactly the same way
by basing the calculations on 1 lb of N, P_2O_5 or K_2O (when using the
short ton) or on UNITS of 1.12 lb (when using the long ton). Table 12
shows unit values for a few fertilizers using all three systems.

FERTILIZER PRICES

Britain. The prices of fertilizers in Britain have been unsettled recently
fixed price lists and recommended prices were not issued by most manu-

TABLE 12

APPROXIMATE PRICES, ANALYSES AND VALUES OF PLANT FOODS IN SIMPLE
FERTILIZERS, SOLD IN BRITAIN IN WINTER 1974/5

	Analysis	Price per* (long) ton) £	Approximate price (new pence) per unit of plant food		
			per kilo-gramme	per pound	per UNIT (1.12 lb)
NITROGEN FERTILIZERS	%N		for N		
Nitrate of soda	16	77.00	47.4	21.6	24.1
Ammonium nitrate	34.5	53.20	15.2	6.9	7.7
'Nitro-Chalk'	25	42.10	16.6	7.5	8.4
Ammonium sulphate	21	42.00	19.7	8.9	10.0
Ammonium sulphate nitrate	25	48.15	19.0	8.6	9.6
Liquid fertilizers					
Nitrogen solution	26	45.00	17.0	7.7	8.6
Aqueous ammonia	28	42.10	14.7	6.7	7.5
PHOSPHATE FERTILIZERS	%P_2O_5		for P_2O_5		
Water-soluble					
Superphosphate (ordinary)	19	41.50	21.5	9.8	10.9
Triple superphosphate	47	90.40	18.9	8.6	9.6
Citric-acid soluble					
Basic slag	15	18.00	11.8	5.4	6.0
Insoluble					
Ground rock phosphate	36	38.00	10.6	4.8	5.2
POTASH FERTILIZERS	% K_2O		for K_2O		
Muriate of potash	60	59.50	9.8	4.4	5.0
Sulphate of potash	50	70.75	13.9	6.3	7.1

*Most prices quoted were recommended prices published in December 1974, some are merchants' quotations for January 1975.

facturers after 1970. Formerly the basis was that *standard prices* were generally quoted for most simple and compound fertilizers for 6-ton lots delivered in the February/June period to the farmer's nearest station. The prices of potash were an exception, as they were usually quoted at the port of arrival and transport to the farm was extra. Recently British prices varied, some were quoted at works or port, some were for materials delivered to the farmer. The quotations in Table 12 are approximate only. Some were published recommended prices in Winter 1974, others were merchants' estimates of prices farmers might have to pay in January 1975.

Table 12 shows that N in sodium nitrate is much more expensive than N in other forms in the Table, which have similar unit prices; aqueous ammonia is the cheapest. P_2O_5 in triple superphosphate is slightly cheaper than in ordinary superphosphate; P_2O_5 in basic slag is much cheaper, the rock phosphate quoted was the cheapest source of P. K_2O in potassium sulphate is much dearer than in the chloride.

Other countries: FAO summarised and discussed fertilizer prices in many countries (ref. 18). They concluded 'many countries do not consider it desirable to allow fertilizer prices paid by farmers to fluctuate at free market rates, and they have adopted programmes of price fixing, price control, or at least price review. Fifty-two countries . . . have such programmes'.

Early delivery rebates are offered by manufacturers to farmers who will take delivery of their compound fertilizers and of some of the simple fertilizers during summer, autumn and early winter. These reductions in price are greatest in late summer and least in January. Savings of £2 per ton or more for a compound fertilizer may often be made by taking delivery in August, while the saving for delivery in January may be quite small. Farmers who use sizeable quantities of fertilizers may save a considerable sum by buying in summer or autumn. Modern granulated fertilizers packed in sealed paper or polythene sacks can be stored in good condition by taking a few simple precautions. The sacks should be handled carefully and should be undamaged when put into store. Piles of sacks should not be too high and they should be raised off a concrete or earth floor on a framework of wooden battens; paper sacks should be protected from rain.

Subsidies

In Britain until 1974 subsidies were paid by the Ministry of Agriculture, Fisheries and Food on inorganic nitrogen and phosphate in simple fertilizers and on these plant foods in compound fertilizers. There was no subsidy on potash or on simple organic fertilizers. The subsidy on nitrogen was at a standard rate per kg of N in both simple and compound fertilizers; for phosphate 1 kg of water-soluble P_2O_5 received a higher rate of subsidy than 1 kg of P_2O_5 in basic slag. The rate for a kilogramme of insoluble P_2O_5 in rock phosphate was lower still. The phosphate subsidies on compound fertilizers were calculated on the basis of their analyses. Water-soluble P_2O_5 ranked for the higher rate of payment, the rate for insoluble P_2O_5 being the lower one applicable to P_2O_5 in rock phosphate.

For example the subsidies paid from April 1971 were (in new pence):

	per kg	per lb	per UNIT (1.12 lb)
N	2.58	1.17	1.31
P_2O_5 : water-soluble	2.27	1.03	1.16
P_2O_5 : in basic slag*	1.60	0.72	0.81
P_2O_5 : in rock phosphate	1.23	0.56	0.62

*Subsidies on phosphate in basic slag depended on concentrations of P_2O_5.

Other countries. FAO (ref. 18) has summarised the arrangements made by the many countries where fertilizers are subsidised to promote their use. Direct payments to farmers or manufacturers, or help with transport costs, are most common. They benefit all farmers uniformly. In some countries only transport to remote areas is subsidised. Some countries use all their resources for subsidies to help crops that are particularly important. For example in Burma rice growing is helped, in Puerto Rico coffee fertilizers are subsidised; many countries give special help with fertilizers for small-holdings, hill farms or other inaccessible areas.

RESIDUAL VALUES OF FERTILIZERS

The outgoing tenant of a farm in Britain is normally paid compensation for the unused residual effects of the fertilizers used during the last years of his tenancy. These payments are based on Tables of Residual Values published by the Ministry of Agriculture, Fisheries and Food. The fraction of the original value assumed to remain after a given time varies with the fertilizer used; all fertilizers are assumed to have no residual effect after four crops have been taken. The latest Regulations were issued in 1969 (ref. 19); unit values were altered in 1972.

Inorganic nitrogen fertilizers, urea and also dried blood, are assumed to have no residual value at all. Organic nitrogen fertilizers, other than blood, are assumed to have half of their original value after one crop is taken and one-quarter after two crops. For *phosphate fertilizers* various residual values are laid down depending on the type of phosphate applied. Phosphate soluble in water or citric acid is assumed to leave the most valuable residues; two-thirds are left after one crop, one-third after two crops and one-sixth after three crops have been taken. Phosphate which is insoluble in water or in citric acid, or which is of unspecified solubility, is assumed to be worth only half as much as soluble phosphate. All phosphates are assumed to be exhausted after four crops have been taken. The total phosphate in basic slag is split; the portion soluble in citric acid gets the larger rate, the remainder has the lower rate of compensation. The amount payable for basic slag residues after one crop has been taken must not exceed two-thirds of the reasonable cost of the slag applied to the land including costs of delivering and spreading it; one-third of the original value is left after two crops and one-sixth after three crops. Potassium fertilizers are assumed to leave residues worth half their original value after one crop, one-quarter after two crops, and to be exhausted after three crops are grown. The residual effects of the different kinds of fertilizers are expressed as money values for use in calculating compensation. Table 13 gives the values which came into force on 4 July 1972; they represent 1% of a ton (22.4 lb) of fertilizer.

TABLE 13

THE RESIDUAL VALUES* OF FERTILIZERS USED FOR TENANTS'
COMPENSATION AS ALTERED IN 1972

Value of 1% of a (long) ton of fertilizer (22.4 lb of plant food).

	After one crop	After two crops	After three crops	After four crops
Nitrogen (N)		new pence		
Inorganic N and urea	0	0	0	0
Dried blood	0	0	0	0
Organic N except N in dried blood and urea	52	26	0	0
Phosphoric acid (P_2O_5)				
Soluble in water or in citric acid	70	35	18	0
Insoluble in water or unspecified solubility	35	18	9	0
Basic slag				
P_2O_5 soluble in citric acid	70	35	18	0
P_2O_5 insoluble in citric acid	35	18	9	0
Bone products				
Total P_2O_5	52	26	13	0
Potash (K_2O)				
Total K_2O	25	12	0	0

*These values include the cost of delivery and spreading.

Liming and chalking residues are compensated. The value of the residue is the reasonable cost of the material plus the cost of delivering and spreading it, reduced by not less than one-eighth and not more than one-quarter for each growing season since the lime was applied. (This means that lime is expected to leave residues for between four and eight years.) There is normally a limit to the dressing for which compensation is paid, this is 5 tonnes/hectare (2 tons/acre) of calcium oxide (CaO) or its equivalent; exceptions are made where it can be shown that a larger dressing was necessary and would benefit the incoming tenant.

Bulky organic materials such as bracken, moss, litter, peat or bean straw that are bought to aid the making of farmyard manure are compensated. Compensation is also given for feeding stuffs used on the farm that have enriched the manure, provided that urine was conserved and that FYM was well made and effectively used. Where slurry is made and applied to land compensation is given for feeding stuffs used by the stock that made the manure (provision for slurry was a new feature in the 1969 Regulations).

CHAPTER FOUR

Nitrogen Fertilizers

There are three forms of inorganic nitrogen in fertilizers:

> *Nitrates* supply NO_3^- ions
> *Ammonium* salts supply NH_4^+ ions
> *Simple amides* are not ionised salts but contain nitrogen in —NH^2 (amide) form or forms derived from this grouping.

Plants take up both ammonium and nitrate ions. Except in very acid soils ammonium-N is quickly converted to nitrate by microbial action; where this conversion is slow due to extreme acidity, plants adapted to the conditions may take up much ammonium. Simple amides (e.g. urea) are hydrolysed quite quickly to ammonium compounds and are then nitrified; if ammonia gas (NH_3) is used as a fertilizer, the N is also converted to the ammonium form in soil and then behaves as an ammonium salt.

There is no recent information on the amounts of different forms of nitrogen sold in Britain but FAO (ref. 18) has given these changes for the whole World:

	1956/7	1961/2	1968/9
	percentage contribution to total N used		
Ammonium sulphate	31	24	15
Ammonium nitrate	27	28	27
Sodium nitrate	4	3	–
Calcium nitrate	5	4	1
Calcium cyanamide	4	2	1
Urea	4	9	16
Other forms			
Solids*	9	12	18
Solutions	15	17	21

*Includes ammonium phosphate.

Ammonium sulphate is only half as important as 12 years ago; 'solids' in other forms are mainly ammonium phosphates, they have become twice as important in 12 years and urea has become a major source of the World's fertilizer nitrogen in this period. The nitrates that dominated sales a century ago now contribute little to world supplies.

A selection of nitrogen fertilizers available to British farmers in 1975 is in Table 11, prices of some are in Table 12.

NITRATES

All nitrates are water-soluble. Differences between their actions on crops depend on the other ions in the fertilizer salt; this may be potassium, sodium, calcium or ammonium. Nitrates are unsuitable for soils that are temporarily or permanently water-logged in any country; they are particularly unsuitable for wet rice growing; under these conditions the process of denitrification (Chapter 1) converts nitrate to nitrogen gas and oxides of nitrogen which are lost to the air. The nitrates of calcium and sodium are hygroscopic and difficult to handle unless packed in air-tight bags or prilled.

Chilean nitrate of soda contains 16% of nitrate-nitrogen and 26% of sodium; it is water-soluble and the sodium makes it a useful fertilizer for sugar beet and mangolds.

Chilean potash nitrate is really a compound fertilizer as it contains 15% N (all as nitrate) and 10% K_2O as well as 20% of sodium.

Potassium nitrate (KNO_3; 13.8% N and 36.5% K (about 44% K_2O)) also supplies two nutrients. It should not be confused with 'Chilean potash nitrate' which is not a pure chemical compound and contains much less K. The chemical compound potassium nitrate is now being made as a fertilizer in some countries. If it becomes cheap enough it will become an important material for direct use and as an ingredient of compound fertilizers.

Calcium nitrate is rarely sold in pure form. The fertilizer-grade 'calcium nitrate' used in some European countries is a double salt ($5Ca(NO_3)_2,NH_4NO_3,10H_2O$) which contains 15.5% N and has satisfactory physical properties when prilled.

AMMONIUM SALTS

All ammonium salts used as fertilizers are water-soluble and all are nitrified quite quickly in slightly acid and neutral soils to form nitrate. Their effects on crops are similar; they differ in their effects on soil acidity.

Ammonium sulphate. With 21% N this was the dominant N-fertilizer for nearly 100 years; it is becoming obsolete and supplies proportionately much less of the N used in Britain than formerly. The sulphate it needs may be costly and its concentration is too small for making modern compound fertilizers. Nevertheless, a considerable amount is still available, much as a by-product from making synthetic fibres. Often ammonium sulphate is preferred in humid tropics where its non-hygroscopic character makes it easier to handle than nitrates or urea. Ammonium sulphate may remain important for many years, but it is certain that mixed fertilizers will be made increasingly from more concentrated materials. Ammonium nitrate and phosphates, urea and anhydrous ammonia are all much more concentrated and with modern technical advances they can be handled cheaply and safely.

Ammonium chloride (26% N) is more concentrated than the sulphate, but the other ion (chloride) is more likely to damage crops. Most of the ammonium chloride used is made in. Japan; it seems very suitable for growing wet rice, particularly on soils where sulphate ion is reduced chemically to the poisonous sulphide. A cheap source of by-product hydrochloric acid is necessary for ammonium chloride to be produced economically.

AMMONIUM NITRATES

These fertilizers provide both ammonium and nitrate; they are all water-soluble.

Ammonium nitrate contains 35% N. The nearly pure salt was little used as a fertilizer in this country until 1965. This was partly because it was difficult to handle, and partly because it increases risks of fires. When ammonium nitrate is mixed with limestone it is safe and easy to handle and in this form has been known to British farmers for 40 years as 'Nitro-Chalk'; such fertilizers contain enough lime to correct the whole or part of the loss of calcium caused by the ammonium present and, if they contain enough limestone, they do not make soils acid. The output of ammonium nitrate in Britain has greatly increased in the last few years as new factories have been opened. Some is now incorporated in granulated compound fertilizers; where ammonium nitrate replaces ammonium sulphate, concentration is raised and there may be advantages from supplying part of the N as nitrate.

A notable advance occurred in Britain in 1965 when prilled ammonium nitrate (with 33-34% N) became available. Since then the quantity used has increased rapidly. There is always a risk that compound fertilizers containing much ammonium nitrate and potassium chloride will decompose if they are heated, and then toxic gases are released, principally chlorine and oxides of nitrogen. Factories have found ways of handling such mixtures safely and it is concluded that ammonium nitrate-containing fertilizers are not, in themselves, a fire hazard; but if they are involved in a fire they make the situation worse than it might have been and may also liberate toxic gases. In the polythene sacks in which it is usually packed ammonium nitrate is no problem on the farm; it cannot burn or ignite itself, either by friction or by a flame. When ammonium nitrate is mixed with organic materials that can burn, the risk of fire or explosion is much greater—as it is with other nitrates. It should *never* be mixed with, or stored near, materials that can burn such as straw, grain, oil or sawdust; this can lead to serious fires. Contact with electrical and heating equipment should be avoided. If these precautions are observed dangers from ammonium nitrate fertilizers on the farm are classed as 'very remote'.

Ammonium nitrate-sulphate (26%N). This is a double salt of ammonium nitrate and sulphate which has been used for many years in Europe and recently in Britain. It handles and stores well and is quite safe; where crops need a sulphur fertilizer its sulphate content is useful.

AMIDES

Urea has the chemical formula $CO(NH_2)_2$, it is also called 'carbamide'. Urea is very concentrated and contains about 46%N, all soluble in water. Ordinary crystalline urea is hygroscopic and difficult to handle, but it can be made in granules or prills that store and spread satisfactorily. Some solid urea has been sold as straight fertilizer in Britain. It is also used as a component of some concentrated solid compound fertilizers and in liquid mixed fertilizers. Small quantities have been applied as sprays.

In soil, urea is rapidly converted by an enzyme (urease) to ammonium carbonate which is unstable and releases free ammonia. If this change occurs on or near the soil surface ammonia may be lost to the air, and the fertilizer is inefficient. If it occurs near germinating seeds or the roots of young plants the crop may be damaged by the high concentration of ammonia.

Calcium cyanamide ($CaCN_2$ with 21-22% N) is hydrolysed in soil to form urea; the process is not quick and some of the intermediate products are toxic to plants. As freshly-applied cyanamide may kill germinating plants it has been used as a weedkiller. These dangerous properties disappear as cyanamide decomposes in the soil to form ammonium nitrogen; if it is to be applied before sowing a crop it should be worked into the soil two or three weeks beforehand. Cyanamide contains lime and does not make the soil acid. Its use has declined in recent years as it tends to be an expensive source of N.

COMPARISONS OF NITROGEN FERTILIZERS

For most purposes a kilogramme of nitrogen has about the same effect on crops whether supplied as ammonium sulphate, as ammonium nitrate supplied alone or in a mixture with limestone, or as 'nitrate of soda' or potassium nitrate. The choice between these materials should be made by considering:

(a) the price charged for 1 kg of nitrogen;
(b) the efficiency of each fertilizer for particular crops and soils;
(c) ease of storage, handling and distribution;
(d) whether the fertilizer causes a loss of lime from the soil.

Prices of British fertilizers quoted by merchants in 1974 are in Table 2.

Ammonium and nitrate

Ammonium and nitrate forms have been compared in many field experiments. The differences have been small in most comparisons on agricultural crops on slightly acid soils. There were no instances of ammonium salts being markedly superior to nitrates unless the nitrate damaged germination, but there are many examples of nitrate giving larger yields than ammonium salts, particularly on very calcareous soils. This suggests that in practice in Britain the reactions that lead to loss of ammonia from ammonium salts may cause more serious loss than occur from leaching or denitrification of nitrates. Nitrates have been superior to ammonium sulphate when used as top-dressings or seedbed dressing for cereals sufficiently consistently for nitrate to be recommended for this purpose in preference to ammonium, provided prices/kg of N in the two forms are similar. For other crops, unless nitrates damage establishment, there seems no reason to prefer either form of N on ordinary soils; nitrates are likely to be superior on highly calcareous soils.

Ammonium salts and fertilizers containing them are sometimes at a disadvantage when used as top-dressings on light-textured soils with a reserve of free chalk or limestone; a portion of the fertilizer may react with the lime and some ammonia is liable to be lost to the air. Under these conditions nitrates will be more efficient. On the other hand ammonium salts are safer than nitrates. Large dressings of nitrate may damage the germination of crops like potatoes and kale if worked into the top-soil just before sowing; they may check the germination of cereals if drilled with the seed; ammonium salts used in such ways are generally safer. If there is good rainfall after fertilizers are applied, nitrates will cause no damage. Excessive rainfall in spring is, however, likely to cause more loss by leaching when the N has been applied as nitrate.

Urea

Urea has been tested in field and laboratory experiments in many countries. It is often regarded as similar to an ammonium source of N but it does differ in being easily leached when first applied to soil. Urea is very soluble and until converted to ammonium compounds through hydrolysis by the enzyme urease it is just as mobile as nitrate. In practice it is usually hydrolysed quickly enough to avoid serious losses.

Commercial urea may contain traces of an impurity (biuret) which damages germination of crops like potatoes, kale or wheat when this fertilizer is applied to the seedbed just before planting. The British Fertiliser Regulations issued in 1973 (ref. 17) define urea as 'commercially pure urea containing not more than 1.5% biuret'; most modern processes give products containing less than this and these materials are quite safe. Urea free from biuret often gives good crops, but may suffer from two disadvantages which are due to its rapid decomposition in the soil or on the soil surface. Free ammonia is formed, and some of this may be lost to the air, so lessening the efficiency of the urea. (Losses of ammonia from urea are most likely when it is used for top-dressing, particularly in dry weather, and on light soils with free limestone, also in showery periods when soil is alternately wet and dry. In other conditions, and particularly when worked well into the soil, no ammonia is lost and urea is fully effective.) The temporary high concentration of ammonia when urea decomposes may also damage germinating seeds or young plants; damage has occurred where heavy dressings of urea have been combine-drilled for cereals, or placed near potato seed, or worked into seedbeds for kale. In most experiments urea has been as effective as ammonium sulphate or ammonium nitrate fertilizers. But in a small proportion of the experiments done (too many to be ignored) biuret-free urea has been much less efficient than other nitrogen fertilizers, either because ammonia was lost, or the crop was damaged, or both effects occurred.

More research is needed to define better the conditions which allow ammonia to be lost from nitrogen fertilizers or where certain fertilizers are liable to damage crops. Special care should be taken if heavy dressings of nitrates or urea are used to replace part or all of the nitrogen which has been generally applied in the past as ammonium sulphate; combine-drilling these fertilizers with cereal seed and contact placement with potato sets should be avoided. These disadvantages of the newer kinds of fertilizers should not be taken as condemning them permanently but rather as present obstacles to their efficient use which further research must overcome. In particular, urea will inevitably become an important fertilizer; it is cheap and easy to make and is highly concentrated and convenient to handle. Safe and efficient ways of using it are being found. Urea used as a top-dressing in wet rice growing always seems to be fully efficient, and much better than fertilizers containing some nitrate. It is also fully efficient if drilled into soil, as the solid or as a solution, at least 8 cm deep and a little distance away from seeds or roots. If used indiscriminately as top-dressing for arable crops or grass in temperate climates 1 kg of N in urea is likely to be, *on average,* only 80% as efficient as 1 kg of N in ammonium nitrate.

Urea is often considered as a component of compound fertilizers rich in nitrogen. A published account describes experiments with compound fertilizers of '2–1–1' or 'High–N' type (types of compound fertilizers are described in Chapter 10) containing from 6% to 34% of urea and tested at rates supplying about 75 kg N/ha (ref. 20). Fertilizers containing up to 12% urea did not damage crops when combine-drilled. With 20 per cent of urea germination and establishment were damaged but yield was not greatly affected. It seems that a proportion (perhaps as much as one-third) of the N in such fertilizers might safely be supplied by urea.

The potential value of urea is best summarised by quoting from the latest review (ref. 21): 'The use of urea as a fertilizer will certainly continue to increase and the compound is probably best regarded, with liquefied ammonia as one of a new generation of nitrogen fertilizers which can be used efficiently, but whose use requires a higher degree of understanding than is the case with simple inorganic salts'.

Soil acidification

Ammonium sulphate is notorious for making soils acid. Roughly 1 kg of ground limestone (or its equivalent) is lost for every kg of ammonium sulphate used. This effect does not matter on limestone soils, but other soils tend to become acid ('sour') after this fertilizer (or compounds containing ammonium salts) has been used for a time, and occasional dressings of lime *must* be added to keep the soil about neutral (that is, 'sweet').

Ammonium chloride has just as large effects as the sulphate because when either sulphate ions or chloride ions are leached out they remove equivalent lime. Other ammonium fertilizers (e.g. ammonium nitrate and phosphate), urea, anhydrous ammonia and nitrogen solutions, all tend to make soil acid. This is because they form nitrate in the soil and if any of this is leached (and some usually is) it removes equivalent lime. On the other hand if *all* the nitrate formed is taken up by crops, no extra calcium is lost because the N-fertilizer was applied. Nitrate fertilizers do not acidify soil, neither do mixtures of ammonium nitrate and limestone if they contain enough lime.

LIQUID NITROGEN FERTILIZERS

Large amounts of nitrogen fertilizer are sold in liquid forms in USA. Much anhydrous ammonia is used in Denmark (29% of the total N used in 1967/8 was ammonia), but most other West European countries use little. The amounts of liquids used in U.S.A. in 1972 were:

	millions of (short) tons of products
Anhydrous ammonia	3.8
Aqua ammonia	0.7
Nitrogen solutions	3.4
Liquid mixed fertilizers	2.9

The total liquids sold (11 million (short) tons) were about 28% of all fertilizers used in U.S.A.

Use in United Kingdom

The liquids used in U.K. in 1973 provided 60,000 (long) tons of N, 10,600 (long) tons of P_2O_5 and 11,500 (long) tons of K_2O. The amounts of all three nutrients used as liquids have nearly doubled in U.K. in the last 4 years, but even the N used in liquid forms was still only 6% of the total N used in 1973.

Forms of liquid fertilizers that supply nitrogen

The common liquid sources of nitrogen are listed below.

Anhydrous ammonia. With 82% N this is the most concentrated nitrogen fertilizer; it is a gas at ordinary temperatures and when liquefied must be transported and handled under pressure. The gas combines with soil but, to avoid loss, it must be placed at least 10 cm below the soil surface for grass and 15 cm for arable crops. The special equipment needed to transport and apply ammonia to soil is costly and is justified only for very large farms (of 200 ha or more in British conditions), or groups of farms, or for use by a contractor.

Aqueous ammonia (21–29% N) is made by dissolving ammonia in water. It is not under pressure and is easier for farmers to handle, but it must be placed below the soil surface to avoid loss of ammonia. Aqueous ammonia, properly used, is as effective as other sources of N, but applied on the surface, or worked shallowly into a seedbed, much of the ammonia may be lost.

Nitrogen solutions are made from ammonium nitrate, urea and ammonia. 'Low pressure solutions' have 30–40% N and usually contain ammonium nitrate, urea and ammonia; they are much easier to handle than anhydrous ammonia. Solutions for direct application that are 'under no pressure' are made mainly from ammonium nitrate and urea (sometimes calcium or sodium nitrates are also added), they usually contain less than 30% N. Often herbicides are mixed with 'no pressure' nitrogen solutions applied to cereals or grass. The two materials seem to be just as effective as when applied separately.

Gas liquor. This was the first liquid nitrogen fertilizer used on a considerable scale in Britain. It is produced when purifying the crude gas from retorts in which coal is distilled. Liquors vary in composition according to the type of purification plant used; they are dilute solutions of various ammonium compounds, together with phenols and tarry matter. Those used on the land contain from 1 to 4% nitrogen, usually as ammonium carbonate or ammonium chloride. Gas liquor concentrations are often given in Britain in terms of 'ounces', but it is better to use the percentage of nitrogen. ('10 oz liquor' (which is often supplied) contains 1.77% N.) 'Scorch' and a check to growth are often caused when liquor is applied to a growing crop. A further disadvantage is that ammonia may be lost by evaporation before the liquor can soak into the soil. By applying gas liquor when there is little growth and under cold and damp conditions, damage due to scorching of cereals and grass may be minimised; trailing tubes fitted to the spray booms reduce contact of liquor with foliage. In experiments liquor used for 'top-dressing' cereals or grass was only about two-thirds as effective as the same amount of nitrogen applied as ammonium sulphate. By applying liquor to arable land and working it in just before sowing, better results can be expected and there is little risk of injury to germination, or of 'scorch'. For kale and sugar beet gas liquor can give as good results as equivalent nitrogen applied by ammonium sulphate, provided that it is worked into the soil to prevent loss by evaporation. Gas liquor can be a useful source of nitrogen if it does not cost too much and if it is used so that it causes no damage to established crops.

Comparisons of liquid fertilizers supplying nitrogen

All liquids under pressure must be injected into the soil before planting, or be applied as side-dressings to established row crops; if placed too close to young seedlings they are toxic. Only non-pressure solution are suitable for applying on the surface, and they may scorch establishe crops and grass. Ammonia, even in dilute aqueous solutions, is unsuitabl for top-dressing grassland, it damages the crop and retards growth, an some ammonia is lost to the air.

Arable crops. Many experiments have been made in Britain to tes anhydrous and aqueous ammonia and 'non-pressure' solutions for arabl crops. In comparisons with solid fertilizers these liquids have either bee injected into soil or sprayed on the surface (non-pressure solutions only for root crops and cereals. Advantage has been taken of the suitability o liquids for deep placement where this was desired, or for injecting besid the rows of crops like sugar beet and kale. Injected liquids must b applied so that no ammonia is lost; liquids sprayed on growing crop should not cause serious 'scorch', nor should they lose ammonia from an urea they may contain. When these conditions have been satisfied, whicl is not difficult to do, liquid fertilizers, and solids applied in conventiona ways have given similar yields when both forms supply equal amounts o N in the same chemical forms. For example anhydrous ammonia has bee as efficient as ammonium nitrate when injected into seedbeds for suga beet or barley, or injected in spring into winter-sown wheat. Aqueou ammonia injected has been equally good, and is easier to inject. Solution of urea injected into barley seedbeds have given yields as good or bette than those from solid nitrogen fertilizers. Solutions containing urea havl often been less efficient when sprayed over land because some ammoni was lost to the air as the urea decomposed near the soil surface. A goo way of making urea a fully efficient nitrogen fertilizer is to inject it a a solution well below the soil surface (but not in contact with seeds).

Grassland. Non-pressure solutions of solid fertilizers sprayed on grass-land are usually as effective as the same materials applied as solids. (Liquids containing much urea may be less effective than ammonium nitrate solutions if ammonia is lost to the air, just as with solid urea used as top-dressing.) No experimenter has resolved the problems of con-sistently injecting anhydrous ammonia into grass swards without loss when using machines suitable for normal farm work. Small scale experi-ments with hand injectors have shown that when ammonia is retained by grassland soils it can be as efficient as top-dressings of solid N-fertilizers but that the grass receiving ammonia grows more slowly. In most com-parisons that have been made, yields of grass have been larger with aqueous than with anhydrous ammonia; it is easier to inject aqueous ammonia without loss. Aqueous ammonia has given as large or larger yields of grass than ammonium nitrate applied at the same time. In recent British experiments aqueous ammonia injected in winter has been more effective than March dressings. Large dressings of aqueous ammonia (250 kg N/ha or more) when injected into grass swards have increased the growth of grass throughout the summer. But to get even production of grass through the year needs more than one dressing; it is usually more difficult to inject ammonia satisfactorily in summer than it is when soils are moister in winter and spring. A good combination on soils which allow liquids to be injected satisfactorily is to inject a large dressing of ammonia in winter or early spring and to apply two later dressings of N during the summer as solid top-dressings or as sprays of a solution. It is worthwhile for farmers to own their own injection equipment if they inject 200 hectares or more of arable land each year, or rather less grassland. Some practical aspects of using liquid fertilizers are discussed in Chapter 18.

Experiments with anhydrous ammonia in Western Europe. An important review has been published (ref. 22) of the experimental work done with anhydrous ammonia in Europe. In experiments on spring-applied ammonia for potatoes, cereals, sugar beet and fodder root crops ammonia has been just as effective as solid nitrogen fertilizer. Because ammonia injection is slower than applying solids, tests have been made of winter dressings that extend the application season. The efficiency of autumn dressings depends on whether winters are dry or wet as this determines leaching. To be successful autumn injection needs cold (average 2°C or less) and dry (less than 150 mm of rain in winter) weather afterwards. Most winters in Western Europe are warmer and wetter than this. Less experiments have been done on grassland; all have shown that the major difficulty is to inject deeply enough to prevent loss of ammonia. Splitting the two amounts of ammonia injected has given larger yields than injecting it all at once, but split dressings of solid N-fertilizers have given larger yields of grass.

Practical advantages and disadvantages of liquid fertilizers have been shown in field experiments. For example one series made by Rothamsted staff used a combined-drilled PK compound fertilizer for barley and tested a broadcast dressing of solid N fertilizer against injected solutions of aqueous ammonia and of urea. The injected liquids gave larger yields than the broadcast solid nitrogen fertilizer. This combination was also superior to combine-drilling the usual 'High-N' compound fertilizer because the double amount of this checked early growth. A liquid fertilizer with roughly a 2–1–1 plant food ratio was also tested; it contained urea and diammonium phosphate. The liquid was less effective than the solid 2–1–1 compound, and killed more plants when drilled; probably when sprayed some ammonia was lost from the surface. The differences in yield found in this experiment from the same amounts of N, P and K, but applied in different ways and forms, amount to nearly 500 kg/ha of grain; they are sufficient to encourage farmers to think hard about the best form of N and the best method of application for their conditions.

Practical difficulties in applying anhydrous ammonia remain to be solved, it must be injected at least 15 cm deep. Injecting liquids deeply needs strong and expensive applicators and a powerful tractor. Figures given in 1968 (ref. 23) quoted in Table 14, show how the real cost of a nitrogen fertilizer applied to a field is made of the cost of the material and the cost of applying it. The cheapest material to buy (in this example anhydrous ammonia) is made expensive when application is costly. New machines will have to be developed, or existing machines used in

different ways, before the cheapness of anhydrous ammonia can become
an advantage to all farmers.

Prices. Some liquid fertilizers which supply nitrogen in Britain are
listed in Table 12. It is difficult to make fair comparisons of prices of
these materials with prices of solid fertilizers supplying N. It is often
assumed that there are constant costs for applying solids, but this is not
really true; to apply 100 kg N/ha as ammonium nitrate involves moving
and applying only half the total weight involved in using sodium nitrate
to supply the same amount of N. Total costs of liquid fertilizers may
involve larger costs for transport in special containers, and larger costs
for buying and using applicators that inject liquids, and which work
more slowly than spreaders used for solids. When valuing liquid fertilizers
by price it seems most relevant to compare *total cost* of a given amount
of N when applied in the field as is done in Table 14.

TABLE 14

APPROXIMATE COSTS OF APPLYING 250 kg N/ha IN DECEMBER 1967

	%N	Cost of fertilizer £	Cost of application per hectare £	Total cost £
'Nitro-Chalk'*	21	19.40	1.80	21.20
'Nitram'*	34	17.00	1.20	18.20
Aqueous ammonia	29	12.00	2.20	14.20
Anhydrous ammonia	82	11.80	4.30	16.10
Liquid N	24	19.80	2.30	22.10

*'Nitram' is straight ammonium nitrate; 'Nitro-Chalk' is a mixture of
ammonium nitrate with limestone (from June 1971, 'Nitro-Chalk'
contained 25%N).

SLOW NITROGEN FERTILIZERS

All of the common nitrogen fertilizers used in agriculture are water-
soluble. Nitrate applied as such, or formed from both ammonium salts
and urea, is liable to be lost by denitrification or leaching. Insoluble
sources that release N either by dissolving slowly, or by being decomposed
slowly by biological processes, have for long been proposed as potentially
more efficient alternatives. But all such materials are more expensive per
kg of N than inorganic N-fertilizers and experiments on agricultural crops
have not shown the much greater efficiency needed to justify the extra
cost. They are briefly discussed here as some are used in horticulture.

Experiments have been done with urea-formaldehyde resins (used in
horticulture in U.S.A), casein and formalised casein (wastes from the
plastics industry) and IBDU (isobutylidene diurea). The last has been
the best in Rothamsted experiments; all the N it contains ultimately
becomes available and it has the advantage that its speed of action can
be controlled by varying granule size. In grass experiments powdered
IBDU and small granules acted as quickly as ammonium nitrate; large

granules were slower and produced less in the first year but more in the second year when residual effects were measured. For the two years together ammonium nitrate gave greater total uptake of N in tests on clay soil. On sandy soil where leaching was more intense, IBDU was inferior to ammonium nitrate on the total of the whole year although large granules gave larger yields at the later cuts. It seems that slow action in a first year, or in the early part of a growing season, is not adequately compensated by prolonged residual effects. In a barley experiment done in a year when spring leaching was severe, powdered IBDU did not provide sufficient N early in the year to give as large grain yields as equivalent ammonium nitrate applied at the same time. Granular IBDU was even slower.

Sulphur-coated urea (SCU) is the latest material offered as a slow-acting fertilizer. It is hoped that the sulphur coating will slow the solution of the urea in soil moisture and that the sulphur may check the action of enzymes and microorganisms which convert urea to nitrate. In experiments described in the Rothamsted Reports for 1973 and 1974 it appeared that the sulphur did check the rate at which urea was released from the granules but it had no subsequent effect on the rate at which the urea was converted to ammonia and then nitrified. In pot experiments SCU did not seem to be genuinely slow-acting as most of the N it supplied was recovered by the first two of nine cuts of grass. In a field experiment nitrate formed from SCU was leached more slowly than nitrate formed from ammonium sulphate but its residual effect was similar to that of urea and both were inferior to ammonium sulphate and calcium nitrate. The conclusion was drawn that SCU had no advantage over ordinary fertilizers and that it shared with uncoated urea a liability to lose part of its nitrogen to the air.

The British experiments done show no justification for introducing slow-acting N fertilizers of the types now available. Better results have been obtained with ordinary inorganic fertilizers *used at correct times and rates*. The slow-acting materials have the disadvantage that, once applied, the supply of N in the soil cannot be controlled and will inevitably depend to some extent on weather. The only advantage that insoluble fertilizers have is that when too much N is applied at one time to sensitive crops, they are much less likely to damage the crops than are soluble forms of N. Because 'organics' cost so much more than inorganic fertilizers it is much more economic to learn to use ordinary fertilizers safely and efficiently than to pay such a large premium for insurance against carelessness.

CHAPTER FIVE

Phosphate Fertilizers

Plants take up their phosphorus mainly as orthophosphate, (the ions are $H_2PO_4^-$, HPO_4^{2-} and PO_4^{3-} — the amounts of each ion depends on the acidity of the solution, but PO_4^{3-} is only significant in very alkaline solutions). There is some evidence that plant roots may absorb unionised compounds of phosphorus that are not orthophosphates and it is possible that they may be incorporated into fertilizers in future. The common fertilizers now used are described in the following sections. Examples are in Table 11.

FORMS OF PHOSPHATE FERTILIZERS

Superphosphate ('super')

This was the first chemically manufactured phosphate fertilizer, introduced first in the 1840s. Present-day products usually contain 8–9½% P, (18–22% P_2O_5) which is soluble in water; it is made by treating ground rock phosphate with sulphuric acid. Besides the water-soluble monocalcium phosphate which supplies the plant nutrient, ordinary superphosphate contains calcium sulphate (or gypsum) as a residue from the reaction between the rock and acid. This substance is of little value on most British soils but the sulphur is useful where this nutrient is deficient. For example in New Zealand where ordinary super is still the most important P fertilizer the sulphur supplied is often essential. Super is suitable for all crops. For arable crops it should be given just before sowing or at sowing-time *if* it is intended to have a direct effect on the next crop. Top-dressings given to a growing crop will be of little value to that crop. For grassland, superphosphate should be applied before growth starts in spring.

Triple superphosphate

'Triple' (also called 'double' or 'concentrated') superphosphate contains about 20% P (46–47% P_2O_5) soluble in water. It is made by treating rock phosphate with phosphoric acid; it is essentially monocalcium phosphate and it differs from ordinary super in containing no calcium sulphate. Triple super is two-and-a-half times as concentrated as ordinary super, and so it must be applied at correspondingly smaller rates. It supplies no sulphur. Apart from this, both triple and ordinary super may be used for the same purposes. Both products are usually bought in granular forms that are easier to handle than powders.

Ammonium phosphates

These compounds also provide water-soluble phosphate (and some nitrogen). In many countries, including Britain, they are the forms of phosphate made by the most modern plants and because they are so concentrated they have displaced ordinary super and, to some extent, triple superphosphate. There are no recent statistics for the forms of phosphate used in compound fertilizers in U.K. (they incorporate 98% of the water-soluble phosphate used here) but figures for U.S.A. for 1967–68 (ref. 24) and for the World for the same year (ref. 18) are:

	% of total P supplied in	
	U.S.A.	World
Ordinary superphosphate	20	37
'Triple' (or 'concentrated') superphosphate	30	15
Ammonium phosphates	35	–
Basic slag	–	10
Other forms	15	38

1968 was the first year in which ammonium phosphates dominated other forms in U.S.A.

Ammonium phosphates are made by reacting ammonia and phosphoric acid; two compounds are important in fertilizers:

Monoammonium phosphate (often abbreviated to MAP) contains 12% N and 26% P (62% P_2O_5) when pure; the salt is completely soluble in water and, like diammonium phosphate, it acts very quickly. Fertilizer-grade monoammonium phosphates containing 11% N and 21% P (48% P_2O_5), or 12.5% N and 22% P (51% P_2O_5), have been marketed in Britain. Most MAP is used to make concentrated compound fertilizer.

Diammonium phosphate (often abbreviated to DAP) contains about 21% of N and about 23% P (54% P_2O_5), the whole salt is completely soluble in water. Diammonium phosphate is not sold as a straight fertilizer in Britain, but is used to make compound fertilizers with large percentages of plant food. The materials often used are intermediate in composition between mono- and di-ammonium phosphate. Grades of DAP commonly used in U.S.A. contain 18% N, 20% P (46% P_2O_5) and 16% N, 21% P (48% P_2O_5), these are mainly diammonium phosphate. *Ammonium phosphate-sulphate* with 16% N and 8.6% P (20% P_2O_5) is made in U.S.A. by neutralising ammonia with a mixture of sulphuric and phosphoric acids.

Ammonium phosphates have become important manufactured intermediates in recent years. They are sold as stable powders or in very small granules which are used by other manufacturers to make compound fertilizers; some is exported. In U.S.A. granulated ammonium phosphates are sold for bulk blending (Chapter 10).

Dicalcium phosphate

This fertilizer is listed in the Schedules to the British Fertiliser and Feeding Stuffs Act Regulations. Little is sold in Britain as a fertiliser, although it is used in some other European countries. This substance contains 23% P but the usual fertilizer grade is only 17% P (40% P_2O_5). Its water solubility is very small, but in good quality samples of dicalcium phosphate practically all the P_2O_5 is soluble in citric acid and citrate solutions.

Dicalcium phosphate is a fine dusty powder that is difficult to distribute by machines, and it is not likely to become a popular fertilizer. Nevertheless it has an indirect interest as it is a major constituent in compound fertilizers made by adding ammonia solution to superphosphate (to make *ammoniated superphosphate*) or by treating rock phosphate with nitric acid to make *nitrophosphates*. Both these processes are used in other countries, and they may be used in Britain as a result of future industrial developments, or the fertilizers may be imported. Although finely powdered dicalcium phosphate is a good fertilizer which may act as quickly and efficiently as water-soluble phosphate, when incorporated in a granulated compound fertilizer its rate of action may be slowed down, so that it is less effective for crops that need a rapid start. The value of the phosphate ions depends as much on the ease with which they leave the granule and enter the plant root, as on their chemical solubility.

Ammoniation of superphosphate (or of solutions resulting from treatment of rock phosphate with nitric acid (to make nitrophosphates)) has the advantage that cheap nitrogen (as ammonia) is introduced. But the real savings that are possible with these processes are limited. If too much ammonia is added all the water-soluble P may be precipitated in forms that act too slowly for quick-growing crops. It is now recognised in most countries where much research has been done with ammoniated superphosphates and nitrophosphates that if these products are to perform, for practical purposes, as well as an equivalent amount of fully water-soluble phosphate, that they must retain about half of their P in water-soluble forms.

Basic slag

These fertilizers (sometimes called Thomas slag in other countries) are by-products of steel-making. Most of the slags sold in Britain have from $4\frac{1}{2}$ to $9\frac{1}{2}$% P (10 to 22% P_2O_5). They contain complex phosphates that are not soluble in water, but in the soil they break down and liberate phosphate. As they are insoluble in water, basic slags must be finely ground so that the phosphate is readily exposed to attack by soil moisture. Coarser products may not act quickly enough; generally 80% or more of the slag should pass a 100-mesh sieve. The value of slags for crops also depends on the quality of the phosphate present, and this is tested by the percentage solubility of the P in a solution of citric acid. If a basic slag is to be *fully* effective 80% or more of all the P it contains should be soluble in the citric acid test. Basic slags with medium solubilities in citric acid (say 50 to 60%) are less useful than those with higher solubilities; slags with small solubilities (40% or less) are of much less value.

The fineness that ensures that basic slags act quickly on crops is a disadvantage in spreading as they do not flow well through some distributors and are unpleasant to apply in windy weather. These difficulties have been overcome in a recent British process where the fine powders are aggregated into quite small granules ('mini-granules' of pin-head size). The granular products are not dusty, they are easy and convenient to spread by machine and do not drift in wind. It seems that the granules break down in soil moisture to the fine powders that release their phosphate ions quickly.

Basic slags, besides being phosphate fertilizers, are also effective liming materials, and 1 tonne of slag may be considered as roughly equivalent in liming value to two-thirds as much ground limestone. Besides having 25–35% Ca, slags contain other plant nutrients, but they are present in varying amounts. These originate in the iron ore which also contains the phosphate that ends in slag. Magnesium is present in most basic slags, some of this is available to crops. Slags also contain a range of micro-nutrients and traces of other elements. The subject of magnesium and trace elements in slags is discussed fully in a recent publication by N.A.A.S. (ref. 25) and is referred to in Chapters 6 and 9. The N.A.A.S. conclude that basic slag may be a useful source of some trace elements. At least some of the magnesium and manganese is available to plants. Too few experiments have been done for firm conclusions on the value of other trace constituents (iron, copper, cobalt, molybdenum, zinc and boron).

Slags are of most use for crops with long growing seasons, and they are less useful for crops that need a stimulus to make rapid growth early in the season. Water-soluble phosphates are best for this. Basic slags have a good reputation for use on grassland and (together with potassium fertilizers where necessary) for improving the growth of clover; slag should be applied to grass in winter, so that it is washed into the soil by rain before growth begins in spring. Some compound fertilizers are made by mixing basic slags and potassium salts; they are especially suitable for grassland. In purchasing basic slags farmers should be guided by

    unit price of P_2O_5 (calculated from the analysis and the
        price)
    solubility in citric acid
    fineness

Generally basic slags are cheaper per unit of P_2O_5 than water-soluble phosphates (Table 12) and some slags offered are cheaper than others. Allowance should be made for the liming value of basic slag; this may be considerable when the lower-grade materials are used.

Ground rock phosphates

These are the raw phosphates as mined, ground finely for direct application to the land. The best rock phosphates used for this purpose in Europe come from North Africa and contain about $12\frac{1}{2}\%$ P (29% P_2O_5). They are quite insoluble in water and there is no satisfactory chemical test for valuing them. To be effective they must be finely ground. The current British Fertiliser Regulations require that the seller should state the total content of P_2O_5 and the fineness of rock phosphate. Generally, rock phosphates are satisfactory if 90% or more of the product passes through the 100-mesh British Standard Sieve. British experiments have shown no gain from grinding more finely than this. Rock phosphates are attacked by acid soils and slowly become available to crops. They do not dissolve in neutral soils and should not be applied where the soil is pH 6.5 or higher. They are at their best when used on moist acid soils for crops of the brassica family. Generally, rock phosphates should only be given to swedes, turnips, rape, kale and other fodder crops grown on acid soils in areas with annual rainfall of 750 mm or more. In these areas they can also be used for grassland on acid soils, but, as they take time to act, they should be applied in autumn or winter so that they are washed into the soil by rain. Where rock phosphates can be used satisfactorily they are good fertilizers, *if* they can be bought *much* cheaper per unit of P_2O_5 than water-soluble phosphates.

Some rock phosphates are very satisfactory fertilizers for perennial crops in warm and wet areas such as the humid tropics. They dissolve satisfactorily in the well-leached acid soils that are common in these areas; they may act for much longer on soils where water-soluble phosphates are 'fixed' to form very insoluble compounds (or under the less common conditions where soluble phosphate is leached from top soil). For example rock phosphate mined from Christmas Island in the Indian Ocean is much used in Malaysia for tree crops, and particularly for rubber and oil palm. In experiments it has been as efficient per unit of P as more soluble phosphates. New laboratory methods of valuing rock phosphate have been described by T.V.A. (ref. 117).

COMPARISONS OF PHOSPHATE FERTILIZERS

The current British Fertiliser Regulations (ref. 17) distinguish three kinds of phosphates in fertilizers:

> Water-soluble P_2O_5
> Citric acid-soluble P_2O_5
> and Insoluble P_2O_5.

Regulations

The total percentage of P_2O_5 should be stated for mineral rock phosphates. For superphosphate and ammonium phosphate and compound fertilizers based on these materials the percentage of P_2O_5 soluble in water must be stated; in addition, the percentage of P_2O_5 insoluble in water must be given for compound fertilizers but not for superphosphate. For basic slag both total P_2O_5 and P_2O_5 soluble in a solution of citric acid must be stated. For dicalcium phosphate, P_2O_5 soluble in citric acid solution must be given. The citric acid test applies to basic slag when it is mixed with potassium chloride or sulphate. Particulars which must be stated for "potassic basic slag" are total P_2O_5, total P_2O_5 soluble in 2% citric acid, amount of K_2O, and amount of slag passing a 0.5 mm sieve.

Chemical forms of phosphate

Water-soluble phosphate has the same high value whatever the source. It should be used when a crop requires a quick start or when the fertilizer is drilled in bands with or near the seed. *Citric-acid-soluble phosphate* is best used when rapid action is not so necessary and when the fertilizer can be applied well before growth starts. Basic slags are at their best on slightly acid and acid soils. *Insoluble phosphate* in compound fertilizers, in superphosphate, and in rock phosphate should not be valued too highly: it acts slowly and is fully effective only on acid soils. That part of the total P_2O_5 in most British compound fertilizers which is insoluble in water is largely unchanged rock phosphate, and it should not be valued more highly than straight rock phosphate.

While there are differences between the immediate effects of different kinds of phosphate fertilizers, and particularly between water-soluble and insoluble P_2O_5, the *residual effect* of a given quantity of P_2O_5 tends to be the same in the years following the application whether the original fertilizer was water-soluble or citric-acid-soluble. Rock phosphate has sometimes given as good results on acid soils as superphosphate in the second and third years after the dressings were applied; sometimes its residual effect has been much less than those of super or basic slag.

In countries where nitrophosphates and ammoniated superphosphates are used widely chemical tests of the quality of the phosphate are made. The solvents usually used are either neutral or alkaline solutions of ammonium citrate and to be satisfactory a large percentage of the total P should be soluble in the reagent. The neutral citrate test is used in U.S.A. The alkaline solution used in some European countries seems a more stringent test of phosphate quality. It is to be preferred to the neutral citrate test when valuing fertilizers containing much P that is not soluble in water such as ammoniated superphosphate and nitrophosphates.

POSSIBLE NEW DEVELOPMENTS WITH PHOSPHATES

Much research has been done recently to make even more concentrated sources of phosphorus. In U.S.A. processes have been developed which lead to 'superphosphoric acid' with 33% P—nearly 50% more than ordinary phosphoric acid (23.5% P). Superphosphoric acid contains some orthophosphoric-acid but also pyrophosphoric, metaphosphoric and polyphosphoric acids; neutralising it with ammonia gives ammonium 'polyphosphate' which has 15% N and 25% P (58% P_2O_5) and is more concentrated than ordinary ammonium phosphate, it is very suitable for making liquid fertilizers. Polyphosphates keep micronutrients in soluble forms and it has been suggested that the small quantities crops need should be mixed with polyphosphates before adding them to compound fertilizers. The non-orthophosphate forms of P in polyphosphates are hydrolysed in soil to orthophosphates and most experiments show they have as good an effect on crops as the same weight of P applied as a water-soluble orthophosphate.

Potassium metaphosphate has been tested in Britain and other countries. The British material contained 25% P and 31% K (58% P_2O_5, 37% K_2O). It is very concentrated and contains no useless ions (such as the chloride in muriate of potash). Being nearly insoluble in water this salt does not damage germinating seeds and the potassium is leached from very sandy soils much slower than the K of potassium chloride. The metaphosphate is hydrolysed in soil and then both P and K are as effective as water-soluble forms of these nutrients.

Magnesium ammonium phosphate ($MgNH_4PO_4$) contains 8% N, 14% Mg and 17% P (40% P_2O_5). It is only very slightly soluble in water and is safe when applied to young plants with tender root systems; all three nutrients become soluble quickly enough to be fully effective. This fertilizer has been used in U.S.A. on a small scale for valuable horticultural crops.

CHAPTER SIX

Potassium, Magnesium and Sodium Fertilizers

These three elements, and calcium, are taken up by plants as positively charged ions (cations). The first two (and calcium) are essential nutrients for all crops; sodium seems essential only to a few kinds of plants.

POTASSIUM FERTILIZERS

Potassium fertilizers are all obtained by mining from the extensive deposits that occur in many countries. (A mine to extract potassium salts found deep underground in Yorkshire is now being sunk.) Most common is potassium chloride (KCl) sold as 'muriate of potash' both for direct use and for making other K fertilizers. Potassium sulphate (K_2SO_4) and potassium nitrate are made from the chloride and so cost more; both are preferred in horticulture and for potatoes and tobacco. Lists of K fertilizers now available are in Tables 11, 12 and 15.

Muriate of potash

Potassium chloride sold for direct application in Britain generally contains 50% K (60% K_2O); occasionally a grade containing 50% K_2O is offered. Potassium fertilizers used to be difficult to handle and spread by machine, as they caked in bags and were hygroscopic. Over 10 years ago 'free-flowing' potash fertilizers were introduced which stored well and flowed more easily through distributors. Granulated forms of potassium chloride are now available.

Sulphate of potash

Potassium sulphate contains 40 to 42% K (48 to 50% K_2O) and, as Table 12 shows, is more expensive than the muriate per kg of K_2O.

Kainit and crude potash salts

These contain from 10–25% K (12 to 30% of K_2O) and they should be bought on their potassium content. These fertilizers often contain sodium salts which make them very useful for sugar beet and mangolds; some kinds also contain magnesium salts. Regulations introduced in

Britain in 1973 under the Fertilisers and Feeding Stuffs Act (ref. 17) give new definitions of two kinds of kainit:

Kainit = 'mineral potassium salt with or without magnesium (Mg)'
Magnesium kainit = 'mineral potassium salt containing at least 3.6% of magnesium (Mg)'

Nitrate of potash or saltpetre

Potassium nitrate contains about 13% of N and 36% K (44% of K_2O). It should not be confused with 'Chilean potash nitrate', which contains a little more nitrogen but much less potash. Nitrate of potash is rather expensive and most of it is used by horticulturists.

TABLE 15

POTASSIUM, MAGNESIUM AND SODIUM FERTILIZERS

POTASSIUM FERTILIZERS

	%K	%K_2O
Muriate of potash	50	60
Sulphate of potash	42	50
Sulphate of potash-magnesium	23.5	28
Kainit*	10–25	12–30
Potassium nitrate	37	44
Potassium metaphosphate	31	37

*With less than 3.6% Mg.

MAGNESIUM FERTILIZERS

	%Mg
Magnesian limestone†	3.1–12
Blast furnace slag	1–6
Ground burnt magnesian lime†	5.6–20
Magnesium sulphate ($MgSO_4.7H_2O$, Epsom salts)	9.7
Calcined magnesium sulphate	18.6
Kieserite ($MgSO_4.H_2O$)	16
Sulphate of potash magnesia	6.5
Magnesite	27
Calcined magnesite	52
Magnesium kainit†	3.6 or more
Examples: magnesia sylvinite kainit	13% K_2O, 4.0% Mg, 13% Na
'Magna Kali'	6–8% K_2O, 11–14% Mg, 4% Na

†The minimum percentages shown here are defined by the 1973 British Regulations (ref. 17).

SODIUM FERTILIZERS

	%Na
Sodium chloride ('agricultural salt')	37–39
Sodium nitrate	27

Other potassium fertilizers

Sulphate-of-potash-magnesia is a mixture of magnesium and potassium sulphates. It contains 21 to 25% K, 25 to 30% of K_2O and $5\frac{1}{2}$ to $7\frac{1}{2}\%$ of magnesium (Mg).

Other sources of potash are only important locally. Wood ashes may contain from 2 to 5% of K. Seaweed ashes or kelp contain considerably larger percentages of potassium. Good dressings of seaweed supply much K. Flue dusts from iron and steel manufacturing processes, and from cement-making, are occasionally sold. They contain from 4 to 12% K, but impurities such as cyanides, sulpho-cyanides and sulphides, which are poisonous to plants, are sometimes present; these toxic substances are destroyed quite rapidly in the soil and it is best to apply flue dust to land a few weeks before sowing a crop.

Comparisons of potassium fertilizers

All potassium fertilizers are water soluble and the K they supply is of equal value; there are no problems corresponding to availability problems of water-insoluble phosphate or differences between forms of N-fertilizers. Differences in agricultural value of K fertilizers depend on the effect of the ion accompanying the potassium and on the value of any sodium or magnesium they contain.

All potassium fertilizers are sold by their content of water-soluble K_2O. So far as the K is concerned the effect of one kg of K_2O in all K-fertilizers will be similar, and normally the cheapest should be used. For special purposes the other materials present may influence the question of which to buy. Many horticulturists, and some potato growers, prefer sulphate of potash to the muriate. Heavy dressings of muriate may lead to high chloride concentrations in the soil, and these can harm some glasshouse crops. In Britain there is no case for *generally* using sulphate of potash instead of the muriate in potato growing, as the two forms of potash have given similar yields, unless the potatoes are grown for special purposes, and higher prices are paid. Sulphate usually gives more, but smaller tubers than the chloride. It is therefore used when growing potatoes for canning where small tubers are needed. Usually special varieties are grown and they are planted closer together, again to encourage many small potatoes to be formed. In Dutch experiments the sulphate gave potatoes containing more dry matter and starch, and of better cooking quality. In recent Scottish experiments, sulphate also gave potatoes containing more dry matter with larger specific gravity that were mealier when cooked. But differences between the quality of potatoes from different sites were very great irrespective of fertilizing. Potassium bicarbonate was also tested in these experiments but the other salts generally gave more yield, so it seems that avoiding the presence of the usual anions did not benefit the plants.

The burning quality of tobacco leaves is greatly damaged when the plants are permitted to take up chloride; muriate of potash or low-grade salts should never be used for this crop; potassium sulphate is widely used, but the nitrate is suitable too when the crop also needs N.

The other cations present in low-grade potash fertilizers are useful for some crops. Sugar beet and mangolds benefit from sodium salts, and these can be supplied by using low-grade potash salts or kainit to provide both potash and sodium. Magnesium is needed on some light soils, particularly in horticulture. This plant food is provided by sulphate of potash magnesia and by some low-grade potash fertilizers such as 'magnesium kainit'.

Potassium is not leached from medium and heavy soils unless far more than is needed is applied for many years. Even on light soils K is only leached from very coarse sands practically devoid of clay. Under these conditions potassium metaphosphate, which is only slowly soluble in water, has been more efficient. This fertilizer (described in Chapter 5) contains no anion that can damage crops; it is worth a trial in wet areas and on light soils where leaching may occur and also in glasshouses where winter flooding is used to remove soluble salts.

MAGNESIUM FERTILIZERS

The amounts of magnesium fertilizers used are increasing, partly because deficiencies in arable crops on light soils are becoming more common, partly because small contents of magnesium in herbage are one cause of grass tetany in animals. Heavy soils are rarely so deficient in magnesium that crop yields are lessened. Estimates of the Mg removed annually by leaching vary between 2 and 30 kg Mg/ha, average crops remove about 10 kg Mg/ha. The total amount lost each year seems to be readily replaced by weathering of the magnesium-containing minerals in the clay of most medium and heavy soils, which may have from 0.2 to 0.8% of total Mg in the topsoil. Sandy soils contain much less magnesium and the rate of release from minerals may be too slow to replace losses. The sandy loam soil at Woburn Experimental Station in Bedfordshire is liable to become deficient in magnesium. Losses of magnesium during the 80-year period from 1888 to 1968 have been followed by analysing the soils. The annual rate of loss has been 4.5 kg/ha of Mg from the top 22 cm of soil. As the rain has supplied about as much Mg as crops have removed, it seems that this light soil has no reserves of magnesium in minerals that can be released and some Mg fertilizer must be used.

The effects of losses of cations by leaching are more serious on light than on heavy soils. When much fertilizers are used, light soils quickly become acid because calcium is lost; but the loss of magnesium which occurs at the same time is often ignored. If the acidity is corrected by liming materials which contain little or no magnesium, the balance of calcium and magnesium in the soil is disturbed further and Mg-deficiency in crops becomes more likely. Because magnesium is not regularly needed as a fertilizer on most land it tends to be ignored; some details are given below of the sources which supply it.

Sources of magnesium

Magnesium is not generally applied as a direct fertilizer in temperate countries but the quantities in other manures are often sufficient to prevent deficiency. Amounts in common liming materials and fertilizers are in Table 15.

Farmyard manure (FYM) made by animals using bought-in feeds provides a net gain of magnesium to a farm, particularly important on arable farms where pigs or poultry are kept. The richest sources are protein cakes from oil seeds with 0.3–0.8 per cent Mg; bran is rich, legume hays are also with 0.3–0.6 per cent Mg, but grass hays usually only half as much. Cereal grains are poor with only 0.09–0.13 per cent Mg. Because the materials used to make farmyard manures vary so much the compositions of manures vary from 0.1 to 0.6 per cent Mg in dry matter. The conclusions from a Scottish survey can be taken as average, 10 tonnes of manure supplied about 12 kg of Mg.

Some kainit contains about 7 per cent Mg, but other batches of kainit may contain little or none. This has been recognised in the 1973 British Fertiliser Regulations. The division is put at 3.6% Mg; materials with more are called 'magnesium kainit'; but the actual %Mg in such materials need not be stated (a farmer should, however attempt to discover the actual content from his supplier).

Basic slags contain up to 7 per cent Mg; some continental workers have concluded this magnesium was fully active, others that it had no more value than one-fifth as much Mg supplied by kieserite. Some English basic slags with 5–6 per cent total Mg can supply magnesium to crops but, even when made by the same process, different batches vary. There is no easy way of determining whether magnesium in slags will be useful to crops, for the total content is little guide. The N.A.A.S. survey (ref. 25) suggests that even if the low estimate that one-fifth of the Mg will be immediately available is accepted, then 1.25 t/ha of slag will supply about 10 kg/ha of useful Mg, as much as is removed by 7½ t/ha of hay. Some of the remaining magnesium may become available later.

Concentrated NPK fertilizers contain practically no magnesium unless it has been added. *If* magnesium is added when a fertilizer is being made, the 1973 Regulations require that %Mg should be stated. An example offered in 1974/75 is a 16–8–8 fertilizer containing 4% Mg.

Liming materials also add much Mg to soil. Poorest in magnesium seems to be ground limestone from Buxton in Derbyshire with only 0.1% Mg. Scottish ground limestones have 0.6–1.2% Mg and ground Chalk from Bedfordshire has about 0.5% Mg. With 1% Mg a tonne of liming material will supply 10 kg of Mg; the ordinary practice of liming every few years does much to maintain magnesium supplies. But only liming materials with large concentrations of magnesium such as dolomitic limestones have much effect on a serious deficiency of magnesium or on per cent Mg in crops; these limestones often contain 10–12% Mg, burnt limes made from them have more (18–20% Mg). More information about magnesium supplied by liming materials is in Chapter 7.

Rain supplies some magnesium in the dust and marine salts it carries down. In many countries the annual gain seems to range from 2 to 10 kg Mg/ha, but in industrial areas the deposit may be larger. Most crops take up rather little magnesium so even the small amounts in rain help in replacing that removed.

Magnesium fertilizers and other sources of Mg are listed in Table 15. Comparisons between various sources have been made. There has been little difference between supplying Mg in various ways and decisions on the form to use should be made from prices and on account of convenience. Dolomitic limestone acts more slowly than soluble magnesium salts. In sugar beet experiments similar yields came from 100 kg Mg/ha as kieserite and 275 kg Mg/ha as dolomitic limestone; Mg in kieserite cost 7.4 p/kg, magnesium in dolomitic limestone cost only half as much and it becomes available slowly as the limestone decomposes in acid soil. Magnesium ammonium phosphate (about 8% N, 14% Mg, 40% P_2O_5) was described in Chapter 5; it is sometimes used in horticulture; crops benefit from the Mg supplied by this slightly soluble salt which does not damage sensitive and valuable plants.

Magnesium deficiencies

Deficiencies of magnesium are most likely on light soils in wet areas. The best diagnosis is likely to be from leaf symptoms. Deficient leaves are light coloured; later they become yellow between the veins; in potatoes the interveinal areas become bronzed. Magnesium deficiency is often observed on potatoes, sugar beet and brassica crops, sometimes it does not appear to lessen yields. Fruit trees often show magnesium deficiency. The trouble also occurs in tomato crops grown under glass, often associated with large potassium manuring.

Both *soil analysis* and *leaf analysis* are used to diagnose deficient soils and crops but neither method is certain. Interpretation of the figures depends much on the type of soil, its pH, other fertilizers used and time of year. It seems however that Mg deficiencies are unlikely unless soils contain 30 ppm of exchangeable Mg or less; some chemists rate soils with 25–40 ppm of exchangeable Mg as 'low', those with $<$25 ppm as 'very low'. There is a great difference between the amounts of Mg in soils that are sufficient to prevent yields being depressed by deficiency and the amounts needed to ensure that crops have large Mg concentrations —which may be desirable in animal nutrition. In Scottish experiments 30 ppm of exchangeable soil Mg was sufficient for maximum yield but 160 ppm was needed to ensure that herbage always contained large %Mg. In experiments on sugar beet 600 kg/ha of kieserite increased yields by 390 kg/ha more sugar. The concentrations of Mg in plants that separated deficiency from satisfactory supply were 0.2–0.4% Mg in leaf dry matter and 0.1–0.2% in petiole dry matter.

Magnesium fertilizing in practice

In general farming there is little risk of deficiency where FYM is used in the rotation, or where dolomitic limestone is the liming material. But in intensive cash cropping systems on light soils without organic manure regular dressings of magnesium salts may be needed in future, either straight or in compound fertilizers.

Herbage, or conserved grassland products, sometimes contain too little magnesium to maintain stock in good health. It is relatively easy to make good the deficiency for dairy stock that receive concentrated feeding stuffs, by giving them a mineral supplement containing magnesium. But for stock that graze continuously and have no other food, farmers may have to fertilize to increase per cent Mg in herbage at the times of the year (principally spring) when stock are prone to hypomagnesaemia (small contents of Mg in blood serum). Where stock suffer from grass tetany, usually the trouble is not due solely to a shortage of magnesium in the soil leading to low magnesium in herbage, and to hypomagnesaemia in the animals. Although low magnesium contents in fodder are a contributory factor, most cases of tetany are due to more complicated physiological effects including those from large K and low Na in herbage, from shock and from bad weather. Using fertilizers containing magnesium, or using dolomitic limestone, are *not certain* means of preventing hypomagnesaemia, but they help to do so. If the *extra* cost of materials containing magnesium is not much greater than that of equivalent fertilizers or liming materials containing little or no magnesium, using such products is a worthwhile precaution. Hypomagnesaemia should not occur where herbage contains 0.20% Mg in dry matter or more; some farmers now dress spring herbage before grazing with calcined magnesite (MgO) to ensure that animals take in enough magnesium.

On acid soils that need regular liming the simplest and cheapest way of avoiding Mg-deficiency is by applying dolomitic limestone. But on neutral soils that need no lime, magnesium salts should be used.

In glasshouses 1–2 t/ha of magnesium sulphate (as kieserite, $MgSO_4$, H_2O) may be needed to *prevent* magnesium deficiency in tomatoes, but the trouble can also be *controlled* by spraying 35 kg/ha of Epsom salt, ($MgSO_4,7H_2O$), dissolved in 400 litres of water on several occasions during the season. (Kieserite is cheaper than Epsom salt, but it cannot be used to make a spray as it dissolves only very slowly.)

On arable land that is acutely deficient in magnesium a massive dose of a 'straight' source will be needed initially and the cheapest effective form will be best. Elsewhere the small doses needed to maintain the magnesium status of soil may conveniently be applied mixed with NPK fertilizers.

SODIUM FERTILIZERS

Sodium does not seem to be essential for survival and growth of any agricultural crop, but some crops give larger yields when they have adequate sodium. Other crops benefit from sodium dressings only when they get too little potassium. Sugar beet and mangolds give large returns from sodium even when they have enough potassium, it seems that for these sodium performs certain functions within the plant better than potassium. Sodium is essential for animals; the amounts in forage crops depends on sodium supplies in soil; sodium fertilizers can therefore affect the value of feeding stuffs produced on the farm.

In British experiments on sugar beet done from 1940–49 applying 600 kg/ha of sodium chloride without potassium chloride gave an extra 460 kg/ha of sugar; giving potassium (150 kg/ha of K_2O) without sodium increased yield by only 280 kg of sugar. Giving sodium chloride when adequate potassium chloride was also applied raised yield by 220 kg/ha; but the gain from potassium when sodium was given was only 40 kg/ha of extra sugar. In a later series of experiments 310 kg/ha of potassium chloride raised sugar yield by 450 kg/ha and sodium gave a greater sugar yield than potassium in most experiments. The use of sodium was profitable whether or not potassium was applied, but when sodium was used, potassium was uneconomic.

In Scottish experiments sodium chloride benefited turnips and kale, but not grassland. This salt increased yield of turnips more than potassium chloride did. Yields of kale have been increased by about $2\frac{1}{2}$ t/ha in both Scottish and English tests. Carrots also benefit from sodium fertilizers and 380 kg/ha of sodium chloride has increased yields nearly as much as 190 kg/ha of potassium chloride.

There is a very clear case for supplying sodium to sugar beet, mangolds and carrots. It is worth making local tests to see if it increases yields of brassicae crops. Sodium is supplied very cheaply by 'agricultural salt' (sodium chloride, with 39% Na) and by sodium nitrate (27% Na). Salt has been incorporated in compound fertilizers but this is not popular because it makes manufacturing more difficult and increases the bulk of the fertilizer, (one fertilizer offered with unspecified sodium content, has 7% N, 7% P_2O_5, 10% K_2O). It seems best to apply sodium chloride separately in winter; if necessary the dressings can be ploughed in for sugar beet and carrots to avoid extra work in spring. Sodium fertilizers are thought to have a bad effect on soil structure but British experiments show no permanent damage of this kind. Annual dressings of several kinds of sodium salts have been applied in the Rothamsted Classical Experiments for over 100 years without damaging the soil.

CHAPTER SEVEN

Calcium Fertilizers and Liming Materials

CALCIUM IN CROPS AND SOILS

Crops

Most crops make little demand on the large supplies of calcium in the soil, the amount taken up usually being less than the potassium removed though more than the magnesium uptake. Cereals, grasses and potatoes take up little calcium, legumes grown for forage, sugar beet and kale take much more. A few data for the cations removed in harvested crops are in Table 2.

Soils

In temperate countries calcium dominates exchangeable cations, even in very acid soils. A heavy clay soil in Britain may contain several thousand kilogrammes per hectare of exchangeable calcium even when it has slightly acid reaction, a sandy loam contains perhaps half as much. In contrast exchangeable potassium rarely amounts to more than a few hundred kilogrammes per hectare. Soils have no mechanism for conserving (i.e. 'fixing') surplus calcium in non-exchangeable but potentially useful forms in the way most can fix potassium and so prevent loss by leaching.

Leaching of bases

Calcium, magnesium, sodium and potassium are retained in soil as positively charged ions which neutralise the negative charges on the colloids of the clay and organic matter fractions. This process protects nutrients from immediate loss but solutions in contact with soil always contain sizeable concentrations of cations, and particularly of calcium. When percolating water moves down the soil it removes the existing soil solution with the cations; but the 'new' soil solution is replenished with Ca^{2+}, Mg^{2+}, Na^+ and K^+ ions from the reserves of exchangeable ions. In this way soil that is leached by rainfall is continually denuded of cations (bases), the loss falling most heavily on the calcium. For the leaching water to remain electrically neutral the cations require equivalent anions, in soil these are chloride, sulphate, nitrate and bicarbonate. Leaching processes are quite complicated. Extra anions entering the soil solution must be neutralised, so the soil solution takes up more cations, again it is principally more calcium. The extra anions may be released from the soil itself (nitrate, bicarbonate and sulphate) or they may be added as fertilizers (nitrate, sulphate, chloride). The rain also contains small concentrations of cations and these may neutralise some of the extra anions. When the exchangeable calcium is replaced by hydrogen ions the soil becomes more acid; but when potassium chloride is added, calcium is exchanged for potassium and is removed, with the chloride in the leaching water; but because there is no net loss of bases, KCl added to soil does not make it acid though the salt increases the loss of calcium.

Wherever the annual rainfall is larger than the water evaporated from crops and soil surface, some water will drain through the soil and calcium will be lost. The actual amount lost depends on amount of rainfall, on supplies of calcium in soil, and on the use of fertilizers. The range of annual losses from agricultural soils is from about 60 kg Ca/ha on very acid sandy soils to 400 kg Ca/ha from soils which have some reserve of calcium carbonate; the average annual loss from British soils is about 150 kg Ca/ha. Table 5 shows average amounts of ions removed annually in drainage from Woburn and Saxmundham Experimental Farms. Saxmundham is a chalky boulder clay soil and exchangeable calcium lost is quickly replaced from natural reserves of calcium carbonate ($CaCO_3$). Woburn soils have no reserve of $CaCO_3$ and they are naturally very acid; they lose less calcium but as much magnesium as Saxmundham soil, the only replacements are from liming materials, manures and fertilizers and the small amounts in rain (Table 3).

Soil acidification

Calcium has a dual function in soil. It is an essential nutrient, but the amounts taken up by most crops are not very large (Table 2). Calcium is also the dominant base and keeps soils neutral in reaction ('sweet'). In soils saturated with calcium, Ca^{2+} ions neutralise most of the negative charges; if calcium ions lost by leaching are not replaced, positively charged hydrogen ions (which are responsible for acidity) take their place and the soil becomes acid ('sour'), i.e. its pH falls below the neutral point (pH 7) and if the process is not stopped by liming, pH may ultimately fall nearly to 4. Soils continue to supply enough calcium as a nutrient for many crops until they became quite acid (pH below 5) but other important changes are caused by the increasing acidity:

  (i) Structure of heavy soils deteriorates.
  (ii) Iron, aluminium and manganese become more soluble and their concentrations in the soil solution may increase until they become toxic to some crops.
  (iii) Phosphates become less soluble.
  (iv) Many soils organisms (eg. insects, worms and some micro-organisms) do not thrive in acid soil, particularly the organisms that convert nitrogen reserves to nitrate and decompose plant remains and mix them

with soil. 'Raw' organic matter accumulates in acid
soils, if they are not disturbed peat may form.

Liming materials correct these conditions because they are the kinds of
calcium (and magnesium) compounds which can neutralise soil acidity
caused by hydrogen ions. Liming is often considered as separate from
fertilizing; but supplying calcium (and magnesium) as their carbonates,
oxides or hydroxides is an essential part of crop nutrition and soil
management. Recent assessments of the loss of lime from British soils have
been published (refs. 104 and 111).

The effects of fertilizers on soil acidity

Ammonium salts displace exchangeable calcium from soil colloids;
this calcium is lost in percolating water, being accompanied by a mobile
anion (bicarbonate, sulphate, chloride and/or nitrate) to maintain electrical neutrality. The ammonium is later nitrified and the nitrate ions formed
neutralise more calcium ions; if nitrate is lost by leaching it removes
more calcium, but if the nitrate is taken up by plants the calcium is
conserved. All ammonium salts behave in this way, but the loss of lime
is greatest when the anion accompanying the ammonium is mobile (usually
sulphate or chloride). In practice using 100 kg of ammonium sulphate is
considered to cause a loss of calcium equivalent to about 100 kg of
calcium carbonate. Ammonium phosphate causes less loss since the phosphate is fixed by the soil (the fertilizer will cause no loss if all the nitrate
formed is taken up by crops). Urea is quickly converted in soil to
ammonium bicarbonate or carbonate, when these are nitrified cations are
needed to neutralise the nitrate; some of the nitrate is inevitably leached,
so using urea continuously can make soil acid, though not as quickly as
ammonium sulphate or chloride does. Even the very alkaline materials,
anhydrous and aqueous ammonia, remove calcium if some of the nitrate
formed from them is leached out.

Potassium fertilizers are often said to have little effect on losses of
calcium; but the large amounts now used provide equivalent amounts of
chloride which are leached from the soil. When the chloride is removed
an equivalent weight of cation must be lost too and using large potassium
dressings increases losses of calcium. Superphosphate is said to have no
effect, though it has an acid reaction. The amounts of calcium in N and
P fertilizers are shown in Table 11. Each tonne of basic slag used
supplies calcium equivalent to about two-thirds of a tonne of calcium
carbonate. Supplying nitrogen as sodium or calcium nitrates conserves
lime in the soil. In light soils that are very acid and have very little
calcium, the amounts added as fertilizers may be quite important in
increasing exchangeable calcium. Ammonium nitrate mixtures with limestone (like the British 'Nitro-Chalk') maintain the calcium supply in
soils.

Farmyard manure and most organic wastes supply some calcium;
poultry manure does not contain enough calcium to counteract the nitrogen
it contains and it tends to acidify soils.

The proportion of total nitrogen fertilizer supplied by ammonium sulphate has diminished greatly in recent years; highly concentrate compound fertilizers usually contain ammonium nitrate and phosphat perhaps urea too. Therefore the *overall* acidifying effect from N ha probably diminished per tonne of fertilizer used. But much more tot fertilizer is now used than formerly; this includes much chloride fro potassium salts and much less calcium which was supplied by super phosphate, now replaced by ammonium phosphate.

Losses of calcium in farm produce

Losses range from 10–20 kg Ca/ha in cereals to 150 kg/ha or mor in a good crop of kale. Legumes remove much Ca, 2 tonnes of clove or lucerne may take from 30–50 kg Ca, 50 tonnes of sugar beet ma remove 80–100 kg Ca. A little is also removed from farms in milk (abou $2\frac{1}{4}$ kg Ca in 1,000 litres). When crops or animals are sold off the farn calcium is lost completely, but where crops are used on the farm muc of the calcium will be returned in farmyard manure, though a proportio will be lost in drainage from sheds. In practice the losses of lime cause by leaching, and by the use of fertilizers that acidify the soil, are muc more important than losses caused by sales of crops and stock.

PRACTICAL LIMING PROGRAMMES

Using regular dressings of lime is a basic principle of good farmin on all soils that have no natural reserve. Even in areas of low rainfa dressings of lime supplying the equivalent of 3–4 t/ha of burnt lim (CaO) or 5–6 t/ha of ground limestone ($CaCO_3$) may be needed onc every five years to correct the losses that occur in intensive arable farm ing through leaching and through using fertilizers, particularly those ric in N and K.

Indications of acidity

On grassland a shortage of lime often leads to a matted and peat layer of unrotted grass debris on the surface of the soil. Absence c clovers is often a sign of lime-deficiency, usually associated with a short age of phosphate. (On certain soils lack of clovers may be due to potas sium deficiency.)

Arable crops such as sugar beet and barley are very sensitive t soil acidity. Often they give the first indication that land is becomin, sour by growing badly or by failing in patches. Other crops that ar affected by sour soils, although less easily, are clover, beans, peas, turnips swedes and wheat. Potatoes, rye and oats are not seriously affected b the degree of acidity usually found in arable land. Certain weeds, such a spurrey or sorrel, indicate that the soil on which they are growing i acid.

A soil test should always be carried out when acidity is suspected usually the results are expressed as pH values. The neutral point is pl 7.0, calcareous soils have values above pH 7, values below pH 7 indicat acid soil. At pH 6.5 all farm crops will succeed, but sugar beet, barle and other sensitive crops may fail at pH 5.5 or below. Many crops are affected at pH 5; at pH 4 soils are very sour and even acid-tolerant crops such as potatoes and oats will fail or grow badly. Soils rich in organic matter, such as peats, usually grow satisfactory crops at smaller pH values than mineral soils do. This is because organic soils have a larger capacity to hold exchangeable bases.

Lime requirements

pH values measure the *intensity* of the acidity and do not indicate the *quantity* of lime needed to bring soils back to the neutral point. To effect the same change in pH value may need twice as much lime on a heavy soil as is needed for a light soil; peats need even heavier dressings to increase their pH values. Measurements of acidity are usually converted into figures for 'lime requirements' (the amounts of lime needed to make soils safe for cropping) by taking account of soil texture and organic matter.

After correcting serious acidity occasional tests should be made and dressings of lime should be given at a suitable stage in the rotation to maintain the soil at a satisfactory pH. Excessive liming without proper advice is wasteful. On some kinds of soils an excess of lime may inactivate trace-elements; excess lime induces manganese deficiency in crops on peat soils. On mineral soils over-liming may cause iron deficiency in fruit trees and the 'heart rot' of sugar beet, which is due to a shortage of boron.

Kinds of liming materials

The materials that are important in liming are listed below.

Calcium oxide (CaO) is sold as *burnt* or *quick lime*.

Calcium hydroxide ($Ca(OH)_2$) is sold as *hydrated lime* or *slaked lime*.

Calcium carbonate ($CaCO_3$) is sold under the names of chalk, ground chalk, screened chalk or ground limestone.

Magnesium oxide (MgO), *magnesium hydroxide* ($Mg(OH)_2$) and *magnesium carbonate* ($MgCO_3$) are also sold together with corresponding calcium compounds in liming materials; they supply magnesium, and where a shortage of this plant food is suspected a magnesium lime product is the easiest way of providing a supply. The commercial materials that supply magnesium are (i) *burnt magnesium lime* (a mixture of magnesium and calcium oxides containing more than 5.5 per cent of magnesium); (ii) *slaked magnesium lime* (made by slaking burnt magnesium lime and consisting of the hydroxides of calcium and magnesium); and (iii) *ground magnesium limestone* (a mixture of calcium and magnesium carbonates containing more than 3% of magnesium).

Various waste products are often sold as liming materials. The most common are carbonates of lime produced by beet sugar factories, paper works and water works. They should be judged by their lime content or 'neutralising value' and also on the ease of spreading them. When purchasing any waste limes that have not been tried before, care should be

taken to see that they do not contain any substances that may injure crops.
Natural liming materials such as lump chalk, marl and calcareous sea
sands are often used.

Purchasing liming materials

The principal liming materials are covered by the provisions of the
current British Regulations (ref. 17) under the Fertilisers and Feeding
Stuffs Act, 1926. The Schedules list and define the main materials sold
and give details of the particulars that must be provided.

Neutralising value. Liming materials should be bought on the basis of
'Neutralising Value', a figure that measures the use of a product for
neutralising soil acidity. It is determined by laboratory tests and
expressed as calcium oxide (CaO). If a product has a neutralising value
of 90, it means that 100 kg has the same effect on soil acidity as 90 kg
of burnt lime (calcium oxide). The 1973 Regulations require that the
Statutory Statement should give the neutralising value for all the
products listed in the Schedules except Chalk (which is simply defined
as 'Cretaceous Limestone').

Fineness. The finer a limestone is the quicker it will neutralise soil
acidity. The British Regulations require that the amount of product
passing through 'a declared British Standard Test Sieve' be stated when
selling screened Chalk. The amount of ground limestone, and ground
magnesian limestone that will pass through a British Standard Sieve
Mesh No. 100 must be stated.

Valuing liming materials. In comparing the values of liming materials
their neutralising values (as delivered to the farm) and the cost of spreading them on the land should be taken into account. The best comparison
for practical purposes is a 'unit cost' for neutralising value. This can be
calculated as:

$$\text{Unit cost} = \frac{(\text{price per tonne of material} + \text{cost of spreading})}{(\text{neutralising value})}$$

In general the *cheapest effective* source of lime should be used. Ease and
convenience of handling, and the cost of the larger dressings that are
often needed with damp and lumpy materials, should be taken into
account as well as the unit value.

Subsidies on lime. The arrangements for subsidising lime in Britain were
described in 'The Agricultural Lime Scheme 1966', Statutory Instruments 1966 No. 794. The contribution from 1966 to March 1970 was up
to 60 per cent of the cost of buying, transporting and spreading liming
materials that were approved (the Regulations list materials for which
contributions are paid). Subsidies are continued in 1974/75 and the current
rate of contributions is about 45 per cent of the total cost of liming land.

Practical comparisons of liming materials

The value of a liming material is simply its effect in neutralising soil
acidity. In field experiments ground carbonate of lime ($CaCO_3$) was

just as effective as a chemically equivalent amount of burnt lime (CaO) or slaked lime (Ca(OH)$_2$), and there was no advantage from using these latter materials that are often more expensive per unit. Most of the important types of liming materials gave similar yields when they were used at equivalent rates; fine grinding of limestone was also found to be unnecessary. Magnesium liming materials were as effective as ordinary burnt lime or limestone when they were applied at equivalent rates, although they are usually considered to be less soluble and to act more slowly than calcium limes.

Burnt limes contain about 85% of lime (CaO), hydrated (i.e. slaked) lime contains about 70% of CaO, and ground limestones about 50% CaO. One tonne of burnt lime is approximately equivalent to 1½ tonnes of slaked lime and to 2 tonnes of ground limestone. Besides being cheaper per unit of lime, ground limestones are more convenient and pleasant to handle and spread than burnt or slaked lime. Limestone stores indefinitely, while burnt lime (which is corrosive) may absorb moisture, and become hot, it often bursts its bags and then sets. Limestones are usually ground so that about 60% goes through the 100-mesh sieve, in this fineness they will act quickly enough and be as satisfactory as burnt lime. Harder limestones should not contain a large proportion of very coarse particles. Generally, ground limestones are satisfactory if all the material passes the 10-mesh screen.

Spreading lime

Coarse liming materials must be used at heavier rates than finer limes, and wet or lumpy products must be applied more heavily still so that all areas of a field receive at least some lime. Spreading should be as even as possible; this is not only because local shortages may result in patchy crops, but also because the thinner the dressing is in some parts of a field, the sooner it will be exhausted so that liming is necessary again. Distributors fitted with 'spinning-disc' spreading mechanisms are often responsible for uneven spreading; these machines give the highest rate of spread near to the disc and the rate tails off towards the outside. Experienced and careful operators can match up their work so that the swathes overlap to produce quite an even spread across the whole field. Often the work is not set out carefully enough and, as varying the speed of the spreader may vary the width of the swathe, overlimed strips alternate with strips that received too little. The results of uneven distribution of lime show in crops and can also be detected by soil tests. The remedy is to set out the runs of the machine carefully so that the swathes of lime overlap correctly; alternatively half of the dressing may be applied in one direction and the other half spread at right angles.

Limestone may be spread on *grassland* at any convenient time of the year. It is best to keep stock off the field until rain has washed the dressing off the herbage.

Where lime is to be applied to very acid soil immediately before growing a sensitive arable crop it is important that it should be distributed through, roughly, the top 10 cm of soil. To do this, lime should be spread *after* ploughing and *before* cultivating. If dressings are spread before ploughing there is a risk that the lime will be ploughed down and that the upper layer of the seedbed will be too acid for a satisfactory crop. Where land is deep-ploughed more lime than the standard dressing may be necessary, since normal recommendations are based on the surface 15 cm of soil. Where the sub-soil is also very acid it may pay to split the recommended dressing, applying half before ploughing and working the remainder into the top-soil. Maintenance dressings intended to keep the soil at a satisfactory pH may be applied at any convenient stage in the rotation; it is best not to apply them before growing potatoes, since the lime may encourage scab.

THE USE OF LIME IN BRITAIN

With the leaching normal in Britain and with crops receiving 100 kg N/ha as ammonium salts annual losses of lime may be equivalent to 400–800 kg/ha of calcium carbonate. Losses may be even larger when account is taken of calcium removed by crops, stock and milk that are sold, and in the handling of excreta. One published estimate is that under *intensive* farming conditions, even in areas of low rainfall, 6½ tonnes per hectare of calcium carbonate are needed once in five years to maintain a satisfactory lime status.

Liming has been subsidised in Britain since 1937; 1¾ million tonnes of liming materials were used in 1939. The amounts increased during the war and in 1948, 4½ million tonnes were applied. In most recent years about 4 million tonnes of liming materials have been used in U.K. (in 1969 4.4 million tonnes). The deficit of lime for the country as a whole seems to be diminishing. Nevertheless we still have much land where acidity prevents crops yielding as much as they could. A recent Survey of Fertilizer Practice showed that about one-quarter of the land cultivated for arable crops or temporary grass needed lime and only 6 per cent was limed each year. Nearly half of the permanent grassland of England and Wales needed liming and the degree of acidity in permanent grass was also more serious. The following figures are from the Survey:

	Tillage land	Temporary grass	Permanent grass	All land in England and Wales
Per cent of total area needing lime	27	29	44	34
Average need (tonnes calcium carbonate/hectare)	3.75	3.65	4.90	4.40

The practical value of lime in Britain

The 1970 Fertilizer Practice Survey in England and Wales showed

hat only 3 to 4 per cent of fields of arable crops or leys surveyed had
eceived lime in that year; for permanent grass the figure was less than
3 per cent. The rates actually used were about 3.1 tonnes/ha on arable
crops, leys and permanent grass. Even when allowance is made for the
areas of land that have natural reserves of calcium carbonate it is difficult
to see that this frequency of liming (one field in 25 each year), or the
amounts used, are sufficient to replace the large losses of calcium that are
inevitable in modern farming.

On some organic soils micronutrient supplies are depressed by over-
liming so that crops suffer. In the west and north of Britain soils contain
more organic matter and there is more risk of over-liming causing micronutrient deficiencies; advisers usually recommend only enough lime to
bring the soil to pH 6.5. But on most mineral soils in England the
balance is in favour of using too much rather than too little lime. Where
crops sensitive to acidity, such as sugar beet or barley, are commonly
grown, much yield may be lost as acidity develops. When lime is needed
it is best given to these crops, so that, on average, soil has a slight reserve
of calcium carbonate. This will result in some waste by extra leaching but
lime is cheap and using more is some insurance against the irregularity in
soil pH common in acid fields and the bad distribution often seen where
lime is spread mechanically.

The effects of lime on crop yields measured in field experiments

H. W. Gardner and H. V. Garner gave a comprehensive account of
liming in Britain and described the results of early experiments (ref. 26).
They showed the long lasting effect of heavy liming in an experiment in
Hertfordshire where light-medium loam cleared from woodland had a
lime requirement of $9\frac{1}{2}$ tonnes/hectare, up to twice this amount was
tested and with this double dressing the soil was nearly neutral 10 years
later:

Chalk applied in 1934 tonnes/hectare	Soil analyses in 1945	
	Exchangeable Ca ppm	pH
0	370	4.3
4.8	630	4.7
9.4	885	5.3
14.1	1230	5.8
18.8	2030	6.5

In recent experiments on clay-loam soil at Rothamsted and on sandy
loam at Woburn, described by J. Bolton (ref. 27), a range of dressings of
ground chalk were applied in 1962. For beans the optimum pH was about
6.8. At both centres yields of barley were largest with soil pH of 6.5
to 7.5; the increases in yields of potatoes caused by changes in soil pH
were not statistically significant. At Rothamsted the largest dressing of
lime (20 t/ha) was still increasing soil pH 6 years after it was applied,
with smaller dressings (5 and 10 t/ha) pH was diminishing slowly. At

Woburn the two largest dressings (12 and 19 t/ha) maintained soil pH at roughly constant values but pH diminished where 5 t/ha had been given. Annual losses of lime from these two soils were estimated to *average* about 300 kg/ha of calcium carbonate; most was lost from the soils with larger pH values as these figures show:

Amount of $CaCO_3$ applied in 1962 t/ha		Annual losses kg/ha of $CaCO_3$		Soil pH in 1967	
Rothamsted	Woburn	Rothamsted	Woburn	Rothamsted	Woburn
0	0	–	–	4.9	5.0
5	5	90	180	5.7	6.3
10	12	140	700	6.7	7.2
20	19	900	1340	7.6	7.5

CHAPTER EIGHT

Fertilizing with Sulphur

The first recorded example of a crop yield being increased in Britain by sulphur fertilizer was in an experiment in 1968 at Wareham in Dorset. Radishes grown on very coarse sandy soil yielded more when supplied with elemental sulphur. Sulphur (S) deficiencies in crops are unlikely where rain supplies annually 12 kg S/ha. At present almost the whole of Britain gets as much as this or more. If, in future, we use much less sulphur-containing fuels (or recover the S from combustion gases), and if the trend to apply less sulphur, as fertilizers become more concentrated, continues, then sulphur deficiency in crops may become more common. This account is intended only as a brief guide to experience in other countries.

Responses to sulphur

Crops have responded to sulphur in parts of Scandinavia (where rain supplies only about 3 kg S/ha each year). In U.S.A. the deficiency is common along the Pacific coast, in the North-West, and in the South-East. Most gains in U.S.A. were recorded with legumes and cotton but yields of cereals have also been increased. In North-Eastern United States, where larger amounts of sulphur-containing fuels are burned, more sulphur is being applied by animal manures, by the rain, and by fertilizers than is lost and crops do not respond to sulphur. In Canada, areas of Alberta, British Columbia and Saskatchewan are sulphur-deficient. Responses in West Indies and in Central and Southern America are reported. The deficiency is widespread in Australia and is often associated with a shortage of soil phosphorus. New Zealand has many areas of soil that are deficient in sulphur and brassica crops and legumes respond to dressings. Ordinary superphosphate is the most important fertilizer in both Australia and New Zealand, some of its benefits are undoubtedly due to the calcium sulphate it contains.

Parts of Africa are sulphur-deficient. In Senegal groundnuts responded to sulphur; responses have also occurred in Ivory Coast, Ghana, Nigeria, other central African states, and in East Africa. Sulphur deficiency is likely where soils have been leached for very long periods, as has happened in Central Africa. Legumes and oil crops are most affected; the deficiency is less common (but does occur) in perennials.

So many examples of gains from sulphur fertilizers are now being reported that any survey of the deficiency is quickly out of date. Gains in yield are often as large as from other fertilizers. For example in one experiment in Kenya sugar cane receiving no sulphur yielded 52 t/ha of cane; with gypsum supplying 42 kg S/ha, the yield was 74 t/ha. Other experiments showed how changes in forms of N and P fertilizers altered yields by varying the sulphur supplied (ref. 28). It can be misleading to compare fertilizers supplying a particular nutrient (for example N or P) when some of the forms tested also supply a second nutrient (for example sulphur). The *full* effects of all the nutrients in a fertilizer on crops must be identified and examined separately and together, if the results of fertilizer experiments are to be interpreted correctly. In territory for which there is no local evidence, it should never be assumed that a nutrient like sulphur has no effect until this has been proved by experimenting. A balance sheet for local conditions should be calculated.

Amounts of sulphur in crops and soils

Some concentrations of sulphur in temperate crops are given in Table 2. Crops such as potatoes, cereals and grasses need about 10 kg/ha or less; legumes and sugar beet take up more. The large demands are made by crops of the brassicae family which often need 40–50 kg S/ha.

Plants obtain most of their sulphur from the soil though some can be absorbed directly from the air. British soils do not contain a permanent reserve of sulphate ions since any surplus is removed by leaching in winter. Sulphate accumulates in soils (usually as calcium sulphate) only in arid districts where rainfall is not sufficient to cause drainage water to pass down the soil. Soil organic matter contains a considerable reserve of sulphur (roughly equal to the organic phosphorus it contains); this sulphur is released only slowly as the organic matter mineralises. (Where the content of soil organic matter does not change, as in most stable agricultural systems, about as much sulphur is locked up in new organic matter each year as is mineralised, so there is no net release of S.) In most industrial countries the air supplies most of the sulphur crops need (it comes originally from the sulphur in coal and oil); some is washed into the soil by rain, some is absorbed directly from the air, both by plants and by the soil. Analyses of sulphur in air and rain are now being made in many parts of the World; for example British figures indicate that much of the country receives, on average, between 16 and 24 kg S/ha but industrial areas and land near large cities receives more than this. Recent Rothamsted work showed that about 28 kg of S/ha was deposited or absorbed directly from the air on to soil, per annum; rain supplied 21 kg S/ha.

The larger crops now commonly grown need more sulphur and make it more likely that natural supplies will not be sufficient. The amounts of sulphur (and other nutrients) in good yields of crops grown in warmer parts of U.S.A. are shown in Table 16. These figures are adapted from a recent publication (ref. 29) which also lists some recent record yields in U.S.A. and shows the amounts of sulphur removed.

TABLE 16

NUTRIENTS IN CROPS GROWN IN U.S.A.

	Yield per hectare	N	P	K	Mg	S
				kg/ha		
Maize	13 tonnes	360	52	230	75	50
Grain sorghum	9 tonnes	290	55	210	40	42
Rice	7 tonnes	150	25	17	17	20
Soybeans	3000 kg	210	25	110	14	11
Groundnuts	3300 kg	250	22	110	30	28
Cotton	6 bales*	140	36	85	18	25
Tobacco	3000 kg	105	12	180	27	24
Onions	50 tonnes	135	27	100	14	28

*1 bale = 215 kg (approximately).

Sulphur fertilizers

Normally sulphur deficiency is rectified cheaply and easily by applying gypsum ($CaSO_4,2H_2O$) but other sources are satisfactory if they are economic. Liquid sulphur dioxide (a gas at normal temperature and pressure) which can be applied by the equipment used to apply liquid ammonia has been tested in U.S.A. and found as effective as gypsum. Elemental sulphur was equally effective, provided the particles were fine enough (less than 0.1 mm diameter) so that they are quickly oxidised by micro-organisms in the soil. In U.S.A. concentrated fertilizers have been coated with sulphur which has also been incorporated in granulated fertilizers. Often the older kinds of dilute fertilizers supplying N, P and K can be used to supply sulphur as well where it is needed. Where the sulphate ion is an essential component of a fertilizer then part of the total cost can be borne by the sulphate. Table 17 gives the amounts of sulphur in some materials used to supply sulphur in fertilizers in U.S.A. (from ref. 29).

TABLE 17

NUTRIENTS IN FERTILIZERS CONTAINING SULPHUR AND SOLD IN U.S.A.

	N	P_2O_5	K_2O	S
		percentage		
Ammonium sulphate	21	0	0	24.2
Ammonia-sulphur solution	74	0	0	10
Ammonium sulphate-nitrate	26	0	0	12.1
Basic slag	0	15.6	0	3
Gypsum	0	0	0	18.6
Kainit	0	0	19	12.9
Magnesium sulphate	0	0	0	13
Potassium sulphate	0	0	50	17.6
Superphosphate (ordinary)	0	20	0	13.9
Urea-gypsum	17.3	0	0	14.8
Urea-sulphur	40	0	0	10

CHAPTER NINE

Fertilizing with Micronutrients

The symptoms in crops caused by deficiencies of micronutrients depend on many factors and correct diagnosis needs a specialist advisory officer who should be consulted if these troubles are suspected. Incorrect dressings of major nutrients, dry or cold weather, and damage by machinery or chemical sprays may all cause crops to have symptoms that may be mistaken for micronutrient deficiencies; these factors also modify the appearance of symptoms actually due to deficiencies. Shortages of micronutrients may become more common in future in areas where few stock are kept. Farmyard manure conserves and redistributes both major and micronutrients and deficiencies are less likely where organic manures return much of the residues from crops and animal foods.

MICRONUTRIENTS IN FERTILIZERS AND MANURES

Most ordinary fertilizers, both older dilute ones *and* the more concentrated materials based on ammonium nitrate and phosphate, do not contain *worthwhile* amounts of micronutrients. Most claims that dilute fertilizers can correct trace element deficiencies are exaggerated. It is unsound to pay more for a certain amount of N, P and K in the hope that the material may also contain some micronutrients; if it does, the quantities in a tonne are usually so small that they could have been bought as chemical salts for a few pence. Nevertheless the quantities applied may help to keep up reserves.

Table 18 gives analyses for micronutrients in fertilizers and FYM taken from Rothamsted stocks. Fertilizers made from naturally-occurring raw materials sometimes contain appreciable quantities of micronutrients. The superphosphate in Table 18 contained significant amounts of copper, zinc, and cobalt, these may have come from the rock phosphate used but they are more likely to have originated in the sulphuric acid used to make the super, it was a by-product of processing metal ores. The ammonium nitrate-limestone fertilizer analysed contained appreciable amounts of manganese, copper and zinc, these may have come from the limestone used. In contrast the two synthetic nitrogen fertilizers (ammonium sulphate and sodium nitrate) contained much smaller quantities of micronutrients. Both potassium fertilizers contained only small amounts of any of the elements, but farmyard manure contained considerable quantities of all.

TABLE 18

THE AMOUNTS OF SOME MICRONUTRIENTS IN FERTILIZERS
AND IN FARMYARD MANURES

Parts per million in dry materials

	B	Mn	Cu	Zn	Co	Ni
'Nitro-Chalk'	–	24	22	15	0	2
Sodium nitrate	–	8	3	1	0	0
Ammonium sulphate	6	6	2	0	0	0
Superphosphate	11	11	44	150	4	13
Potassium chloride	14	8	3	3	1	0
Potassium sulphate	4	6	4	2	0	0
Farmyard manure	20	410	62	120	6	10

Fertilizers containing a small concentration (about 10 ppm) of a micronutrient and applied at a common rate (about 500 kg/ha), will make little contribution to the micronutrient status of soils and will apply considerably less of the elements than may be removed by an annual crop. Where elements are present at higher concentrations, normal dressings of fertilizers may apply amounts which are of the order of those removed each year by crops, but which are much smaller than the amounts given to correct a deficiency. Materials applied at rates of many tonnes/hectare, such as farmyard manure, provide much greater quantities of micronutrients, approaching the amounts applied as sprays to correct deficiencies in crops. Common dressings of the materials shown in Table 18 would supply approximately the amounts of micronutrients shown in Table 19, which also shows the amounts removed by 5 average arable crops at Rothamsted (ref. 30). Most of the amounts of micro-

TABLE 19

AMOUNTS OF MICRONUTRIENTS SUPPLIED BY FERTILIZERS
AND FYM

	Cu	Mn	Mo	Zn
		grammes/hectare		
	Supplied by FYM and fertilizers			
FYM (45 t/ha)	560	3360	11	1120
'Nitro-Chalk' (560 kg/ha)	11	11	0.5	11
Superphosphate (450 kg/ha)	22	5	0.8	67
Potassium sulphate (220 kg/ha)	2	2	0.02	2
	Removed by five arable crops grown in rotation			
	300	2500	10	1800

nutrients supplied by dressings of the ordinary fertilizers are much less than are needed to replace soil reserves of the elements. But a dressing of FYM of this quality once in a rotation, would replace as much copper, manganese and molybdenum as is removed by four or five common crops, and nearly as much zinc.

Farmyard manure

The amounts of trace elements that would be supplied by 10 tonnes of FYM of average composition in the West of Scotland were calculated (ref. 5) in terms of the common salts used to correct trace element deficiencies; they were approximately: 1,450 grammes of manganese sulphate, 420 grammes of borax, 170 grammes of copper sulphate, 14 grammes of cobalt sulphate and 14 grammes of sodium molybdate. FYM is the only regular source of extra trace elements on most land; although the amounts in 10 tonnes are generally much less than the equivalent quantities of salts recommended for preventing deficiencies, they must often be useful.

Basic slag

The N.A.A.S. (ref. 25) have summarised information on micronutrients in basic slags. Average analyses from 6 steel works in U.K. showed that 1.25 t/ha of basic slag would provide the amounts of elements which are shown below together with the amounts removed by 5 t/ha of hay:

	In 1.25 tonnes of slag	In 5 tonnes of hay
Mg	50 kg	7.5 kg
Mn	39 kg	0.57 kg
Cu	163 g	30 g
Mo	24 g	4 g
Zn	24 g	200 g
Co	4 g	0.4 g

Assuming that the total amount of each of the trace elements in basic slag is available to crops, the dressing of slag would supply all the magnesium and micronutrients, except for zinc, removed by 5 t/ha of hay. The ranges of total amounts of some elements in European basic slags and samples from U.S.A. were:

		European samples	U.S.A. samples
Manganese	%	2.3–5.3	1.9–2.5
Magnesium	%	0.6–3.0	2.6–3.8
Copper	ppm	5–200	9–35
Cobalt	ppm	2–8	1–3
Boron	ppm	10–1,000	10–30
Zinc	ppm	<30	10–29
Molybdenum	ppm	5	9–10

The total *amounts* of trace elements in basic slags are impressive but there is little experimental evidence, except for magnesium and manganese, to show that the elements do in fact become available. A proportion of the Mg and Mn present can be useful; the N.A.A.S. Report (ref. 25) emphasises the need for more experimental work but says basic slag could play a useful part in maintaining trace element supplies to soil.

CURING DEFICIENCES IN FIELD CROPS IN BRITAIN

Boron (B).
In some areas soils are deficient in boron, causing heart-rot in sugar beet, mangolds and table beet, brown-heart ('raan') in turnips and swedes, internal cork in apples and pears and hollow stem in cauliflowers. Cereals and potatoes are not easily affected. Over-liming may cause boron deficiency, but the trouble may be bad in some years and not appear in others. In areas where boron deficiency is common, borax is often included in compound fertilizers sold for turnips, swedes or sugar beet. For example one compound fertilizer offered for beet in 1970 contained 20% N, 10% P_2O_5, 10% K_2O and 0.32% B. The deficiency may also be controlled by working 20 kg of borax (equivalent to 2 kg/ha of boron) *very uniformly* into the seedbed. If too much borax is applied, and this may occur with uneven spreading, the crop may be poisoned, since there is only a small range between toxic amounts and supplies that are too little for crops. (Some crops, barley for example, are damaged easily by boronated fertilizer.) Sprays of soluble boron compounds are now usually preferred as they can be spread more uniformly than solid borax. An example is 'Solubor' containing 20% B; it is readily soluble and applied as a solution in water containing 0.2–2.5% of the compound.

Copper (Cu).
Copper deficiency lessens yields of cereals on some of the reclaimed fen peats and glacial sands of Eastern England, it also occurs on some Chalk soils in Southern England. Sugar beet and carrots have shown no symptoms but have yielded more with copper sprays. Potatoes have not responded (the fungicides traditionally used to control potato blight have supplied much copper). Liming increases copper deficiency. It is controlled by spraying the crop at a suitable time with 1 kg/ha of copper sulphate dissolved in 1,000 litres of water. In experiments on copper-deficient soils where wheat yielded only 200–300 kg/ha, these sprays increased the yields to 4 t/ha. Dressings of 20 kg/ha of solid copper sulphate applied to soil have lasted for several years and have been more effective than sprays alone; sometimes best results have been obtained from both sprays and soil dressings used on the same field. Solutions of copper sulphate sometimes damage the leaves of crops; sprays containing copper oxychloride or cuprous oxide avoid this damage and cure the deficiency as well as copper sulphate does. In recently published A.D.A.S. experiments (ref. 112) copper did not increase yields.

Iron (Fe).

Fruit trees and horticultural crops grown on calcareous soils often show iron deficiency, which can be controlled by spraying ferrous sulphate as a 5% solution at 400 litres/ha. This salt is also effective when injected into trunks of affected trees. Often it is difficult to get deficient crops to absorb enough iron; when this occurs chelated forms of iron (discussed later) are often more effective.

Manganese (Mn)

Manganese deficiency usually occurs only on neutral and alkaline soils that are rich in organic matter and have a high water-table. The two most common diseases due to manganese deficiency are 'grey-speck' in oats and 'marsh spot' in peas; both are most common on wet organic soils that are naturally calcareous or which have been over-limed. Wheat also suffers from 'grey-speck', but barley is less easily affected. In sugar beet and mangolds manganese deficiency causes 'speckled yellows'— the name describes the symptoms. Peas affected by 'marsh spot' often show no symptoms on the leaves, but the peas in the pods have brown spots of decayed tissue that lessen their value; the disease occurs on low-lying soils containing free lime, such as those in Romney Marsh in Southern England. Potatoes also suffer from manganese deficiency and show characteristic spots on the leaves.

Manganese deficiency is controlled by broadcasting 40–100 kg/ha of manganese sulphate over the soil before sowing, but it is often cheaper and more effective to apply 10 kg/ha of manganese sulphate dissolved in 200–1,000 litres of water as a spray as soon as the deficiency is diagnosed. Treating soil with manganese usually costs more than spraying, and only a little of the dressing will be taken up by the crop, the rest may be 'fixed' by the soil in a useless form. Experimental work in Ireland showed that manganese deficiency in cereals can be controlled by drilling 14 kg/ha of manganese sulphate with the seed. This treatment was better than normal spraying. Similar results with 20 kg/ha of combine-drilled manganese sulphate have been obtained in Scotland where this method was considered the most effective and economic treatment. Soil dressings or combine-drilling cannot, of course, be used when the deficiency is not diagnosed until the crop is growing; only sprays can then give a satisfactory crop.

Molybdenum (Mo).

Molybdenum deficiency causes 'whiptail' in brassica crops (principally cauliflower and broccoli) and, occasionally, failures of legumes. The availability of molybdenum is increased by liming and deficiencies are most common on acid land. In experiments on acid soils at pH 4.8 in the Welsh Hills the yields of reseeded pastures dressed with sodium molybdate were increased by 15% in the first year. When lime was given there was no response to molybdate. Further work is needed to see whether molybdate will improve yields generally on hill pastures as dressings commonly do in Australia and New Zealand where molybdates are often essential for good growth of clovers in pasture.

Acid soils should be limed, but if molybdenum deficiency persists after liming, it will be controlled by spraying 1 kg/ha of sodium molybdate as a solution in water.

On the alkaline Lower Lias Clays of Somerset the soil contains an excess of molybdenum which is taken up by clover; animals feeding on the clover suffer from scouring, a condition called 'teart'.

Zinc (Zn).

Zinc deficiency affects fruit trees in many parts of the world, but rarely in Britain. A severe case in apples has occurred in Surrey; in Romney Marsh, potatoes and cereals responded to zinc sprays. In Western U.S.A. sugar beet is sensitive to zinc deficiency on some alkaline soils, other farm crops are not. Soil treatment with zinc sulphate, oxide or carbonate is said to be better than spraying the crop. As with manganese, copper and boron an excess of zinc is toxic to crops. Poisoning of this type has been caused by industrial wastes or by sewage sludge containing zinc.

INTERACTIONS OF MICRONUTRIENTS WITH EACH OTHER AND WITH MAJOR NUTRIENTS

Spectacular increases in yield have been recorded in experiments with micronutrients in many countries and particularly in Australia and New Zealand. Examples are given in a good review which has been published (ref. 31) of the need for trace elements for pastures and fodder crops in Australia. Often there have been large interactions with major nutrients, sometimes between two micronutrients. Sometimes when one crop responds well, a similar crop may respond less or not at all. For example Table 20 gives results of field experiments on cereals in

TABLE 20

RESPONSES OF CEREALS TO MOLYBDENUM IN WESTERN AUSTRALIA

(from ref. 31)

(all plots received superphosphate)

Fertilizer applied per hectare		Yields of grain, kg/ha	
N	Mo*	Wheat	Oats
None	None	630	680
None	75 g	770	770
25 kg	None	800	540
25 kg	75 g	1020	790

*Applied as roasted molybdenum oxide containing 55% Mo.

Western Australia where molybdenum and nitrogen both increased yields of wheat and the effects of N and Mo used together were greater than when the two were given separately. Only molybdenum was needed for oats, nitrogen alone depressed yield. Quite complicated experiments are often needed to investigate multiple deficiencies; Table 21 shows yields of pasture in South Australia affected by the triple interaction of phosphate, copper and zinc.

TABLE 21

EFFECTS OF PHOSPHATE APPLIED WHEN SOWING, AND OF
COPPER AND ZINC ON YIELD OF PASTURE IN
SOUTH AUSTRALIA

(from ref. 31)

	Yield in second year kg/ha of dry matter	
	Without P at sowing	With 125 kg/ha of superphosphate
Without Cu or Zn	763	1413
With Zn	1416	2082
With Cu	1305	1608
With Cu and Zn	1534	2571

Total yield was raised by each of the nutrients, P, Cu and Zn, but the maximum yield, which was three and a half times the untreated yield, was only obtained when all three nutrients were applied together. Sometimes effects on both soil and crop occur at the same time. Table 22 shows that in an experiment in New South Wales subterranean clover virtually failed without molybdenum or lime. Practically the maximum yields could be obtained either with a very small amount of lime (250 kg/ha) plus molybdenum (applied as 140 grammes/hectare of molybdenum trioxide) or with 1,000 kg/ha of lime alone.

TABLE 22

EFFECT OF MOLYBDENUM AND LIME ON YIELD OF SUBTERRANEAN
CLOVER IN NEW SOUTH WALES

(ref. 31)

Yields of dry matter, kg/ha

Lime applied kg/ha	Without Mo	With Mo
0	439	402
63	678	1205
250	2384	4681
500	2748	4782
1000	4493	4794

The very small dressings of lime were necessary for the crop to respond to Mo at all. The much larger dressings of lime released molybdenum from soil reserves and made the micronutrient dressing unnecessary.

MICRONUTRIENT FERTILIZERS

Experience and practice with micronutrient fertilizers has been well summarised by V. Sauchelli (ref. 32); he listed the recommendations made for using micronutrient fertilizers on crops grown in different parts of U.S.A. K. H. Schütte (ref. 33) discussed biological aspects of trace elements and their role in nutrition.

Deficiencies of micronutrients are usually corrected by sprays or soil dressings of simple salts containing the deficient element. Table 23 lists the salts which are usually used, their compositions, and the amounts commonly applied in Britain and North America.

TABLE 23

FERTILIZERS COMMONLY USED TO SUPPLY MICRONUTRIENTS

Micronutrient	Salt used as fertilizer Name	Formula	Percentage* of element	Amount of element applied per hectare as spray	to soil
Boron (B)	Borax	$Na_2B_4O_7,10H_2O$	10.6 (B)	500 g	2–3 kg
Copper (Cu)	Copper sulphate	$CuSO_4,5H_2O$	25(Cu)	250 g	5–25 kg
Iron (Fe)	Ferrous sulphate	$FeSO_4,7H_2O$	20(Fe)	1–3 kg	–
Manganese (Mn)	Manganese sulphate	$MnSO_4,4H_2O$	24(Mn)	2.5 kg	10–25 kg
Molybdenum (Mo)	Sodium molybdate	$Na_2MoO_4,2H_2O$	39(Mo)	40–200 g	40–200 g
Zinc (Zn)	Zinc sulphate	$ZnSO_4,H_2O$	36(Zn)	1–8 kg	2–25 kg
	Zinc chelate	Zn EDTA	6(Zn)	0.5–1 kg	2–10 kg

*Approximate analyses to allow for use of commercial salts.

Enriched NPK fertilizers

In countries where micronutrient deficiencies are common the elements likely to be needed are often added (as a soluble salt) to ordinary fertilizers so that the recommended NPK dressing will supply sufficient micronutrients as well. The only element commonly supplied in this way in Britain is boron which is incorporated into NPK fertilizers intended for sugar beet and brassicae crops. Examples sold in 1975 are:

	% N	% P_2O_5	% K_2O	% B
A	20	10	10	0.32
B	15	10	10	0.3
C	9	23	18	0.45

In U.S.A., recommendations for an area often include micronutrients in mixed fertilizers where experiments have shown the elements are often needed. Some farmers favour the 'shotgun' method where small amounts of several micronutrients are added to fertilizers to insure against the risk of deficiencies that have not been proved (ref. 32).

Frits

Some micro-nutrients are easily leached from soils (boron for example); others combine with soil (particularly with deficient soils) in very insoluble forms (for example, copper becomes very insoluble in some organic soils). Two ways of making such elements more efficient have been tried. One way is by fusing salts of boron, or other elements, with glass which is then shattered and applied to soil. The element is released as the glass slowly dissolves. These materials, called *frits,* have been successful in curing deficiencies where normal methods have failed.

Chelates

These are complexes of the metallic micronutrients with organic substances; they do not ionise, so they protect the element from wasteful reactions with the soil, but they are soluble and allow the metal element to be taken up by plants. Sometimes the substances used to chelate metals are called *sequestering agents.* A very common chelating agent is ethylenediamine tetraacetic acid (EDTA), but others are used. The metals usually protected as chelates are copper, iron, manganese and zinc. The metal chelates are water soluble and are often applied as sprays on foliage to correct deficiencies quickly, but they may also be applied to soil. Chelated materials are often very much more efficient than ordinary salts of micronutrients and much less may be needed, particularly where soil precipitates the element from ordinary salts in very insoluble forms. An example from U.S.A. shows the large savings that are possible: In a citrus orchard where iron chlorosis was severe 1,200 kg of iron *per tree* improved the appearance of the foliage, but there were larger improvements from only 10 to 50 grammes of iron in chelated form applied to each tree. Chelate preparations are available in Britain to supply calcium, cobalt, copper, iron, magnesium, manganese and zinc.

TOXICITIES OF MICRONUTRIENTS

Crops or livestock may be damaged by heavy metals present in, or added to soil. The amounts that may cause damage are quite small and therefore micronutrient fertilizers should not be used unless they are likely to be needed. Damage may be caused by application of materials such as copper salts that are applied regularly as fungicide, by unnecessary spraying with micronutrients, by applying town wastes or sewage sludge containing heavy metals, or it may be due to a naturally-occurring high concentration of the element in soluble form in the soil. For example *manganese* toxicity is quite common on acid soils and the effects are often confused with a shortage of calcium or the direct effect of high acidity. It is prevented by liming.

In countries where large areas of deficient soils have similar origins, compositions, and are used for the same crops, general recommendations can be made and fertilizers enriched with micronutrients can be supplied. This happens in parts of Australia and Russia, large areas may be deficient in molybdenum, cobalt, boron or copper. In U.S.A., although the need to fertilize with micronutrients is increasing, it is still considered unwise to use trace element fertilizers as insurance dressings because if used too often they could become toxic. In Britain, geology is very varied and soils usually vary from field to field and it is not justified to make general recommendations for micronutrient fertilizing in an area where only a few fields have been found to be deficient. Advice should always be obtained before using micronutrient dressings to avoid the waste of money and risk of toxicity that 'blunderbus' dressings entail. Where one crop may be deficient, others may secure enough micronutrients. For example boronated fertilizers for turnips, sugar beet and brassicae crops may be recommended in areas where the need is proved but these fertilizers must not be used for other crops as they can damage them. The risks of unnecessary dressings must be stressed because it is generally more difficult to deal with toxicity than deficiency. For example if molybdenum is added to fertilizers indiscriminately, the stock of this element in the soil will be raised until there is a risk of increased uptake of herbage causing disorders in animals, such as now occur in the 'teart' areas where molybdenum is naturally rich in the underlying clay.

CHAPTER TEN

Compound or Mixed Fertilizers

Compound fertilizers (usually called mixed fertilizers in U.S.A.) contain any two or all three of the nutrients nitrogen, phosphorus and potassium. Some are made by simply mixing straight fertilizers. Other compounds are made by more complex chemical processes in which rock phosphate is first attacked by acid, ammonia is added next, and finally potassium salts. Such processes give concentrated fertilizers that are usually cheaper per kg of plant food than ordinary manufactured compounds made by mixing the straight fertilizers that farmers can buy.

GRANULATION OF COMPOUND FERTILIZERS

For the last 25 years most compound fertilizers sold in Britain have been granulated or prilled and we have been ahead of some other countries where only in the last 10 years have fertilizers commonly been granulated (ref. 24). Granulated materials are usually made by drying a wet slurry in a heated rotating drum. *Prills* are made by allowing a molten fertilizer or very concentrated solution to fall through air (or sometimes through a liquid such as oil). The droplets become round and solid as they fall. Granulated fertilizers are easier to handle, store and spread than corresponding home-made or factory-made powdered mixtures. Experimental work comparing granulated and powdered fertilizers has shown that when the same quantities of N, P_2O_5 and K_2O are supplied in similar chemical forms the two types of fertilizer give the same crop yields. Any residue left in the soil at the end of the season contains no plant food as the water-soluble N, P and K have diffused away. Where placement drills are used for cereals and other crops, granulated fertilizers should be used so that the dressings are applied evenly at the correct rate; checks to drilling caused by blockages in delivery tubes are much more common with powdered than with granulated fertilizers. A further disadvantage of powders is that they are blown away by high winds, while granules fall directly on the soil. The only case where granular products are at a disadvantage is when they contain much insoluble phosphate; granulation impedes the action of soil moisture on the insoluble material and also hinders the diffusion of phosphate away into the soil. Basic slag contains phosphate that is insoluble in water and must be finely ground. Recent developments in processing have given slags in very small granules that do not hinder the phosphate becoming active but which spread satisfactorily and have little dust.

ANALYSES OF COMPOUND FERTILIZERS

When compound fertilizers are sold the British Regulations (ref. 17) require that the 'amounts, if any, of nitrogen (N), potash (K_2O), phosphoric acid (P_2O_5) soluble in water, and phosphoric acid insoluble in water' should be stated. In addition to this the name must be stated of any pesticide or herbicide, or of the elements boron, cobalt, copper, iron, magnesium, manganese and molybdenum where these have been added during manufacture or preparation for sale and the total amount present must be stated as a percentage (if 0.1% or above) or as parts per million (if below 0.1%). (If these elements happen to be included in a conditioner or filler that improves the handling of the fertilizer, they need not be declared.)

Grades. The total of the percentages of N, P_2O_5 and K_2O is called the *grade* (or concentration) of the fertilizer. Variations in grades are obtained by using different sources of nitrogen and phosphorus. Triple superphosphate gives higher grade compounds than are made with ordinary superphosphate. Most of the concentrated fertilizers sold in U.K. contain mono-ammonium and di-ammonium phosphates, and ammonium nitrate and/or urea. Highly concentrated fertilizers can be cheaper to the farmer because costs of handling during manufacture, packaging and transport to the farm are less, a constant sum for these items per tonne of product being spread over a greater amount of plant food. The agricultural performance of a fertilizer does not depend on grade, and a kilogramme of plant food in high-grade and low-grade materials will give the same crop yields. Farmers often find that with high-grade fertilizers a given quantity of plant food can be applied to the land more cheaply, especially if large areas are covered.

Other countries require other kinds of information about fertilizers. In most countries phosphorus and potassium are expressed as P_2O_5 and K_2O respectively, but a few countries (Ireland, Norway, New Zealand and South Africa are examples) use the more logical %P and %K. But as P_2O_5 and K_2O are still used in Britain, for international trade in fertilizers, and in international statistics they are used here in practical discussions of fertilizer use. The supplementary information to describe forms of N, P and K vary too. In U.S.A. procedure between States varies, but generally the percentages of total N, 'available P_2O_5' (measured by solubility in neutral ammonium citrate solution) and soluble K_2O must be stated. Some American firms print guarantees in terms of both P and K and P_2O_5 and K_2O on their bags.

Brief descriptions of compound fertilizers are conveniently based on their analyses. A fertilizer with 20% N, 10% P_2O_5, 10% K_2O is described as '20–10–10'; one without nitrogen, but containing 20% P_2O_5 and 20% K_2O is called '0–20–20'. This shorthand is very convenient in advisory work, in informal discussions and in writing. It is used throughout this book.

PROPORTIONS OF STRAIGHT AND COMPOUND FERTILIZERS

Compound fertilizers are very important sources of plant nutrients in most developed countries and particularly so in Britain. Thirty years ago they supplied about half of the N, P and K used in U.K. During the last fifteen years the amounts of 'straight' fertilizers (containing only one nutrient) sold have diminished. In 1973 49 per cent of the N was sold straight, and 8 per cent of the K. Very few farmers now mix their own fertilizers, while 20–30 years ago many did. Only 3 per cent of the total water-soluble phosphate was used as straight fertilizer in 1973, but 74 per cent of all the water-insoluble phosphate was sold 'straight' (mostly as basic slag).

Similar statistics are not readily available for most other countries but some figures for U.S.A. are:

	N	P	K
	percentages of total nutrients used that were applied in mixed fertilizers		
1950	49	69	93
1964	32	81	84
1972	25	82	64

F.A.O. figures (ref. 18) cannot be compared with these British and American data. F.A.O. report the amounts of 'complex' fertilizer used in many countries but 'complex' means that N, P and K are combined in manufacture; mechanical mixtures of straight fertilizers are excluded. For Europe as a whole F.A.O. reports that in 1968/9 33% of N, 40% of P and 34% of K was in 'complex' fertilizers; five years earlier the percentages were 24, 26 and 23.

CHANGES IN CONCENTRATIONS OF COMPOUND FERTILIZERS

Fertilizers have become much more concentrated in most countries during the last 20 years. In 1950 some British fertilizers were based on ammonium phosphate, but most were made from ammonium sulphate (21% N), ordinary superphosphate (18% P_2O_5) and potassium chloride (50% K_2O). Now much of the N and P is supplied by ammonium nitrate (34% N) and ammonium phosphate (11% N + 48% P_2O_5); 'triple' superphosphate (47% P_2O_5) and some urea (46% N) are also used. These changes have increased the percentages of plant nutrients in compound fertilizers. In 1948 the total of %N + %P_2O_5 + %K_2O in British compounds averaged about 24, in 1959 it was 32.4, in 1963 35.9, in 1967 40.2 and in 1973 41.8

Total percentages of nutrients (the *grade*) of mixed fertilizers sold in U.S.A. have changed similarly from 19.7 in 1940, to 23.2 in 1950, 31.6 in 1960 and to 41.2 in 1972. Increasing concentration lessens bulk and therefore cheapens costs of transport, handling of bags, and lessens the work in spreading.

Many compound fertilizers are nearly as concentrated as is possible from the materials used to make them. The increases in *average* concentration will, no doubt, continue as less ammonium sulphate and more ammonium nitrate and urea are made. Ammonium phosphates will replace more of the single and triple superphosphate now used. If more use is made of purer grades of diammonium phosphate, and if urea is acceptable agriculturally as the main source of nitrogen, we may expect to see even more concentrated materials; fertilizers like 19% N, 19% P_2O_5, 19% K_2O or 26% N, 13% P_2O_5, 13% K_2O can be made. If polyphosphates are used instead of orthophosphates, concentrations may increase still further but this change seems unlikely in Britain during the next few years.

CLASSIFICATION OF COMPOUND FERTILIZERS

Most compound fertilizers sold are designed for particular crops and soils. To use compounds efficiently some easy way of seeing if a particular product suits an advisory recommendation is needed. Unless there is a general method of classifying compound fertilizers, it is often difficult for farmers to choose between the alternatives that are available. The analysis of a compound fertilizer shows what proportions of plant foods it contains, and some manufacturers state '*plant food ratios*' on their lists. These figures express the ratio of the percentages of N, P_2O_5 and K_2O, usually by taking the percentage of N (but sometimes the percentage of P_2O_5) as being equal to 1. Any one of a group of fertilizers with similar plant food ratios will be suitable for a particular purpose, but the amount applied must be varied according to the strength of the fertilizer. For example four fertilizers made by three quite different processes are:

	%N	%P_2O_5	%K_2O
Low-analysis product based on ordinary superphosphate	7	7	7
Medium-analysis product based on triple super	10	10	10
High-analysis products based on ammonium phosphate {	15 17	15 17	15 17

These four fertilizers have the same plant food ratio,

$$\%N \; : \; \%P_2O_5 \; : \; \%K_2O$$

being 1:1:1 for each. They are equally suitable for any particular crop, *provided* the overall rate of application of each is chosen correctly. Thus if 100 kg of N, 100 kg of P_2O_5 and 100 kg of K_2O are required the low-analysis compound (containing 7 kg N per 100 kg) will supply these quantities in $\frac{100}{7} \times 100$ kg = 1.43 tonnes. 100 kg of N from the medium analysis fertilizer will require $\frac{100}{10} \times 100$ kg = 1 tonne; the highest analysis fertilizer (17–17–17) will supply the same quantity of N in only 0.59 tonne of product. If a fertilizer with the correct plant food ratio has been chosen, all that is necessary is to divide the number of kilogrammes of N (or P_2O_5) required on a hectare by the % of N (or P_2O_5) in the fertilizer and multiply by 100 to get the weight of compound fertilizer needed in kilogrammes/hectare. The analytical figures on bags or invoices in $\%N$, $\%P_2O_5$, $\%K_2O$ represent the number of kilogrammes of these plant foods in a 100 kg bag (or the number of pounds in a 100 lb bag, or the number of UNITS of 1.12 pounds in a one hundredweight bag of 112 lb).

Classification that aids selection of compounds

Any particular field and crop has characteristics that assist in choosing a suitable compound fertilizer. The fertilizers also have characteristics (their plant food ratios) that assist in naming them. To help the farmer to get the right fertilizer to the right field the following scheme of names was proposed over 10 years ago. It is now used by many manufacturers and by most farmers and advisors. The terms 'phosphate' and 'potash' are widely used and so are used here. Examples only are given in Table 24 of the large range now available in Britain; where possible two fertilizers are listed to show the ranges of concentrations.

General Purpose (1:1:1) NPK compounds with equal percentages of plant foods are often needed for grassland, fodder crops and horticultural crops; they suit soils of average fertility. Grades range greatly; 17–17–17 seems the strongest. At the other extreme is 7–7–7, the first standard compound introduced in 1942 under the war-time rationing scheme and still listed in 1970 (at no higher price than in 1942!)

High Nitrogen ($2\frac{1}{2}$:1:1, 2:1:1, and $1\frac{1}{2}$:1:1) *NPK* compounds are used for spring cereals (which need more nitrogen than phosphate and potash) and for kale and other leafy crops. They also suit grassland managed intensively, root crops grown on soils well supplied with phosphate and potash, and sugar beet which receives salt or kainit. The choice of $\%N$ to P_2O_5 and K_2O in this group depends on crop and soil; actual nutrient ratios and grades vary greatly, roughly from 14–9–9 to 25–9–9; some grassland fertilizers are even richer in N, e.g. 29–5–5.

Low Nitrogen NPK compounds are equally correctly called *High Phosphate, High Potash* (1:2:2 and 1:$2\frac{1}{2}$:$2\frac{1}{2}$) *NPK* compounds. They suit crops grown on soils that are well-supplied with nitrogen; examples are peats, land ploughed from good leys and the arable land in wet

TABLE 24 A SELECTION OF COMPOUND FERTILIZERS SOLD IN U.K. IN 1974/5

(Liquid fertilizers listed are marked 'L' and are in parenthesis)

Type of compound fertilizer	Approximate Plant Food Ratio N : P₂O₅ : K₂O				Grade (%N + %P₂O₅ + %K₂O)					
	N	P₂O₅	K₂O	(L)	to 36 N %	to 36 P₂O₅ %	to 36 K₂O %	over 36 N %	over 36 P₂O₅ %	over 36 K₂O %
General Purpose NPK	1	1	1		10	10	10	15	15	15
	1	1	1	(L	(9	9	9)	17	17	17
High-nitrogen NPK	2½	1	1		14	6	—	25	10	10
	2	1	1	(L	14	9	8	20	10	10
	1½	1	1		6	9	9	21	14	14
Low-nitrogen NPK	1	2½	2½	(L	4	10	15	10	25	25
High-phosphate NPK	1	1½	1		10	15	10	12	18	12
	1	1½	1	(L	8	12	8	16	24	16
Low-phosphate NPK	1½	1	3	(L	12	6	18	18	6	20
	2	1	3	(L	8	4	10	14	8	24
High-potash NPK	1	1	1½		7	7	18	12	12	18
	1	1	2	(L	—	—	10)	15	15	23
	1	1	2½	(L	—	—	—	10	10	18
Low-potash NPK	1	1	⅔		—	—	—	12	12	30
					—	—	—	17	17	12
Nitrogen-potash NK High-nitrogen	1	0	⅓	(L	16	0	8)	25	0	16
Nitrogen-phosphate NP High-nitrogen	2	1	0		20	10	0	30	13	0
High-phosphate	1	2	0		12	24	0	—	—	1
Phosphate-potash PK General purpose	0	1	1		0	12*	10	0	20	20
High-potash	0	1	2		—	—	1	0	14	28

*Based on basic slag.

areas of the North and West where responses to nitrogen fertilizers are often smaller than in the East and South. They suit soils that are deficient in both phosphate and potash. Also this class of compound is often used for combine-drilling with the seed of autumn cereals, because it supplies the small amount of N needed in autumn, whilst providing the crop with enough P and K. Grades (and ratios) from individual manufacturers vary, for example from 6–15–15 to 10–24–24 and 9–25–25.

High Phosphate (1:2:1 and 1:1½:1) *NPK* compounds suit crops grown on soils that are very deficient in phosphate, for example potatoes on fen peat soils, and crops such as swedes and turnips that need much P.

Low Phosphate (or *High Nitrogen, High Potash*) (1½:1:2, 2:1:3 and 2:1:4) *NPK* compounds are needed on soils well supplied with phosphate but where crops need full dressings of N and K. Crops grown on well-farmed land in the traditional arable areas are often in this category; compounds of the 2:1:3 group are ideal for sugar beet in most areas.

High Potash (1:1:2 and 1:1:1½) *NPK* compounds are needed for most row crops grown on potash-deficient soils in the drier parts of the country. They suit crops that require heavy dressings of K such as potatoes, sugar beet and many horticultural crops.

Low Potash NPK compounds are not often made.

Nitrogen-Phosphate NP compounds (2:1:0 and 1:1:0) are used where soils contain sufficient potassium for the coming crop. They suit many crops, and especially cereals on Chalky Boulder Clays which are rich in K but poor in P; they suit sugar beet on heavy land if sodium chloride is given.

Nitrogen-Potash (1½:0:1) *NK* compound fertilizers, which contain no phosphate are not common. They are useful for root crops grown on soils well supplied with phosphate or where basic slag or ground rock phosphate has been given. They are also used to supply the N and K needed by intensively-managed grassland, a dressing being given before cutting or grazing.

Phosphate-Potash (0:1:1 and 0:1:2) *PK* compounds contain no nitrogen. Those based on water-soluble phosphate are suitable for combine-drilling with autumn-sown cereals where no nitrogen is needed. They are also used where straight N fertilizer (e.g. a liquid) is used. Legumes such as peas, beans, clover and lucerne fix their own nitrogen from the air and need no nitrogen fertilizer, but they need much K; *High-Potash* (0:1:2) types of *PK* compound will generally suit them. Compounds of this group which contain basic slag are very suitable for grassland.

Changes in classes of compound fertilizers used

Before 1940 compound fertilizers were mostly rich in phosphate and poor in potassium and nitrogen. Experiments showed that much more nitrogen should be used; therefore during the War, and for some years after, the compositions of the mixed fertilizers that could be made were regulated, three types only being important: High-K, High-N and High-P. After restrictions were stopped, in about 1950, many hundreds of compounds were made, but the most important were 'High-K' types, which were intended for potatoes and sugar beet, but were also used for many other crops. This tendency continued until about 10 years ago when 'High-N' mixed fertilizers became the most important.

Table 25 shows large changes in the kinds of mixed fertilizers produced in Britain. Twelve years ago, only 22% were rich in N but more than 40% of compounds were rich in K, whereas now 48% are rich in N but only 19% are rich in K. These changes (towards nitrogen-rich compounds) have been made so that a single dressing of compound

TABLE 25

CHANGES IN THE TYPES OF COMPOUND FERTILIZER USED IN GREAT BRITAIN

Type of compound		Typical ratio	1962	1964	1966	1973
			Percentage of all compounds used			
NPK compound	High-N	2–1–1	22	34	42	48
	Low-N	1–2–2	–	11	10	11
	High-P	1–2–1	7	6	4	4
	Low-P	2–1–2	–	3.4	3	3
	High-K	1–1–1$\frac{1}{2}$	41	35	26	19
	Low-K	1–1–$\frac{1}{2}$	–	0.4	0.3	0.3
	General Purpose	1–1–1	4	6	9	6
NP, NK, PK		–	–	4	5	8

fertilizer can supply enough N for spring cereals, for grass, and for animal feed crops (such as kale); they have been responsible for the amount of K used changing little in the last few years while the N used has increased greatly. There is little reason to expect further dramatic changes in the types of compounds used, but the total numbers offered to British farmers may diminish as manufacturers rationalise their production.

MIXING COMPOUND FERTILIZERS LOCALLY

Most farmers use large amounts of fertilizer and because granulated compounds are so convenient and labour is scarce, few will now mix straights themselves. These notes are to help the very few farmers, but larger number of market gardeners, who wish to make their own mixtures.

By mixing together ammonium sulphate, ordinary or triple superphosphate, and muriate of potash, a compound fertilizer can be made that will contain the same amounts of plant foods as a purchased com pound, it may be cheaper. The home-made mixture may, however, be more difficult to handle and distribute than a factory-made granulated mixture. Mixtures of powdered fertilizers are liable to set if they stand for more than a few days, so it is best to make up only as much as can be applied straight away. Conditioning materials, such as ground rock phosphate, steamed bone flour or fine peat, may be added to lessen set- ting, but they raise the cost of the mixture so that some of the advan- tage of home-mixing is lost.

Straight fertilizers that have caked should be broken down and sieved before they are mixed. The best way of mixing is to put ammonium sulphate, superphosphate and muriate of potash into a dry concrete mixer and to run the machine for a few minutes. As an alternative, appropriate amounts of straight fertilizers may be mixed with shovels on a dry floor.

Farm mixing is most satisfactory if the materials used are restricted to ammonium sulphate, superphosphate and muriate of potash; these straight fertilizers will give no trouble, except that the mixture may set if it is stored. Other straight fertilizers give trouble in mixtures. *Alka- line* materials like lime or basic slag should not be mixed with ammonium sulphate, since ammonia will be lost. Both lime and basic slag mixed with superphosphate cause 'reversion' of water-soluble phosphate to an insoluble state. *Nitrates* absorb moisture and make powdered mixtures sticky. They must *never* be mixed with organic materials as they increase risk from fire. Chemical deterioration of the mixture will be avoided if superphosphate, ammonium sulphate, ammonium phosphate or triple superphosphate are *not* mixed with the following fertilizers: basic slag, ammonium-nitrate-limestone mixtures, liming materials or wood ashes. If sticky and deliquescent materials, like potash salts, kainit or nitrate of soda, are mixed with other fertilizers, the mixture should be used immediately, or it may become very difficult to spread.

Bulk blending is the name given in U.S.A. to the common practice of mixing together *granulated* fertilizers; the mixtures are made to parti- cular farmers' orders or are specified for an area by local advice based on soil analyses. Because well-conditioned granulated 'straights' are used difficulties in storage and handling disappear. The mixtures are sold in bulk, often being taken immediately and spread by contracting ser- vices. Some bulk blends are sold in bags. There were over 4000 bulk blending plants in U.S.A. in 1968; these supplied over a quarter of all the mixed fertilizers used (ref. 24). Success in U.S.A. of bulk blending seems to be due to the ability of the system to supply the special needs of individual farmers ('prescription mixing') in the form of a fertilizer with good physical properties. The method has not been widely used outside North America. Most British farmers (and farmers in other European countries) find it possible to meet their special needs from the range of compound fertilizers offered—such as is given in Table 24.

LIQUID MIXED FERTILIZERS

Use in Britain

These figures show how the total use of liquid fertilizers has increased. About 6% of the total N used in 1973 was sold in liquid forms, most as straight fertilizer.

	N	P_2O_5	K_2O
Thousands of (long) tons used as liquids in U.K.			
1965	11	2.6	3.3
1967	20	3.9	4.9
1970	38	7.5	8.7
1973	60	11	12

All the P and K was in mixed fertilizers. Although the quantities have more than doubled in 4 years they are still only about 3 per cent of the total P and K used in compound fertilizers.

Use in U.S.A.

In U.S.A. $2\frac{1}{2}$ million short tons of liquid mixed fertilizers were sold in 1967—about 12% of all mixed fertilizers. Much of their success seems due to the same reasons as the success of bulk blending in U.S.A.—it allows small factories to meet farmers' special needs with an economic and handleable product; if desired, micronutrients can be added, sometimes pesticides too. Liquid mixtures are now mostly made from ammonium polyphosphate solution (10–34–0 or 11–37–0), urea-ammonium nitrate solution and muriate of potash. These mixtures tend to be more expensive than equivalent bulk blends in U.S.A.; they are also more dilute, the average analysis of liquid compounds was 7.5–16.2–6.1 in 1967, as compared with 8.6–16.6–12.7 for all mixed fertilizers. Recently 'suspension' fertilizers have been introduced—liquids containing saturated solutions of solids and stabilised to minimise settling by a special clay; these materials may be as concentrated as solids (for example 15–15–15 is made) but they are not as easy to handle as clear liquids.

Practical application and comparisons of liquid fertilizers

The advantages of liquids are:

(1) They are easily and cheaply moved by pumping.
(2) They can be distributed more accurately than solids.
(3) They can be placed where needed in the soil by being injected under pressure.

While anhydrous and aqueous ammonia are much cheaper than equivalent solids, liquid mixed fertilizers are usually more expensive than equivalent solids per unit of plant nutrient. Easy handling was important in expanding the use of liquids in the U.S.A.; they suit bulky handling and spreading, as they can be moved by pumping. Liquid mixtures containing N, P and K save some manufacturing and transport costs, but need special equipment for distribution; they are corrosive and more expensive to store, and they are more dilute than modern concentrated solid compounds. In British experiments liquid fertilizers supplying N, P and K have been as effective as equivalent solids. Increases in the use of liquids will depend on the factors listed and especially on the handling advantages to farmers who employ few workers but who can engage efficient contractors. A few examples of liquid compound fertilizers are in Table 24.

VALUING COMPOUND FERTILIZERS

Since compounds contain two or three plant foods in varying proportions their values cannot be compared by simply comparing costs per tonne of fertilizer. The amounts of plant foods they contain must be valued in terms of standard prices for kilogrammes of plant foods supplied by straight fertilizers. A few examples will make this clear. Prices and analyses are recent quotations in 1974.

Prices of nutrients in straight (simple) fertilizers assumed in these calculations are similar to those for fertilizers sold in Britain in 1974 (Table 12) and were:

N	:	15 pence per kg
P_2O_5 (water-soluble)	:	20 pence per kg
P_2O_5 (insoluble)	:	10 pence per kg
K_2O	:	10 pence per kg

Compound Fertilizer A was a High-N NPK compound with analysis 20% N, 10% P_2O_5, 10% K_2O (9.5% of the P_2O_5 was water-soluble, 0.5% P_2O_5 was insoluble) it cost £69.60 per tonne. One tonne contains:

		pence
200 kg N (at 15p/kg)	=	3000
95 kg of soluble P_2O_5 (at 20p/kg)	=	1900
5 kg of insoluble P_2O_5 (at 10p/kg)	=	50
100 kg of K_2O (at 10p/kg)	=	1000
Total cost		5950

The nutrients, if bought as straight fertilizers, would have cost £59.50 for amounts equivalent to 1 tonne of 20–10–10 compound. Therefore the compound fertilizer is a dearer source (by about 17%) of plant foods than straights at the prices assumed. (An allowance for mixing and handling the straights should be made.)

Compound Fertilizer B was a High-potash NPK compound with analysis 12% N, 12% P_2O_5, 18% K_2O (of the P_2O_5 11% was water-soluble and 1% was insoluble). It cost £68.00 per tonne. The value of the nutrients in 1 tonne of fertilizer is:

		pence
120 kg of N (at 15p/kg)	=	1800
110 kg of soluble P_2O_5 (at 20p/kg)	=	2200
10 kg of insoluble P_2O_5 (at 10p/kg)	=	100
180 kg of K_2O (at 10p/kg)	=	1800
	Total cost	5900

Again the cost of nutrients in the compound fertilizer was greater (by about 15%) than the cost of equivalent straight fertilizers.

Compound Fertilizer C was a High-N NPK compound in liquid form. Its analysis was 14% N, 6% P_2O_5, 8% K_2O (all the P_2O_5 was, of course, soluble in water). It cost £50 per tonne. The value of the nutrients in a tonne in terms of straights was:

		pence
140 kg N (at 15p)	=	2100
60 kg of P_2O_5 (at 20p)	=	1200
80 kg of K_2O (at 10p)	=	800
	Total cost	4100

Straight fertilizers could have supplied these nutrients in a tonne of this compound for £41, so the liquid was about 22 per cent more expensive. Its extra cost would have to be balanced against the costs incurred in mixing straights, handling and spreading the mixture. The comparison would be more appropriately made with a solid compound fertilizer of '*High-N*' NPK type; the decision which to buy would have been made after considering convenience and labour costs involved in using solid or liquid forms of NPK compounds as well as their prices.

Cost of extra nutrients in NPK fertilizers. Similar calculations can be made to discover the extra cost of a fertilizer containing magnesium, sodium or a micronutrient. For example *Compound Fertilizer D* was a High-N NPK mixture containing 15% N, 10% P_2O_5 and 10% K_2O + 0.3% B; it cost £66.30 per tonne. The value of the N, P and K contained in 1 tonne, in terms of straights, was:

		pence
150 kg of N (at 15p/kg)	=	2250
93 kg of soluble P_2O_5 (at 20p/kg)	=	1860
7 kg of insoluble P_2O_5 (at 10p/kg)	=	70
100 kg of K_2O (at 10p/kg)	=	1000
	Total cost	5180

The compound therefore cost about 28 per cent more than equivalent straights for the N, P and K alone. The other High-N compound (A) supplied N, P and K for about 17% more than the assumed cost of straights. Therefore Compound D is relatively more expensive than Compound A and some of the extra cost must be accounted for by the boron it supplies. The worth of the additive must be assessed by knowing the cost of the same amount of boron purchased as borax and estimating the cost of applying it as a powder or a solution. Costs of applying boronated fertilizers are, of course, no greater than the cost of applying an equivalent material containing no boron.

Sometimes the comparison can be made more simply by obtaining quotations for similar NPK fertilizers with and without an extra nutrient. For example a 20–10–10 fertilizer containing boron cost £75.10 per tonne; the similar material (Compound Fertilizer A above) cost £69.60 per tonne. So having boron present increased the fertilizer's price by £5.50 per tonne.

CHAPTER ELEVEN

Planning Fertilizing for Maximum Yields

When farmers first use fertilizers they are satisfied if yields are increased and the money spent returns a profit. As farming becomes more intensive, costs increase and maximum yields are needed to keep the system profitable. In most intensive agriculture fertilizers are responsible for only a small part of total costs (10% in U.K. in 1973) and it will usually pay to use sufficient to grow the largest yields that are possible. Too much emphasis on cutting costs by lessening fertilizer use impedes progress by preventing full returns from other advances in agricultural science; larger crops, however achieved, need more nutrients. Economic analyses of fertilizer effects based on results of old experiments where maximum yields were not achieved cannot help to build more productive farming systems. Advances in agriculture are now achieved so quickly that some kinds of experimental work are soon out of date; advisers and farmers must be prepared to modify fertilizer practice as new results become available. This Chapter discusses how fertilizer dressings are chosen and shows how yields are built up by exploiting the interactions between factors that affect yield.

Whether the fertilizers are needed to increase production in a developing system, or to maintain large yields in a system that is already intensive, the problems are the same. A farmer must choose:

> the *right amount*
> of the *right kind* of fertilizer
> and apply this dressing
> in the *right place* and
> at the *right time.*

The next few chapters discuss the scientific information needed to decide on *kind* and *amount* of fertilizer. Timing and placing are discussed in Chapter 18.

The first decisions that have to be made are about the nutrients that must be applied because the soil cannot supply enough; ways of deciding which are needed are discussed in the section below.

DIAGNOSING NUTRIENT DEFICIENCIES

Symptoms of plant food deficiencies

When soils are *very* deficient in one or more of the major plant nutrients, crops grown on the land often show what is wrong by symptoms on their leaves or by their habit of growth. When such symptoms appear they indicate such a serious deficiency that yields of the current crop will be lessened. If nitrogen deficiency is detected at an early stage it can usually be corrected by fertilizer dressings, since nitrogen is washed into the soil by rain. If symptoms of phosphate or potassium deficiency appear, it will probably be too late to secure a full crop by giving P or K fertilizers, as these are only useful when they are mixed with soil round the plant roots.

Nitrogen deficiency usually gives poor and thin plants with pale-green or yellowish-green leaves. Later in the season some nitrogen-deficient crops develop other colours: for instance kale, cabbage and swedes have shades of purple, blue, or red on their leaves.

Phosphate deficiency in cereals leads to young plants with bluish-purple stems and leaves. With most other crops a shortage of phosphate causes the plants to be stunted or spindly, but without any unusual colouring of the leaves.

Potassium deficiency is usually seen as a discoloration of part of the leaf; afterwards the affected part may die. In cereals the tips of the leaves become yellow and may wither. K-deficient potatoes have dark-green leaves which, later, become a purplish-bronze. Patches of the leaves may die and crack away. Severe potassium deficiency in sugar beet, mangolds and kale leads to 'scorching' of the leaves, followed by the death of the affected parts. Clover, lucerne, peas, beans and other legumes are particularly sensitive to K shortage; acute deficiencies lead to yellow spots on the leaves. Later these spots turn brown and run together and the leaf dies. Tree fruit and soft fruit quickly show the effects of K deficiency by 'scorching' on their leaves.

If symptoms of these sorts appear on crops, the diagnosis should be confirmed by analysing plant tissue. The manuring for future crops must be planned so that the deficiency of plant food in the soil is not allowed to interfere with crop growth.

Simple fertilizer tests

Simple fertilizer tests are often made in developing areas to determine which nutrients are deficient in local soils. The name 'subtractive method' is sometimes given to this way of diagnosing deficiencies because a basic fertilizer mixture is used and on each plot of the 'experiment' one nutrient is omitted; plots with the complete mixture and with no fertilizer at all are included. This method has been widely used for diagnosing deficiencies in tropical soils. Results of three such tests on cotton in Africa (given by R. Chaminade in ref. 34) are below:

Yield of cotton with NPKS fertilizers kg/ha	Yield of cotton grown without one or more nutrients as percentage of yield with full fertilizing				
	without fertilizer	N	without P	K	S
893	43	43	87	92	35
1087	53	79	82	97	72
2014	69	90	77	93	96

The first experiment shows that both sulphur and nitrogen are seriously deficient and that some phosphorus is needed too. The second shows that N, P and S are all needed; at the third site, which gave best yields, phosphorus was the most serious deficiency. These trials show that the soils contain sufficient potassium—this is often so in developing tropical agriculture.

Large numbers of experiments of simple patterns have been made under F.A.O. auspices; often the intention was only to demonstrate overall benefits from a fertilizer mixture and the tests have been of N, NP, and NPK fertilizers against none. Other experiments of 8 plots have tested all combinations of none versus single dressings of each of the nutrients N, P and K. Some results of these experiments are given in Chapter 12.

While deficiency symptoms and these simple field tests draw attention to serious shortages of plant foods and show the *kind* of fertilizer needed, they give no information on the *amount* that must be applied to obtain a full yield. Besides, crops that appear quite normal often give profitable returns from fertilizer dressings. Planning fertilizing by waiting until deficiency symptoms appear can never be better than an attempt to control a disaster, for no annual crop that has been so severely checked early in its growth as to show leaf symptoms can be expected to give maximum yield. In newly developing areas where fertilizers have been little used, leaf symptoms may help in deciding what fertilizers should be tested. But in developed intensive agriculture, if symptoms of a deficiency of a major nutrient appear, the methods used to control fertility by fertilizers should be reviewed. A watch for leaf symptoms should however be kept in intensive farming because the first warning of a micronutrient deficiency may be obtained in this way.

CHOOSING AMOUNTS OF FERTILIZERS

Optimum dressings from field experiments

Only field experiments that test several increasing dressings of each nutrient can show how fertilizers affect crop yields and form a basis for deciding the right amount, that is the 'optimum dressing'. But it must be recognised that choosing dressings by studying experimental results is not a simple business and it becomes more difficult to make correct decisions as agriculture is intensified. Responses to fertilizers are very variable and depend on crop, type of soil and past use of the land even when these factors are the same, they still depend greatly on local weather and the characteristics of the whole season. A single experiment or even a group of experiments in one small area, cannot give reliable conclusions on the effects of fertilizers over much larger areas, nor can they identify and analyse all the factors that alter responses. Averages of responses in individual experiments can be a basis for advice to a whole district; to give satisfactory advice for a single field—which is how *all advice has to be given*—regional advice *must* be modified to take account of all that is known about local conditions.

When farmers used much less than published optimum dressings, as happened 25 years ago, the optima were useful targets to increase fertilizer use. The present situation in Britain and many developed countries is quite different. Farmers use *on average* as much or more fertilizer as is generally recommended for many arable crops, but often they do not differentiate between poor and rich land. On some rich fields part of the fertilizer is wasted, sometimes yield is lost by using too much. On other fields that are poorer, average optima are too small for crops in continuous arable farming. Recent British experiments on root crops and cereals show that previous cropping, and particularly a period under a legume or grass, affects the amount of nitrogen fertilizer needed for maximum yields; previous cropping and manuring, and soil type, affect the responses to both phosphorus and potassium.

It is useful to ask whether we have enough evidence from field experiments alone to give advice on using fertilizers in Britain:

Nitrogen fertilizers. Adequate advice on using *nitrogen* for all our important arable crops and for grass *can* be given for 'average' conditions, though weather introduces uncertainties. Most farmers use less fertilizers on grassland than we know to be justified; as more fertilizers are used on grass, more information will be needed on the response to nitrogen of different types of herbage used in different ways.

Phosphorus and potassium fertilizers. Enough field experiments have been done on high-value arable crops—potatoes and sugar beet, and on swedes, to calculate average optimum dressings of *phosphorus* and *potassium* fertilizers. Other arable crops (including cereals), and grass, only give small *average* responses to P and K; often these fertilizers have no immediate effect. Some basis other than responses in field experiments has to be used to make recommendations for these crops, which get about four-fifths of the phosphate and potassium bought by farmers. Scientific information on soils and nutrient cycles must be used to avoid both waste and loss of potential yield.

Auxiliary information that helps in deciding amounts of fertilizers

Besides a general knowledge of local soils, farming conditions and fertilizer experiment results, an adviser's decisions on *amounts* of fertilizer may depend on some or all of the following kinds of information:

(1) *Nutrient cycle* on the farm—particularly important for planning to use potassium on both arable crops and grass, but important too for planning N-manuring.

(2) *Interaction effects* associated with methods and intensity of farming. (Interactions that are unimportant when yields are small may become very important as yields are raised by applying other nutrients or water, or by growing the crop differently.)

(3) *Past manuring* with FYM and fertilizers leaves residues of N, P and K in soil that benefit following crops. The residues of inorganic N fertilizers usually last only for a season, but the residual effects of continued manuring with P and K may last for many years.

(4) *Farming system:* Crop rotations, methods of using crops, systems of keeping stock, and the use of FYM all affect the supply of nutrients to crops.

(5) *Weather* affects fertilizer responses. No allowance can yet be made for the season that is to come, but useful adjustments can be made each spring to allow for the effect of the past winter's rainfall on the nitrogen needed by crops. Leaching also occurs when heavy rain falls in spring; the risks of loss of seedbed dressings must be assessed so that further top-dressing of N can be given if they are needed.

(6) *Soil analysis* is the only way of assessing the need for calcium fertilizers (liming materials). Repeated analyses on the same land also assess the changes in soluble P and K caused by the cropping and manuring used. Soil analyses show how P and K dressings should be adjusted; large fertilizer dressings are needed to produce immediate responses to P and K on poor soils, on richer land smaller dressings will maintain satisfactory amounts of soluble P and K in the soil. Soil analyses cannot yet be used as a satisfactory guide for nitrogen manuring. Measurements of soluble magnesium (and sometimes of micronutrients) are useful.

(7) *Plant analysis* often shows when crops contain too little nutrients for satisfactory yields. Measurements of nitrate in plant tissue can indicate the need for extra N fertilizer while there is still time for it to be effective. Similarly micronutrient deficiencies may be detected by leaf analysis and immediately corrected by a spray, the small quantity that is essential being taken in by the leaves. Measuring other elements in plants often helps in investigating nutritional problems. In some countries plant analysis is widely used to aid advice on fertilizing. It is particularly useful for tree and other perennial crops where field experiments are often costly and difficult to do and are slow to yield results. The nutrition of crops such as sugar cane, pineapple and oil palm is often controlled by leaf analysis.

BUILDING MAXIMUM YIELDS

Targets

All work on crop improvement needs targets by which improvement may be judged. These are rarely available except where permanent field experiments are done. Usually improvement is step by step, factors that limit yields are identified and then removed. At Rothamsted we have several experiments where this is done, notably our 'Ley-Arable' rotation experiments where test crops of wheat, barley and potatoes are grown in rotations that avoid root diseases. Average 'best' yields from these experiments in Table 26 are compared with national average yields since 1952. For long periods best yields of cereals at Rothamsted have been nearly twice the national average yield, recently best potato yields have been more than twice the national average. Comparisons of 1962-64 and 1965–69 national averages suggest that cereals yields are not now increasing, but potato yields are rising slightly.

TABLE 26

COMPARISONS OF BEST YIELDS OF WHEAT, BARLEY, AND POTATOES GROWN IN THE ROTHAMSTED LEY-ARABLE EXPERIMENTS WITH AVERAGE YIELDS FOR U.K.

| | Wheat (grain) | | Barley (grain) | | Potatoes (tubers) | |
	Rothamsted*	U.K. average	Rothamsted*	U.K. average	Rothamsted*	U.K. average
	Yields in tonnes/hectare					
1951-3	4.9	2.9	4.2	2.7	34	21
1954-5	5.1	3.1	6.0	3.0	28	19
1956–8	5.3	3.1	5.3	2.9	39	19
1959-61	6.6	3.6	6.4	3.2	43	22
1962-64	7.3	4.2	6.2	3.6	47	22
1965	6.7	4.1	6.3	3.8	58	25
1966	7.4	3.8	7.0	3.5	60	24
1967	8.4	4.2	6.9	3.8	60	25
1968	5.4	3.6	5.4	3.5	44	25
1969	9.3	4.1	7.0	3.6	57	25
Average 1965-9	7.4	4.0	6.5	3.6	56	25

*From 1959 to 1968 the wheat variety was Cappelle Desprez, in 1969 Joss Cambier was grown; from 1954 to 1964 the barley grown was Proctor, from 1965 Maris Badger was used. Potato varieties at Rothamsted were: 1951-67 Majestic, 1968 and 1969 King Edward.

Increases in yields of cereals in these Rothamsted experiments have been from introducing improved varieties, using more correct fertilizer dressings and better herbicides, and from improving cultural practices. Marked increases in yield occurred when modern stiff-strawed varieties were introduced in the experiments, for barley (Proctor) in 1954, for wheat (Cappelle Desprez) in 1959. (A mildew resistant barley (Maris Badger) introduced in 1964 raised yields immediately, but it had lost its resistance by 1968.) Similar improvements in yields are shown by national averages. Potato yields depend much more on season, particularly on the incidence both of dry summer weather and of leaf disease. Both the experimental and the national average yields show steady improvements.

Interactions with fertilizer effects

Experiments that test one factor alone, such as increasing amounts of nitrogen, soon show how much N is needed for maximum yield under these conditions. Separate experiments may test amounts of other factors —such as phosphate, or irrigation water. But experiments on one crop should be planned to test all the factors that can be varied and when this is done it is often found that the gain from correcting one factor that normally limits yield is greater when another limiting factor is corrected at the same time. Thus the maximum benefit from N will only be obtained when crops have sufficient P, K and water. Responses to N will be larger in the presence of P than without P, and *vice versa*. These effects, when the gain from factors A and B tested together are greater than the sum of gains from factor A + factor B tested separately, are called *interactions*. It is by exploiting the interactions of nutrients with each other, and with improved practices such as better varieties, irrigation, weed control, and disease control that yields have been raised dramatically in many countries in the last 30 years. In most older experiments small amounts of each fertilizer were tested singly; responses recorded were less than they would have been on crops where other factors that increase fertility were tested. Results of experiments with fertilizers done more than 20 years ago are mostly out of date and fresh experiments must be made continuously so that the effects on fertilizer responses of other ways of improving crops can be tested. Further progress towards maximum yields in developed agriculture will depend largely on exploiting interactions.

Nutrient interactions

In experiments done in the last 20 years interactions between nutrients have become increasingly important as the amounts of fertilizers tested have increased. As the basal manuring with PK fertilizer for potatoes, or K for sugar beet, was increased, responses to nitrogen increased. Often large amounts of nitrogen are justified only when much PK fertilizer is used.

Potatoes are the only important *agricultural* crop grown in Britain for which all three nutrients must always be considered together in planning to manure individual fields. Responses to each nutrient depend on supplies of other nutrients more than with any other crop and interaction effects must be exploited to secure maximum yield. Table 27 shows yields derived from a group of experiments on potatoes on peat soils done by N.A.A.S. (ref. 35); maximum response to N (or to PK) depended on using enough PK (or N).

TABLE 27

NITROGEN RESPONSE CURVES FOR CONTRASTED LEVELS OF PK

USED ON POTATOES GROWN ON FEN PEAT SOILS

Mean yields, t/ha

Level of N	Level of PK		
	00	11	22
0	(26.9)	(28.3)	–
1	33.3	34.6	32.6
2	32.6	38.5	39.0
3	29.4	35.4	39.4

Potatoes often show large gains from farmyard manure (partly due to increased supplies of potassium, partly from physical soil improvements) which are only realised when extra nitrogen is applied. Fig. 1 shows 5 years of results on clay-loam at Rothamsted and sandy-loam at Woburn in experiments that test NPK fertilizers with and without farmyard manure (FYM). In both experiments there were large responses to nitrogen, but the whole level of yield was raised at each site when FYM was applied as well as fertilizer; surprisingly the two response lines were parallel and maximum yields were not achieved. The effects of FYM in these experiments come from direct effects of fresh dressings plus the residual effects of dressings in previous years, they are thought to result from interaction of nutrients supplied by FYM with increasing amounts of fertilizer nitrogen.

Grass. Supplies of nutrients in soil that are sufficient for small yields are often not enough when larger yields become possible. The results of an experiment at Rothamsted that has tested N and K fertilizers on grass for 10 years which are illustrated in Fig. 2 are a good example. Potassium from soil was practically sufficient for the yield that 37 kg N/ha per cut could give. When twice as much N was used 32 kg K/cut was needed and there was no gain from using more K. But with the largest amount of N, there was a gain from the larger dressing of K.

Maize. The results of an experiment in Illinois (U.S.A.) testing N and K fertilizers on maize are in Fig. 3. They show well how yield is built up by the two nutrients acting together. They also show how the optimum dressing of N depends on the amount of K used. With no potassium fertilizer 160 kg N/ha was enough, with 45 kg K/ha, 200 kg N was justified; when 135 kg K/ha was applied the most N tested, 270 kg N/ha, was not enough for maximum yield.

Micronutrients. Examples have been discussed in Chapter 9 of interactions between micronutrients and also of multiple interactions between a major nutrient and micronutrients.

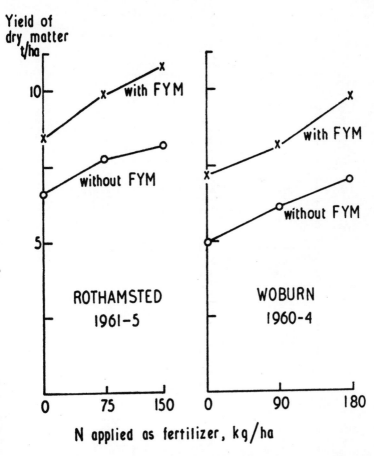

Fig. 1   Yields of potatoes grown in Reference Experiments at Rothamsted
and Woburn. Farmyard manure (FYM) and N-fertilizer interact to
produce larger yields and the amount of N-fertilizer justified is not
diminished by giving FYM.

Water-nitrogen interactions

When yield potentials can be raised by irrigation, more fertilizer
will be justified to produce the extra crop. Fig. 4 gives yields averaged
over 3 years in an experiment with spring wheat on sandy loam at
Woburn where 4 amounts of N were applied with and without irrigation
to make good the moisture deficit. Irrigation increased the maximum
yield measured in the experiment by 825 kg/ha. The largest N dressing
justified was 100 kg N/ha without irrigation but 150 kg when the crop
was watered.

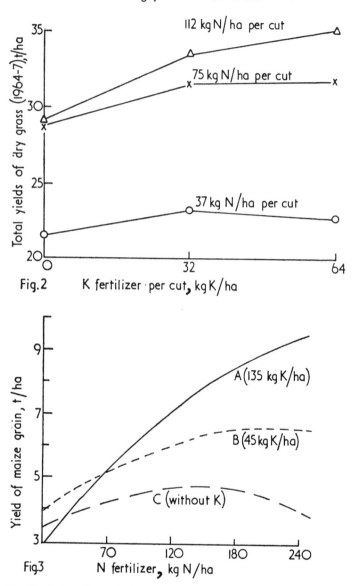

Fig.2

Fig.3

Fig. 2   The interactions of nitrogen and potassium fertilizers used on grass the amount of K needed depends on the amount of N used.

Fig. 3   The effects of N-fertilizer on yields of maize. The amount of N fertilizer needed for maximum yields depends on the amount of K fertilizer applied.

Interactions with methods of growing crops

When more plants can be grown on an acre, more nutrients are needed to produce the extra yields that are possible. A typical example is the demonstration by American workers that close-planted stands of hybrid maize justify larger dressings of nitrogen than crops with smaller plant populations (examples are given in Chapter 22).

Interactions of new crop varieties and fertilizers

An excellent example of how plant breeding produces new crops that have larger potential yield and justify much more fertilizer is given in the section on rice in Chapter 22.

Nutrients other than N, P and K

As larger yields are obtained more of the other major nutrients are needed (and more micronutrients too). It is quite possible for yields, and responses to N, P and K, to be restricted by shortages of other nutrients and for some of the potential yield to be lost. Supplies of Mg, S, Ca and micronutrients from soil and natural sources that suffice for one set of yields may be too small for much larger yields. Experiments in many countries have shown that very large yields are possible when well-managed crops are grown on good soils. Recently-published examples from U.S.A. are in Table 28. Very large amounts of calcium, magnesium and sulphur were removed by these record crops which are much larger than the average American farmer now gets, but which may become common in future.

TABLE 28

CALCIUM, MAGNESIUM AND SULPHUR TAKEN
UP BY VERY LARGE YIELDS OF CROPS IN U.S.A.

	Lucerne (California)	Maize (Missouri)	Wheat (Washington State)
Crop yield tonnes/ha	36 (dry hay)	20 (grain)	14 (grain)
		Nutrients in crops, kg/ha	
Ca	726	120	47
Mg	123	110	48
S	109	74	50

Fertilizing for Maximum Yield

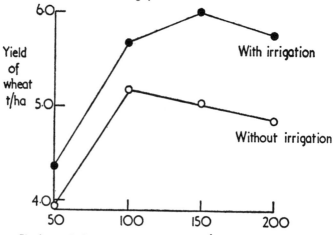

Fig.4   N fertilizer applied, Kg N/ha

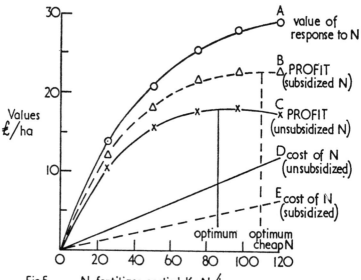

Fig.5   N fertilizer applied, Kg N/ha

Fig. 4   The effects of N-fertilizer on yields of spring wheat grown with and without irrigation at Woburn. The amount of N needed for maximum yield depended on whether the crop was irrigated.

Fig. 5   The relationships between the value of extra wheat grain produced by nitrogen fertilizer, the cost of the dressing and the profit obtained. Prices assumed are: wheat £28/tonne
N in fertilizer: 10 np per kg N without subsidy
                            5.4 np per kg N after deducting subsidy

CHAPTER TWELVE

The Responses of Crops to Fertilizers

The *proof* that fertilizers are profitable, by giving increases in yield worth more than the dressings cost, can only be got from field experiments. Deficiency symptoms, crop analyses, and soil analyses can all indicate that certain nutrients *may* be deficient and that appropriate fertilizers are worth testing. However, fertilizer programmes for producing maximum yields from intensive cropping systems must be devised against a background of information from modern field experiments using appropriate cropping systems on local soils. This Chapter discusses the relationships between fertilizer dressings and crop yields, that is, the response 'curve'. Older methods of dealing with experimental results are described first and this is followed by some modern ideas that are important for obtaining maximum yield at least cost.

METHODS FOR DERIVING OPTIMUM DRESSINGS

The usual way of finding out *how much* fertilizer should be used for a crop is from the results of field experiments that test different quantities of plant foods and measure the extra yields given by the dressings. From the results of a sufficient number of trials of this sort practical recommendations can be made on the 'most profitable' (or 'optimum') dressings that will give the greatest return from the outlay on fertilizers. The dressings will be appropriate to the average of conditions in the area where the trials were.

The classical way of calculating optimum dressings is best understood from Fig. 5, based on the average results of many old experiments on stiff-strawed varieties of winter wheat. The value of the response (extra yield given by fertilizer), the cost of the fertilizer and the profit are all plotted on a hectare basis against the amount of nitrogen applied. The average responses to nitrogen were not directly proportional to the amounts of fertilizer used. When the amounts of N increased regularly by equal steps, the extra crop given by each additional step decreased regularly as in line OA. The value of the extra grain produced by the first steps of the nitrogen dressing, as shown by OA in Fig. 5, is much greater than the cost of the dressing. As the dressings are increased their cost (line OD) becomes more important and a stage is reached (at about 85 kg of N per ha) where the last step of the fertilizer dressing costs as much as the extra grain it produces is worth. The amount of fertilizer that produces the greatest profit per acre is called the '*optimum dressing*'. If more than the optimum is given, then profit will fall again; if less than the optimum is used, some profit will be sacrificed.

In Fig. 5 the greater part of the total profit is obtained from the first steps of a fertilizer dressing. The first 25 kg of N, costing about £2.50 per hectare, produce an extra crop worth £13.20 per hectare leading to a profit of £10.70 which is more than half of the total profit (£19 per hectare) given by the optimum dressing. In such circumstances it is more profitable, if money for buying fertilizer is limited, that all fields should receive some fertilizer than that some fields should receive the optimum dressing and the rest none. If the likely returns from the crop are uncertain, either because prices are not known in advance, or because the crop grown will be used for feeding livestock on the farm, then it is better to use rather less than theoretical optimum dressings.

Effect of change in fertilizer prices

This discussion assumes the price of N in ammonium nitrate current in Britain in 1971. Lines OB and OE in Fig. 5 show corresponding data for a cheaper form of nitrogen—either cheaper to buy (as N in anhydrous ammonia is) or subsidised at the rate current in spring 1971. While the general nature of the effects shown is similar at both prices, the profit curve is displaced so that, when calculations are based on the smaller prices, the optimum dressing is about 110 kg of N per hectare instead of 85 kg.

With relationships of this kind optimum dressings are not sharply defined. Line OC in Fig. 5 shows that while 85 kg of nitrogen brings in most profit (£19/ha) using only 60 kg produces nearly as much profit (£17.80) and if 110 kg N are used the profit is still only 38 np less than the maximum that can be obtained. In this particular example the profit obtained from the range of dressings between 60 and 120 kg of N did not vary by more than £1.25 per hectare. Using *exactly* the right amount of fertilizer is not very important; provided that an amount near to the optimum quantity is used, a farmer may expect to get quite near to the maximum possible return. Also, there is *no need* to change fertilizer dressings continually to try to allow for small changes in soil fertility or prices. The season will always have a large effect on the fertilizer responses and, as weather cannot be forecast when the crop is sown, the exact profit cannot be forecast either.

Exponential response curves

It is quite satisfactory to derive optimum dressings of fertilizers by a graphical method like that in Fig. 5. But when *many* sets of experimental results have to be handled tedious plotting can be avoided by using an equation that fits the response curve.

Curve OA in Fig. 5 has an *exponential* form as it approaches more and more closely to the maximum but does not reach it, or diminish. The equation for such curves was developed many years ago by Mitscherlich in Germany and has been much used in Britain for assembling experimental results and for calculating optimum dressings. Its form is:

$$y = y_0 + d(1 - 10^{-kx})$$

where y is the yield with a fertilizer dressing of x kg N, P or K per hectare, y_0 is the yield without fertilizer, d is the limiting response, and k a value assumed to be constant for each of the three principal classes of fertilizer. The form of the curve implies that the response ($y_1 - y_0$) to a unit dressing of amount p of a fertilizer bears a constant ratio, 10^{kp}, to the additional response to a second unit. The latter response bears the same ratio to the further additional response obtained by the application of a third unit. British experiments done before 1940 showed the following average values: for nitrogen $k_N = 1.1$, for phosphorus $k_P = 0.8$, for potassium $k_K = 0.8$, and for farmyard manure $k_{FYM} = 0.04$.

E. M. Crowther and F. Yates (ref. 36) showed how relative responses to different dressings varied in relation to the response to a standard dressing. They also showed how to calculate optimum fertilizer dressings. If v is the value per hectare of the response to the standard dressing, and c is the cost per hectare of that dressing, Table 29 shows the optimal dressings corresponding to a series of value/cost (v/c) ratios, when the relationship between yield and fertilizer has an exponential form.

TABLE 29

OPTIMAL DRESSINGS FOR VARYING VALUES OF CROPS AND FERTILIZERS

From E. M. Crowther and F. Yates' paper (ref. 36)

v/c*	Optimal dressing in kg/ha of N	P or K
1	15	28
1.5	35	57
2	49	76
3	69	104
4	84	122
6	102	150
8	118	170
10	128	186

*v = value per hectare of the response to a standard dressing, and c is the cost per hectare of that dressing.

This method of treating field experiment results was developed by Crowther and Yates to rationalise advice on the use of fertilizers in the 1939–45 War and it remains the basis of much modern advice. They found their results applied to much of Northern Europe. The exponential curve was adequate while farmers used (and most experimenters tested) amounts of fertilizers that were less than optimum. Many farmers now use as much or more fertilizer than is indicated by calculations of optimum dressings based on the exponential equation for each nutrient considered separately, and this treatment has therefore become inadequate for advising on practical conditions.

Parabolic response curves

Many recent experiments have been planned to find the full forms of response curves for large rates of dressings and to show how response to one nutrient is affected by the amounts of other nutrients used. In experiments done since 1950, average response curves have often been parabolic, yield rising to a maximum and then falling as more fertilizer is given. Fig. 3, which illustrates the results of an experiment on maize in U.S.A., has already given a good illustration of the way that fertilizer response curves vary in shape according to the conditions of the test. Where no potassium was given the response curve (C) for N is parabolic, yield rising to a maximum and then falling again as more N is given. The response to N with the moderate dressing of K (45 kg/ha), curve B, is already diminishing with the largest rate of N. With the large dressing of K, maximum yield had not been reached with the largest dressing of N. The form of the response curve A suggests that the exponential equation would fit these results, but curves B and C suggest that had the test been continued to much larger amounts of N, a parabolic curve might also have been found. These differences in response to one nutrient caused by differences in amounts of a second nutrient show how impossible it is to forecast the form of the response curve before doing the experiment, and to decide how the results should be treated mathematically.

It is very important, (1) to establish the full shape of the response curve in properly designed and well-conducted experiments (paying attention both to the slope of the steeply rising part with the smaller dressings, *and* to the effect of excessive dressings), and (2) to define as precisely as possible the soil, cultural and weather conditions under which the test was made. These aspects of field experimentation with fertilizers are much more important than elaborate calculations on the results, perhaps using a standardised equation which is inappropriate for these data. An old saying is that 'silk purses cannot be made from sows ears', and satisfactory fertilizer recommendations cannot be derived from results of inadequate experiments, however good the mathematical treatment may be.

Other mathematical relationships

When experimenters tested large fertilizer dressings, and found that too much often depressed yields, a variety of methods of treating the data mathematically were developed. No standard equation is widely acceptable, which is an acknowledgement of the fact that fertilizer

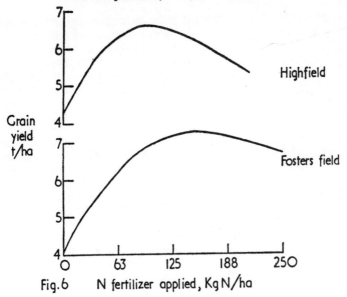

Fig. 6 N fertilizer applied, Kg N/ha

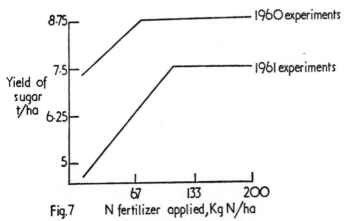

Fig. 7 N fertilizer applied, Kg N/ha

Fig. 6 The effects of nitrogen fertilizer on yields of wheat in two similar experiments at Rothamsted. Until 1948 Highfield was very old pasture, Fosters Field was old arable land. The shapes of the response curves and the N needed for maximum yield depend on previous history of the land. (Ref. 44)

Fig. 7 Two-part linear relationships between the N-fertilizer used on sugar beet and the yields of sugar. (Ref. 39)

response curves have no standard form, their shape depends not only on crop and nutrient but on all the local conditions of soil, weather and farming methods. Fig. 6 is an example; it shows the responses of wheat to nitrogen fertilizer in two identical experiments on the same soil type on Rothamsted Farm. The two fields are cultivated identically and the same varieties are grown; they are 1 km apart, but differ in their old history. Highfield was permanent grass until 1948, Fosters Field has always been arable. Even 20 years after ploughing Highfield, the soil still contains much more nitrogen than Fosters Field as a residue of its time under grass. This extra N is released in greater amount and at different times from the N released from Fosters soil. The differently shaped response curves to N used on the same crop are the result of these differences, the greater slope of the downward part of the curve on Highfield being due to the extra nitrogen liberated by the soil.

F.A.O. (ref. 37) has produced an excellent account, which should be studied, of the different ways of treating modern experimental results to determine how much fertilizer should be used. Apart from the Mitscherlich exponential equation given above, two others were found widely satisfactory:

Quadratic curve $y = a + bx + cx^2$

y = yield obtained with x units of fertilizer,
a, b, c are constants to be determined (c measures the curvature of the response)

Square root formula $Y = a + b \sqrt{x} + cx$

Other curves have been used, sometimes because they are easy to handle, for example:

$$\frac{1}{y} = a + \frac{b}{x + c}$$

F.A.O. (ref. 37) reported that much data from modern experiments in India were suited by the quadratic equation which gives a good fit. They described, with examples, how the equation is fitted to a series of field experiment results, how to calculate the constants a, b and c and how to calculate the optimum dressing x_0, which is derived from the expression

$$x_0 = \frac{b - (q \div p)}{2c}$$

where q is the cost of a unit of fertilizer
p is the price of a unit of yield
b and c are, again, constants.

An example of an English comparison of equations was made by Levington Research Station in interpreting results of experiments on maincrop potatoes (ref. 38). The Mitscherlich equation was unsatisfactory because much fertilizer depressed yields. Two other equations were tested:

$$\text{the square root function, } y = ax + b\sqrt{x} + c$$
$$\text{and a squared-term function, } y = ax + bx^2 + c$$

The square-root function gave the best fit with the data and was used to calculate economic optima. Nevertheless these workers advised caution in using the results because (1) there is a wide tolerance around the calculated optimum in which profit is little affected by amount of fertilizer used; (2) experiments are subject to considerable errors which make the shape of the curves uncertain.

The F.A.O. manual also discusses the much more difficult treatment needed to calculate response *surfaces* (i.e. curved *areas*, not simply the curved *lines* for one nutrient). Surface must be calculated whenever there are interactions between two nutrients because the optimum amount of one nutrient depends on the amount applied of the second.

Response curves are useful for standardising responses in series of experiments using different amounts, for comparing the relative efficiencies of two sources of one nutrient or different methods of application, and for calculating residual effects in terms of direct effects of fresh fertilizer. It may, however, be misleading to assume that the constants in any equation for response curves have fixed values and erroneous conclusions can be derived from using one set of constants to interpret experiments made under quite different conditions.

Recent developments in interpreting field experiment results makes it doubtful whether study of the actual *form* of the response curve derived from the averages of numbers of experiments will help in raising yields in highly developed farming systems. The results of fewer experiments closely related to the local conditions of a particular farmer's field are likely to be more useful than the more precise *averages* derived from experiments over much larger areas. Average results of series of experiments apply only to the 'average' field; expressions of average fertilizer effects are made up of results from fields where fertilizers were less effective and more effective than the average. Average advice is rarely applicable, it *must* be improved by adjusting fertilizing to the local supply of nutrients in each field and to the crop's potential yield.

LINEAR RELATIONSHIPS BETWEEN YIELD AND FERTILIZER DRESSINGS

Recent work on the results of field experiments testing several amounts of fertilizer rising to rates above the optimum has questioned whether the curve', for nitrogen at least, is a curve at all. Two linear relationships on either side of a point of inflexion fit the field results better than a curve.

Sugar beet and nitrogen

Responses in individual experiments, and in groups of related experiments, on sugar beet made in the last 10 years have been examined by staff of Rothamsted and Broom's Barn Experimental Stations (ref. 39). They found a practically linear relationship (illustrated in Fig. 7) between yield and amount of N needed for maximum yield; with larger dressings there was a sharp transition to a linear and almost horizontal part of the curve. Their results are confirmed by experiments in the Netherlands on sugar beet; as in the British work responses to N-fertilizer were not large, but most experiments showed a sharp point of inflexion above the rising part of the curve, followed by practically no change in yields as dressings increased.

Cereals and nitrogen

Most experiments on cereals have tested too few rates of nitrogen to show the shape of the response relationship. A recent review (by D. A. Boyd in ref. 34) of the experimental results that explore the 'curve' more thoroughly shows that nitrogen increases yields rapidly and more or less linearly up to a transition point beyond which yield may change little, or may decrease slowly and linearly. One example of the effect of nitrogen on cereals is in Fig. 4, another given by D. A. Boyd is in Fig. 8.

Grass and nitrogen

Many experiments testing N on grass show a two-part or three-part response relationship, yield rising steeply with increasing N to a sharp point of inflexion above which yields were again linearly related to N but the line was less steep. Work with very large dressings of N on grass shows that dry matter yields do not continue to rise indefinitely (but 'protein' yields often do) up to the largest N dressings tested. Fig. 9 shows an example from an experiment testing very large amounts of N on a tropical grass in Peurto Rico. Very few British experiments on grass have tested many dressings of N rising to amounts larger than are likely to be used in practice. The little information there is suggests that the relationship is in three parts: In the first part N increases yields linearly up to total amounts of about 350 kg N/ha used in a year (but split into several dressings). Above this transition point the rate of increase is less although N still increases yield. When annual total dressings exceed about 500 kg N/ha, the extra fertilizer has little effect on yield, although it increases yield of 'crude protein' further.

General form of nitrogen response curves

When the errors of individual field experiments are taken into account there is often no sound reason for using non-linear (i.e. curved) relationships between yield and fertilizer dressings. Fig. 10 shows this simply by

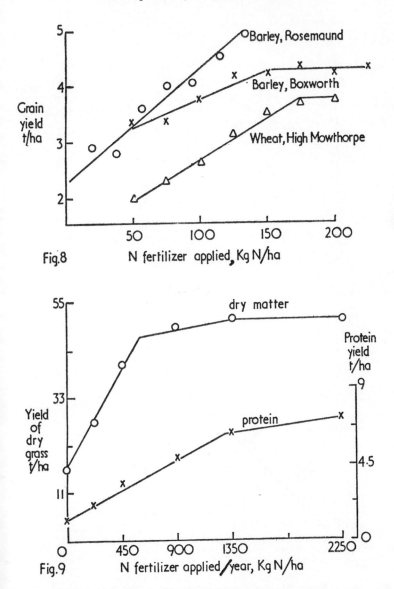

Fig. 8 Linear relationships between yields of cereals and the amounts of N-fertilizer applied. (Ref. 34)

Fig. 9 Two-part linear relationships between yields of dry matter and of protein and the amounts of N-fertilizer used on grass in Puerto Rico.

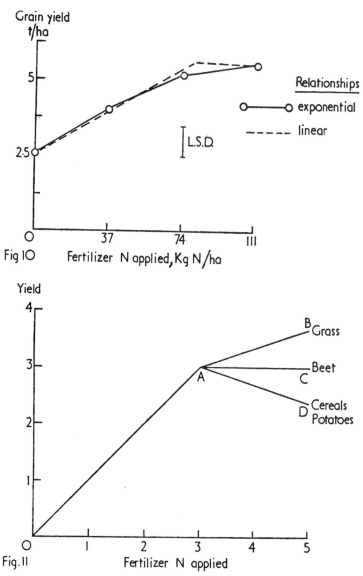

Fig. 10   Exponential and linear relationships between the amounts of nitrogen fertilizer used and the yields of barley at Rothamsted in 1967.

Fig. 11   Three possible relationships between the effects of nitrogen fertilizer on the yields of different kinds of crops.

plotting results of a recent experiment on barley and showing the 'least significant difference' (LSD) associated with each point on the response relationship. Differences less than the LSD may arise by chance (that is they are not due to experimental treatment but to variability in soil or crop) more often than once in twenty tests. Although it would be customary to draw an exponential curve through such points, the two straight lines shown are equally justified by the experimental evidence. Three possible relationships between yield and nitrogen are shown diagrammatically in Fig. 11; they differ in the ways the crops behave above the point of inflexion (A):

Grass: Using more N increases yields further as expressed by AB, but the increase/kg of N is much less than in section OA. The practical decision whether to use more N than is needed to reach point A depends on the value of the grass in relation to cost of N from section AB of the response relationship.

Sugar beet: Line AC is typical of the British experiments reported (ref. 39); yield of sugar being virtually unaffected by increasing N fertilizer. Differences between individual sites could not be explained in terms of effects of weather, soil or crop husbandry on N responses; so a recommendation was made to apply more than the optimum indicated by A because losses from applying too little fertilizer (on OA) were much greater than from applying too much (on AC)—the latter is simply the cost of the unnecessary fertilizer.

Potatoes: Yields diminish sharply when too much N is used (as in AD). With such valuable crops which have this pattern of response it is most important (but very difficult) to identify point A for individual fields. Cereals behave like potatoes, too much N depresses yield, particularly if lodging occurs.

Cotton: There is much less information on the relationship between amounts of fertilizer dressings and yields of tropical crops than exists for temperate crops. An account of the responses of cotton to N-fertilizer has been given recently by Indian workers (V. Ranganathan and others, Fertilité, No. 38, pp. 23–28). They showed that yield was directly proportional to amount of N applied up to a very sharply-defined point of inflexion. Giving more N than this depressed yield, and the depression was directly proportional to the surplus N supplied.

Responses to phosphorus, potassium and other nutrients

We have little evidence on the form of the response relationships of most crops with P and K fertilizers and even less on response to other nutrients. In most experiments only two amounts of fertilizer have been tested, these are too few to define the shape of the 'curve'. The few experiments made with several rates of P and K show that within the limits of experimental error a linear relationship may often be inferred. An example in the upper part of Fig. 12 is taken from experiments testing superphosphate on swedes in 1942; there is no justification for assuming a curved relationship in preference to one composed of two straight lines. D. A. Boyd (ref. 34) has given other examples of straight-

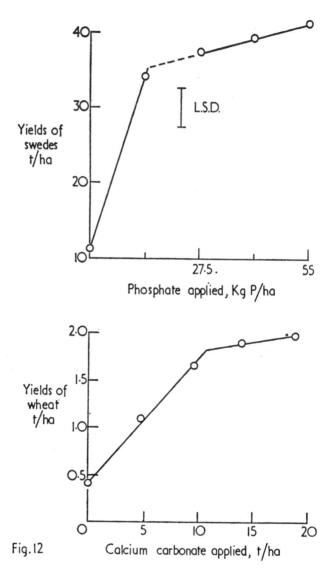

Fig.12

Fig. 12 **Upper graph:** Effects of superphosphate on the yields of swedes;
the results of 4 experiments on acid soils were averaged.
Lower graph: Effects of lime dressings on yields of wheat grown
on a very acid soil.

line relationships with phosphate fertilizers. Many experiments with phosphate done in other countries can also be interpreted in the same way. The lower part of Fig. 12 shows an example of a two-part linear relationship in a liming experiment on wheat.

Implications of recent work on response relationships

A linear relationship between fertilizer applied and crop yield is expected if the nutrient is mobile and other factors do not affect its use by the plant; uptake by the crop continues until the total is sufficient, then a sharp transition occurs. The slope of the linear portion is a measure of the efficiency of uptake; when a plot is made of nutrient applied and nutrient in the crop (both in kg/ha and on the same scale), the slope should be $45°$ if uptake is 100 per cent efficient; less steep slopes indicate waste of fertilizer, perhaps because factors such as root disease, leaching, or fixation by soil, limit uptake. With less mobile nutrients, response curves may not be linear. Phosphate and potassium ions reach roots by diffusion processes as mass flow does not provide enough (Chapter 1); where response per unit P (or K) applied diminishes with increasing dressings (as with the exponential curve) the reason may be that large concentrations of phosphate or potassium ions are needed to increase diffusion rates sufficiently to increase uptake by the crop.

When the effect of a nutrient changes sharply from deficiency to sufficiency, and further supplies have no effect (or may depress yields), the relationship is that of the 'Law of Limiting Factors', proposed by the German agricultural chemist Liebig over a century ago. The transition point where the relationship changes from the rising to the flat (or nearly flat) part occurs with different amounts of fertilizer in different fields, the position of the point may also change from year to year in any one field. When many experiments are grouped together the *average* response relationship *will* lie on a curve. Curved relationships may occur if there are large interactions of the tested nutrient with supplies of other nutrients from soil, or from fertilizer (as in Fig. 3).

When experimental results are used for advising on fertilizing, the sites *must* be arranged in groups that separate different soils, previous cropping systems, and seasons. When this is done response 'curves' are likely to be in two or three straight lines and there is no need for complicated calculations of optimum dressings. D. A. Boyd (ref. 34) pointed out that rectilinear response curves could make advice to farmers simpler and more precise. Except near the point of inflexion fertilizer recommendations will be little affected by small changes in crop and fertilizer prices; thus in one set of cereal experiments quoted by Boyd doubling or halving the crop: fertilizer price ratio would not change the optimum dressing of N by more than 10%. Yield of sugar from beet is not diminished by excess nitrogen so growers may use more than average optima with no larger penalty than wasting the cost of the fertilizer. But on cereals excess nitrogen may diminish yields through lodging and it is very important to determine the point of inflexion accurately.

A SUMMARY OF THE EFFECTS OF FERTILIZERS

This section contrasts the average responses recorded in series of field experiments made in countries where the use of fertilizers is only now developing with results in Britain and North America.

Experiments in developing countries

In developing countries of the tropics most soils are poor because they are deeply weathered and have been leached for very long, and because little fertilizer and farmyard manure has been used. Results from F.A.O's Freedom from Hunger programme of experiments (ref. 40) in Table 30 show that food crops give large responses to fertilizers on very poor soils; there were similar average increases in yields from the dressings of N, P and K tested and increases were obtained at most sites.

TABLE 30

AVERAGE RESPONSES TO FERTILIZERS IN 5 YEARS OF TRIALS
UNDER THE F.A.O. F.F.H. CAMPAIGN

	Responses (in kg/ha) to		
Near East and North Africa	N	P	K
Maize	546	369	259
Wheat	235	225	65
Potatoes	1368	780	587
West Africa			
Maize	178	173	150
Rice	224	226	158
Millett	137	110	102
Yams	895	839	946
Northern Latin America			
Maize	427	401	175
Rice	303	159	−20
Beans	189	120	68

These were simple experiments, only one rate of each fertilizer was tested and as interactions were not examined their value is limited. The work does not show *how much* N, P and K should be used and *what combinations* are needed to secure maximum returns from interaction effects.

In some developing countries quite large programmes of field experiments have been organised; advanced experimental designs have been used where several rates of nutrients have been tested in factorial combinations. These experiments have shown how much fertilizer should be used; the interaction effects calculated have shown whether good benefits could be obtained from nutrients applied singly or whether the gain from one nutrient was only realised when a second was also applied. A good example is a review of the responses to fertilizer by crops grown in Northern Nigeria, published by R. M. Meredith (ref. 40a). Over 600 experiments were done between 1952 and 1961. Many crops responded to N and P fertilizers, but large interactions were common as the following example of average results of experiments on sorghum showed. The unfertilized crop yielded 726 kg/ha; average responses to N and P were:

Ammonium sulphate applied (kg/ha)	Amount of superphosphate applied, (kg/ha)		
	0 (P_0)	63 (P_1)	126 (P_2)
	Increase in grain yield, kg/ha		
0 No	–	143	193
63 N1	84	293	359
126 N2	75	344	470

Without phosphate the return from N was small and the large dressing gave less yield than the small. With phosphate there was a large return from N and the relationship suggests that much more N would have been justified. As these results are a good example of interaction effects the formal calculation is given below:

$$\text{Interaction N} \times \text{P} = (NP - P) - (N - O) = NP - N - P + O$$

$$\text{or in this example } N_2P_2 - N_2P_0 - P_2N_0 + N_0P_0$$
$$= 470 - 75 - 193 + 0 = 202$$

This large value may be compared with the overall (or main) effects of N and P calculated in this way:

Main effect of N:
$$N_2P_0 + N_2P_2 - N_0P_0 - N_0P_2$$
$$= 75 + 470 - 0 - 193 = 352$$

Main effect of P:
$$N_0P_2 + N_2P_2 - N_0P_0 - N_2P_0$$
$$= 193 + 470 - 0 - 75 = 588$$

For statistical reasons the interaction calculated as shown above is divided by 2; main effects and interactions can then be added together.)

Experiments in Britain

Most highly-developed agricultural systems in which much fertilizer has been used for many years are in temperate parts of the World, north and south of the Tropics. Soils are richer in nutrients, partly because they have been less weathered than tropical soils, partly because they often contain fertilizer and manure residues. Average responses in series of experiments made in Britain and recently summarised in a number of papers are in Table 31.

TABLE 31

AVERAGE RESPONSES (PER HECTARE) BY BRITISH CROPS TO FERTILIZERS

No. of expts.		N	P₂O₅	K₂O
			Response per ha to	
124	POTATOES (kg of nutrient tested)	(100 N)	(125 P$_2$O$_5$)	(188 K$_2$O)
	Response, tonnes of roots	4.4	3.6	3.0
219	SUGAR BEET (kg of nutrient tested)	(100 N)	(125 P$_2$O$_5$)	(150 K$_2$O)
	Response, kg of sugar	530	134	276
	CEREALS (kg of nutrient tested)	(75 N)	(about 40 P$_2$O$_5$)	(about 40 K$_2$O)
–	Response, kg of grain Wheat	680	100	100
–	Barley	730	55	–16
260*	GRASS (kg of nutrient tested)	(31 N)	(60 P$_2$O$_5$)	(60 K$_2$O)
	Response, kg of dry hay	1456	302	302

*207 experiments provided P responses, and 191 experiments K response.

Notes Responses by sugar beet and potatoes are from large series of similar experiments. Data for cereals are from several series of modern experiments, numbers of tests of P and K were less than those of N. The responses by grass were measured on hay crops in many old experiments, responses by modern grass are likely to be larger for N and less for P and K.

Only potatoes give roughly equal responses to the three nutrients (and the experiments show responses to N, P and K are also equally frequent). Sugar beet gave much larger responses to N than to K, and even less response to P. Grass responded much more to N than to P or K. The average results of the cereal experiments also showed large responses to N, but only small responses to P and K.

Experiments in U.S.A.

Similar information for U.S.A., but on a less comprehensive scale, has been published by T.V.A. workers (ref. 41). Their results for cotton and maize showed that nitrogen increased yields on three-quarters of sites, potassium on a tenth to a quarter of sites and phosphorus on fewer still. The average responses to N were also much larger than responses to P and K.

Results from many countries show that in developed agricultural systems most crops give much larger and more frequent responses to nitrogen than to phosphorus or potassium. Most effort must go into making nitrogen fertilizing efficient while developing systems of manuring that avoid losses of yield by using too little P, K, Ca, Mg and S.

CHAPTER THIRTEEN

Variability in Responses of Crops to Fertilizers

E. M. Crowther and F. Yates (ref. 36) made the first attempts to use averages of large series of field experiment results (such as are in Table 31) to formulate general advice on manuring. There is a large amount of unexplained variability in the results of all groups of field experiments which make up such averages. For example, the table below summarises the results of 21 experiments, made on potatoes (usually grown after cereals) between 1953 and 1964 in north-east Scotland (ref. 42).

OPTIMUM NUTRIENTS NEEDED BY POTATOES IN NORTH-EAST SCOTLAND

	N	P_2O_5	K_2O
		kg/ha	
Average	95	176	217
Range	0–179	0–428	0–484
Number of sites where no fertilizer was needed (out of 21)	5	2	3

The optimum dressings calculated from individual experiments ranged widely. The data shows that it may be impossible to forecast correct dressings for individual fields on which experiments are *not* made from the results of a series of experiments on the same crop in the same area. Using the fertilizer dressing appropriate to each site increased overall profit by £20 per hectare as compared with using the average recommendation on all sites.

Causes of variability

Responses to fertilizers vary greatly from year to year on many fields, even when the same crops are grown. There are many reasons. The results of a field experiment are subject to many errors. Variability due to irregularities in soil, site and crop are examined statistically in modern experimental designs and analysis of variance is used to assess the validity of the results. A field experiment also samples biological variability caused by the unpredictable, and often unidentifiable, effects of weather, both on total yield, and on the chemical availability and physical accessibility of nutrients in soil. Temperature and rain alter growth of roots, and therefore their efficiency in feeding the crop.

On most agricultural land total supplies of nutrients are determined at least as much by soil type as by fertilizer. The total supply of *nitrogen* is lessened by leaching and increased by micro-biological activity, both effects depend on season. Total supplies of immobile nutrients, measured by chemical extraction, are little altered by season; but, because P and K are not mobile, roots have to grow to these nutrients which also move a little by diffusion (Chapter 1). Size and arrangement of pores determines whether roots gain access to much or little of total soluble reserves. Structure (measured by pore space) depends much on cultivations and cropping in previous years; the effects of cultivations on soil also depend on weather.

Responses to fertilizers measured in experiments of similar pattern are usually consistent from year to year on very poor soils (where much fertilizer is always essential), and on very rich soils (where fertilizers have little or no effect). Soils with intermediate fertility are those where responses vary annually. This is because the value to crops of stocks of nutrients in the soil, *and* the effects of fertilizer dressings on yield, are *both* altered by weather, soil conditions and cultivations. Responses are also altered by nutrient balance; they tend to diminish when fertilizer residues increase and to increase as nutrient reserves are exhausted. These effects are discussed in the following sections.

VARIABILITY IN RESPONSE TO NITROGEN

Effect of leaching

Nitrogen fertilizers are all converted to nitrate and this, together with nitrate formed by mineralisation of organic materials, is leached by water that passes through the soil since there is no chemical mechanism to retain it. (Some nitrate may escape leaching when held in pores that are not in the main channels for drainage water moving down the soil profile.) In badly-drained soils nitrate may not be lost by leaching, but it will be removed even more certainly by denitrification. Because a readily-available reserve of inorganic N cannot accumulate in soil, the supply of N in soils which are leached each year is always too small for maximum yield of cultivated crops grown on old arable land, or of grasses.

Winter rainfall leaches nitrate in most years in Britain. The amount of leaching depends partly on how dry soil is in autumn (this alters the amount of autumn and winter rain needed to bring the soil to field capacity); it also depends on amount of winter rainfall. Winter leaching varies much from year to year. At Rothamsted practically none occurred in 1964/5 when a dry autumn was followed by a drier than average winter; the following year (1965/6) a wet autumn followed by a wet winter left 250 mm of surplus water to cause leaching. Many investiga-

tions have been made of the effect of winter rainfall on response c
following crops to N fertilizers, but no standard rules have been derivec
In early work on sugar beet 25 mm per month more rain than averag
in the winter increased response to fertilizer by 300 kg/ha of sugar
correspondingly, drier weather than average decreased response by
similar amount. In Holland recommended dressings of nitrogen fertilize
in spring allow for the previous winter's rainfall; between 0.25 and 1.2
kg/ha of extra N is needed to compensate for the effect of 1 mm of rai
above average in winter. A.D.A.S. (ref. 43) suggest that after drier-than
average winters 25 kg/ha less N than the value calculated for averag
seasons should be given; after wet winters 25 kg/ha more N should b
given.

Previous cropping and manuring affect both total supply of N anc
leaching losses. Amount of winter rainfall alters the need for nitrogen i
the following spring only when soils themselves supply much N. Ol
arable soils growing cereals continuously usually release only 20–30 k
N/ha annually and the need for 100 kg N/ha as fertilizer to grov
maximum yields is little affected by winter weather. But when soil
contain large reserves of N combined with soil organic matter or organi
manures they will supply much to a following crop, the actual amoun
supplied varies greatly with variations in autumn and winter rainfall
This is true whether the reserves come from residues of fertilizer usec
on the previous crops of roots, from residues of leguminous crops, ley
or organic manures, or from soil organic matter itself. Previous crop
ping and manuring should always be examined to see if large N residue
in soil are likely to be present; if they are the fertilizer N dressing shoul
always be adjusted (1) by using less N (that is less than for continuou
cereal growing) and (2) to allow for winter rainfall. If this is done mucl
N will be saved, losses of yield due to too much or too little N will b
avoided, and risk of lodging in cereals will be less.

Spring rainfall increases the amount of N-fertilizer that must be give
as top-dressing to established crops when leaching removes part of a
dressing already given. The quantities of nitrate lost depend on botl
amount and intensity of rain. Losses may be very different in two sprin
months with similar total rainfall. Well-distributed small falls may caus
no leaching at all. Large falls (20-30 mm or more in 24 hours or less
have caused large losses of nitrate even when previous weather was dry
In a wet spring responses to early dressings of N that suffer leaching
will often be smaller than responses to top-dressings of N given when
crops are well established.

Summer weather causes further changes in N-supply. In warm and
wet seasons more N is mineralised than in cold dry weather; this causes
differences in the nitrogen fertilizer needed. Differences due to summe
weather cannot be forecast in spring when fertilizer-N is applied anc
therefore allowances cannot be made. In warm and wet summers when
extra nitrate is released from soil reserves, the spring dressing may prove
to be excessive and will make cereals lodge.

Changes in nitrogen supply caused by previous cropping

These differences in nitrogen supplied by soil which are caused by weather, are superimposed on differences caused by previous treatment of land. Soils which have grown non-leguminous arable crops for many years contain the least reserves of organic nitrogen and mineralisation produces least nitrate. Dressings of organic manures, or ploughed annual legumes, leave some reserves of organic N, but nitrogen reserves are largest where old grassland is ploughed. Past land use alters N-reserves and therefore the responses of crops to fresh dressings of N fertilizers; the actual value of the reserves in a particular season depends on the weather which alters *both* the amounts of N mineralised, and also the amounts of nitrate that are leached.

Table 32 gives ranges of amounts of N-fertilizer needed by wheat for maximum yields between 1961 and 1966 in the two 'Ley-Arable' experiments at Rothamsted (ref. 44). Fosters field has been arable for a century or more, Highfield was ploughed from very old grassland in 1948 and still releases more N than the old arable soil. The largest differences

TABLE 32

NITROGEN NEEDED FOR MAXIMUM YIELDS BY WHEAT IN THE ROTHAMSTED LEY-ARABLE EXPERIMENTS, 1961-1966

Preceding crops	Fosters	Highfield
	kg N/ha	
3 years lucerne	50 to 100	38 to 75
3 years clover ley	50 to 150 or >	38 to 113
3 years of grass receiving much N fertilizer	50 to 150 or >	38 to 113
3 years of arable crops (hay, sugar beet, oats)	100 to 200 or >	38 to 113
Reseeded grassland (ca 15 years old)	50 to 100	0 to 75

between the N-fertilizer needed for maximum yield are caused by previous cropping or the history of the land before the experiments started (old arable or old grassland). But for any one cropping sequence on each field the N needed has varied two- or three-fold during the 6 years shown in the Table. Since maximum yields of wheat in this period varied little (ranging between 6 and 7.5 t/ha) these changes in N needed are due to weather affecting either the N released from the soil, or that leached by rainfall, or to both causes.

No other results of long-term experiments of this kind have been published in Britain, but there is similar evidence from other countries. Two experiments made in Iowa (U.S.A.) (ref. 45) measured responses by maize to nitrogen fertilizer; the test crops were grown in several different rotations. A common response curve fitted all the rotations and differences in yields were due only to nitrogen supply. Taking continuous maize as zero, the amounts of N equivalents available in different cropping systems from 1958 to 1964 were:

	kg/ha of N
Continuous *maize**	0
Maize, maize, oats	45
Maize, *maize*, oats, 1 year meadow†	67
Maize grown directly after meadow†	112–145

(*Maize crop in italic is that which tested the N;
†meadow = 'ley' in English usage, the herbage used was *Medicago sativa*, *Bromus enermis* and *Trifolium repens*)

The errors associated with these averaged data were caused by seasonal differences in nitrogen supply. The nitrogen supplied to maize grown after maize, or after oats, varied little from year to year, but for maize after meadow the N supplied by soil varied greatly.

Adjustments to nitrogen fertilizing in England

In planning fertilizer dressings allowance must always be made for previous cropping and manuring and for weather. A.D.A.S. (ref. 43) have devised an index system which uses previous cropping to assess, empirically, the amounts of N the soil may supply. Examples given by L. J. Hooper (ref. 46) are in Table 33. The indices allow for residues from legumes, leys, organic manures and for much N used for root crops; they also allow for farming system. For example, the standard recom-

TABLE 33

SOME N.A.A.S. RECOMMENDATIONS FOR USING N ON POTATOES AND BARLEY

	Previous crop					
	Cereal or cut grass ley	Annual legume or root crop		Leguminous 3-year ley or roots with FYM		Ley treated with FYM or slurry
		In arable system	In a ley system	In arable system	In a ley system	
Index	0	1	2	2	3	4
Soil type			kg/ha of N			
POTATOES						
General grouping of sandy loams and other textures	226	188	151	151	100	75
Newport series	226	188	151	151	100	75
Adventurers light and loamy peat	126	126	100	100	75	75
Fenland silts	226	188	151	151	100	75
BARLEY						
Sandy soils	151	113	75	75	38	0
Silty loams	100	63	38	38	0	0
'Other textures'	126	88	63	63	38	0
Adventurers series light and loam peat	0	0	0	0	0	0
peaty loam	50	50	0	0	0	0

mendation for barley grown as a second cereal in an arable system on sandy loam soil is 150 kg N/ha. Grown on the same soil in a ley rotation after ploughing lucerne or a good clovery ley, the barley would need only 38 kg N/ha; if the ley preceding the barley had received FYM or slurry, no N at all would be needed. The A.D.A.S. Tables also show how adjustments are made to allow for soil type. For example the dressings recommended for barley as a second cereal in an arable system range from 150 kg N/ha on sands and sandy loams, to 100 kg N on silty loams, to 50 kg on some fenland soils; on some peats barley needs no N-fertilizer. These Tables (refs. 43 and 46) should be consulted; they establish a basis for advice on most crops in cool temperate regions. The tables allow for crop varieties and soil depth and for areas with summer rainfall above or below 400 mm. Crops grown in wet areas need less fertilizer N than those in areas with dry summers as, other things being equal, soils contain more N the wetter the climate.

Where rain or irrigation water is likely to leach nitrate during the growing season a 'starter' quantity of N should be applied at planting, the amount being varied according to the severity of leaching since the previous crop was harvested. Subsequent dressings of N should be planned to supplement the supply from soil; split dressings may be more efficient where summer leaching occurs. To achieve scientific control of N fertilizing during the growing season we may, in future, have to depend on periodic measurements of nitrate in plant tissue.

Adjustments to nitrogen fertilizing in Ireland

The Soils Division of An Foras Talúntais has published recommendations in their 'Fertilizer Manual' (ref. 47) which take account of previous history of land and summer rainfall. The advice is appropriate to a wetter climate than central and eastern England; because grass dominates more of the cropping system, soils are richer in nitrogen (as they are in western Britain).

Cereals. Recommendations range from 55 kg N/ha for wheat grown in mainly arable rotations in areas with less than 500 mm of summer rainfall to 18 kg N/ha for wheat following pasture in a wetter area or where much FYM or slurry was used.

Root crops. Recommended dressings for root crops varied in the same way according to previous cropping. After arable crops, hay or silage, potatoes need 100 kg N/ha and sugar beet 90 kg. After good pasture, a legume crop, grazed roots, or where FYM is much used, the amounts needed are only 55 and 45 kg N/ha respectively.

VARIABILITY IN RESPONSES TO PHOSPHORUS AND POTASSIUM FOUND IN FIELD EXPERIMENTS

Annual experiments. N.A.A.S. made 124 experiments on potatoes in England and Wales from 1955 to 1961 (ref. 35). Differences between both soils and seasons contributed largely to variance per site; of the factors identified, soil type and season had the largest effects on variability. At the smaller rates tested, responses to P and K were nearly as much affected by season as by soil type; responses to the second dressing of P were more influenced (but response to K less influenced) by season than by soil. (In contrast, response to nitrogen at both levels of dressing was influenced much more by season than by soil type.)

Long-term experiments at Rothamsted. In some Rothamsted experiments fertilizers have been supplied each year for very long periods, much more P and K has been applied than the crops have removed; therefore soluble P and K in the soils are now much larger than crops need. But in other experiments smaller dressings were used, or they were discontinued (ref. 48). Reserves of P and K in the plots of these latter experiments are in the 'intermediate' or sensitive range and there are large seasonal variations in response. In favourable years reserves are enough for crops and so fertilizer has no effect, but in unfavourable years they are too little and large responses to fertilizers are measured. An example of variability in the amount of phosphate needed in successive years for maximum yields was obtained in the Agdell experiment (ref. 48). On a plot which had received an average 125 kg/ha of superphosphate for each of 100 years the amounts of P needed for maximum yield varied greatly in the three years 1959, 1960 and 1961 when tests were made. Potato responses were most constant, maximum response needing between 50 and 75 kg P/ha. In 1959 responses by both sugar beet and barley were very small and practically no fertilizer-P was justified for these crops. In 1960 responses were large and in 1961 larger still. In the last year sugar beet needed 150 kg P/ha and barley 75 kg P/ha to obtain maximum response. These changes reflected differences in the usefulness of soil P in three successive years; they cannot be explained but show how difficult it is to interpret responses in a single-year experiment when no test was made on the soil in earlier years and little is known of its history.

Effect of weather on response to P and K fertilizers

There are no clear relationships between weather and response to P and K but some general principles are known.

Phosphate is the classical example of an 'inefficient' fertilizer, only between 10% and 30% of a fresh dressing being used by the crop to which it was applied. Because soluble phosphates combine with soil constituents to give materials with small solubility products, the diffusion of phosphate ions in soil is slow, and because roots do not explore the whole soil, much more P has to be supplied as fertilizer than the crop takes up. The amounts of rain, and the soil temperature, alter speed and extent of growth of roots—they grow further and faster when soil is moister and warmer. Increases in soil moisture increase the amount of P in the soil solution, *and* the ease with which it can diffuse to roots, they therefore diminish the need for extra P as fertilizer and responses to P dressings are smaller when the weather favours maximum use of soil phosphate.

Potassium is retained by soils with clay minerals of the kinds which fix' K ions in their lattices. But, since diffusion of potassium may be oo times faster than diffusion of phosphate, larger proportions of a K-ertilizer dressing are taken up by crops and some experiments record omplete recovery of K fertilizer by single crops of kale or grass—these ake up large amounts. Other things being equal weather probably has ess effect on response to potassium than to phosphate.

When weather is *very* dry fresh dressings of fertilizer mixed in the urface soil may be much less effective than residues of old dressings nixed with all the cultivated soil. This happened in experiments in Eastern England in 1970. In very dry weather fresh dressings of super-phosphate had practically no effect on yields of potatoes, sugar beet and parley grown on very poor soils, but there were large responses to esidues of past dressings remaining in the soil.

THE ACCUMULATION OF FERTILIZER RESIDUES IN SOILS

The common assumptions that inorganic nitrogen fertilizers have no esidual value, and that inorganic phosphorus and potassium fertilizers eave residues that last for only a few seasons, are often wrong. Residues of fertilizers left in the soil often raise yields in ways that are difficult to mitate with fresh fertilizer dressings, sometimes responses to fresh Iressings (and therefore optimum dressings) are unaffected by residues of previous dressings, but usually residues lessen the size of the fresh Iressings needed (ref. 49).

Nitrogen fertilizers
Recent dressings of organic manures, or recent periods under herbage crops leave large active residues that cannot be ignored in planning manuring. The large rates of inorganic nitrogen fertilizers now commonly used often leave residues that increase the following crop.

Phosphorus and potassium fertilizers
Manuring with P and K fertilizers, and with FYM, leaves residues in the soil that benefit following crops and such reserves must be allowed for in planning to use fresh fertilizer. The residual effect of one *single* dressing of phosphorus or potassium is usually much smaller than the direct effect the year before, it may be too small to measure accurately in experiments and can usually be ignored in planning. But the cumulative residual effects of many annual dressings are large and cannot be ignored; they may be sufficient for normal yields of crops without applying fresh fertilizer. In long-term experiments continuous annual manuring has accumulated P and K residues in soils which are measured by chemical analyses; the value of these residues to present-day crops has been proved on several kinds of soil.

It seems wise to build up reserves of nutrients as better crops are often obtained from soils containing residues of fertilizers or FYM than can be obtained from poorer land that has no residues, even when very large dressings of fresh nutrients are applied. Often to obtain yield from new fertilizer dressings used on poor soils, equal to those obtained from soils containing phosphorus and potassium fertilizer residues, the fresh P and K must be thoroughly mixed with all the cultivated soil. Widely recommended 'average optimum' dressings of P and K may be far too little for maximum yields on poor soils, particularly when the land is in bad physical condition; much of a fresh fertilizer dressing may then be inaccessible because the crop roots cannot grow to the nutrients.

EXHAUSTING NUTRIENT RESERVES IN SOILS

When soils which contain no reserve of calcium carbonate are cropped and much nitrogen fertilizers are used, calcium supplies are depleted and soil pH diminishes until crops will respond to lime. In the same way continuous cropping without sufficient P and K fertilizer depletes soil reserves so that crops begin to respond to fertilizer where they did not previously. At first responses are small on average but very variable; as deficiency increases they become larger and more consistent. As crops remove only a few kilogrammes per hectare of phosphorus depleting soil P reserves is very slow and there are few examples of this happening during the life of one experiment. By contrast the exhaustion of soil potassium can occur quite quickly when crops which take up much K are grown and removed. In many experiments potassium fertilizers have given no response at first, but after several years crops have begun to respond regularly. An example is given below:

A five course experiment at Woburn Experimental Station was started on sandy loam which had nearly enough soluble potassium for the crops grown. There were only small responses in the first two years; the third year crops responded to potassium and by the fifth year responses were very large indeed. The potassium removed in the crops fell dramatically during the five years. The figures below illustrate the rapid exhaustion that occurred:

	1960	1962	1964
	Response to K-fertilizer, kg/ha of dry matter		
Potatoes	730	1620	2970
Sugar beet roots	−290	1350	5720
	Potassium taken up by crops, kg/ha		
Potatoes	109	52	29
Sugar beet	244	132	23

East Africa. A series of experiments in Buganda and Western Uganda, described by D. Stephens (ref. 50), compared responses by several crops in two successive two-year cycles. Responses by maize, cotton, sweet potatoes, millet and sorghum were all larger in the second than the first rop cycle. There was no similar clear pattern in responses to N and P. Potassium responses are summarised in the table below:

	Response to K fertilizer, kg/ha	
	First cycle	Second cycle
Maize	−30	138
Cotton	10	46
Sweet potatoes	1050	1800
Beans	13	12
Millet and sorghum	−84	85

There were 8 experiments and the trend for response to K fertilizer to increase occurred in seven of these. There were similar increases in response to farmyard manure which probably acted as a source of potassium. This work suggested that potassium deficiency may have limited yields in Uganda more than had been suspected, especially after continuous cropping.

Similar changes in responses to fertilizers in experiments in *West Africa* have been reported (ref. 16).

CHAPTER FOURTEEN

Soil Type, Soil Analyses and Fertilizer Responses

Soils vary in their capabilities for supplying plant nutrients according
to their parent materials and the processes by which they were formed
and with past manuring and cropping. Fertilizer recommendations
should allow for the fertility of each field. Field experiments with
fertilizers cannot be done on every field, but information about the soil
can be obtained to help advisers recommend the dressings that are best
for local conditions.

INFORMATION THAT IS USEFUL IN ADVISING ON
FERTILIZING

Soil is formed from superficial geological deposits by the combined
action of chemical and biological processes. Slow chemical changes
(weathering) are caused by percolating water; the speed of these changes
depends on temperature, they are fastest in the wet tropics. Plants growing in the soil add organic remains in their roots and the 'tops' that
fall back on soil; these are decomposed by micro-organisms, insects and
small animals to make humus. The composition of the original geological
deposit determines the texture of the soil (amount of sand and clay); the
weathering and biological processes of soil formation determine how
much of the original endowment of plant nutrients in the rock remains
in the soil that is formed; climate and soil texture determine how much
of the nitrogen fixed from air by micro-organisms is retained in the
soil. The result of these processes is a 'natural' soil; some of its physical
and chemical properties are quickly altered when used for agriculture.

Soil classification
A soil surveyor usually describes the 'profile' of a soil (that is the
characteristics of the 'horizons' (layers) from surface down to parent
material). Profiles of soils that have not been greatly modified by farming
reflect their origin and development and serve as the basis of classification. Soil classes ('series') group together soils formed in similar ways

rom similar material; within one series the soils tend to behave simi-
arly and to supply similar amounts of plant nutrients unless different
arming systems have had large residual effects. For convenience in using
oil survey information the distribution of each kind of soil in a district
s recorded on a map. If an advisor knows that two fields are on the
ame soil series, he can extend knowledge gained on one field to the
ther, being cautious however to allow for differences in the way the two
ields have been farmed.

Soil texture

'Texture' expresses the proportions of coarse sand (the particles are
–0.2 mm in diameter), fine sand (0.2–0.02 mm), silt (0.02–0.002 mm)
nd clay (less than 0.002 mm). Texture affects the way the soil behaves
physically and determines how easy it is to plough and cultivate, it also
ffects nutrient supply. Sandy soils contain little clay and are intrin-
ically poorer than clays in all nutrient cations, often in micronutrients
oo. They become acid quite quickly. Sands tend to contain smaller
eserves of organic matter and nitrogen because organic materials oxidise
nore quickly in the open texture of a coarse soil. Clays have the reverse
f these properties; they usually have larger reserves of cations in the
lay fraction that can be released by weathering for use by plants, and
hey contain more nitrogen and organic matter. Clays may be naturally
cid just as sandy soils are, but when limed to make them neutral they
ontain larger reserves of lime so that acid-forming fertilizers do not
nake them acid again as quickly as happens in sands.

Textural groupings are one kind of soil classification and knowing the
exture of a soil is of much help in planning its fertilizing.

Available nutrients

The amounts of nutrients that agricultural soils can supply depend on
both the origin of the soil and the residues of previous cropping and
manuring. Soils are analysed to estimate these reserves and the informa-
ion is used to modify standard recommendations.

The commonest analyses are to determine whether lime is needed.
nterpreting the results of tests for lime requirements is much simpler
han interpreting tests for other nutrients because the neutral point (pH
) in the pH scale is a 'target' in planning liming. Although it may not
e necessary to lime to pH 7 to get good crops (and it may be undesir-
ble and uneconomic to do so), there is usually no benefit from giving
nore lime than will bring the soil to this pH. No corresponding 'target'
igures have been agreed for interpreting analyses for other nutrients.

Some laboratories use tests for assessing the potentially useful nitrogen
n soils, but they are often unsatisfactory. Nitrogen supply usually limits
production in the agricultural systems of developed temperate countries,
nd N-fertilizers give large returns. In settled arable agriculture nitrogen
s generally needed for all non-legumes and the results of field experi-
nents with nitrogen fertilizer are a better guide to manuring than are
oil test results. Methods of measuring the potentially useful micro-

nutrient reserves in soils are now being developed but these analyses a▮ not made as a routine in most laboratories, being used only for invest gating special cropping problems. Most experience in the developmeᵀ of soil analysis methods, and in their interpretation, has been with pho▮ phorus and potassium, but the general principles apply to other nutrient▮

SOIL TEXTURE AND SOIL GROUPS

Several recent investigations have shown the value of knowing so▮ texture and, where possible soil series, in grouping together experimenᵀ where fertilizers have tended to give similar responses. The large seri◦ of 124 potato experiments done by N.A.A.S. (ref. 35) provided th▮ examples in the following sections.

Soil texture

Average responses after grouping soils into 'light' and 'heavy' were▮

	Response to		
	N	P	K
	tonnes of potatoes per hectare		
All sands and sandy loams (56 sites)	6.28	2.59	3.36
All loams, silts and clays (40 sites)	2.31	3.39	0.43

These groupings show clearly that responses to N and K are, on averag◦ much larger on light than on heavy soils, but heavy soils are usuall▮ more deficient in phosphate.

Drainage

Soil surveys record drainage characteristics and these have been close▮ related to phosphate responses. These groupings of soils from Englan▮ used in the N.A.A.S. potato experiments show this:

	Response to superphosphate tonnes of potatoes per hectare
Drainage free to below 45 cm	1.58
Drainage impeded at 30-45 cm	2.91
Drainage impeded at 30 cm or less	4.44

Soil classification

The soils were arranged in groups based on texture and Soil Surve▮ classification and the effect of these groupings on the variability i▮ response to P and K fertilizers was tested statistically. Taking accour▮ of soil analysis removed 37 per cent of the variance in responses to ▮ and 27 per cent of variance in responses to K. Variance caused by so▮

groups accounted for 19 per cent of total variance in P responses but 39 per cent in K responses. In other words, knowing the soil analysis was most useful for interpreting phosphate responses, knowing the soil classification was most use with potassium. When account was taken of both soil groupings and of soluble P and K in the soils, variance in fertilizer responses was less than half of the original amount for both P and K.

Recommendations for potatoes and other root crops and for cereals grown in England and Wales (ref. 43) are varied by soil texture and where possible by soil series.

SOIL ACIDITY

The clay and organic matter in soils have negative electrical charges which attract and retain the positively charged cations of hydrogen, aluminium, calcium, potassium, magnesium and sodium. These ions can be displaced by treating the soil with a strong solution of other cations and are therefore called exchangeable. In very acid soils most of the negatively charged sites are occupied by hydrogen, (and some by aluminium). As more of the hydrogen is displaced by calcium the acidity is diminished and pH rises. The pH of a soil suspended in water indicates the saturation of the exchange complex with hydrogen; titrating soil with added lime determines the amount of base needed to bring the pH to the neutral point.

The background of recommendations for liming acid soils is usually a laboratory titration of the lime needed to bring soil up to a given pH value (often pH 7 in the arable areas of the East and South of England, but often pH 6.5 in the west and north of the country). Most recommendations are now made after a simple test of soil pH, the adviser knowing from type of soil and its texture how much lime will actually be needed to achieve the desired pH in the field. For example laboratory tests showed 1.25 t/ha of calcium carbonate would raise the pH of an unlimed sandy soil to neutrality; in the field twice as much was needed. Some advisers, instead of measuring pH, estimate the exchangeable calcium and any free calcium carbonate in the soil; most take 0.20 per cent of exchangeable Ca as the safe limit above which no lime is recommended.

TESTS FOR 'USEFUL' NITROGEN

Measuring potentially useful soil nitrogen needs different methods from those used for measuring the stocks of soluble (and potentially useful) phosphate and nutrient cations; most of the reserves of N are combined with organic materials and must be transformed to inorganic-N (ammonium and nitrate) before being used by plants. Laboratory methods have the difficult task of forecasting chemically the amount of organic-N likely to be transformed to inorganic-N by a biological process which depends on temperature, moisture, and the other inorganic and organic conditions in soil that govern microbial life, when neither they, nor the weather of the season to come can be forecast accurately. The amount of nitrogen to be released in a year may vary from perhaps only 30 kg N/ha to more than 100 kg/ha from a *total* stock that normally ranges from 1000 to 3000 kg N/ha in the top 15 cm of soil. Both the amount of mineral-N produced by soil on incubation in the laboratory, and the amount actually provided for a field crop vary according to season.

Although laboratory methods of measuring 'available' N have been related to responses to fertilizers, the methods seem too uncertain to be applied to the wide range of conditions met in advisory work. They are not extensively used in any country. The amount of nitrogen in soil that is useful to crops usually depends on the small fraction of organic matter coming from recent additions of organic manures and from roots and other plant residues. For the present advisors must therefore use their knowledge of the history of the land, the results of local fertilizer tests, the crop rotation, previous manuring, and weather, in guiding farmers in use of nitrogen on arable crops.

MEASURING SOLUBLE PHOSPHORUS AND POTASSIUM

The greater part of the large total amounts of P in soils (often 2000 kg/ha of P and 10 times as much K) are so tightly combined with soil constituents that they are useless to crops. The very small amounts dissolved in soil water are too small to feed crops for long. About 100 years ago the famous German chemist Liebig tried dissolving P and K from soil with dilute acids and showed that more was extracted from continuously manured soils in the Rothamsted experiments than from unmanured soil. Seventy-five years ago an English chemist, B. Dyer, established 1 per cent citric acid solution as a soil extractant because it had about the same acidity as the sap of many common plants; he calibrated his method using the Classical experiments at Rothamsted.

Development of analytical methods

Developments after Dyer's time were mostly of methods designed to simplify and quicken the analysis and with the hope that the P or K extracted might be more closely related to the fraction used by crops. Most of the solvents used were acid and the methods adopted generally distinguished satisfactorily between soils rich and poor in P and K. 'Acid' methods usually failed to dissolve the correct fraction of P from calcareous soils; when a small quantity of weak acid is neutralised by the excess lime, too little phosphate is dissolved, and using more acid may dissolve large particles of calcium carbonate containing P that is useless to crops.

The early workers hoped that analyses would measure the nutrient the soil could supply, and that, by knowing how much the crop contained, the difference could be supplied by fertilizer. This simple view, now mostly abandoned, was never adequate. Field experiment results showed that on some soils with 'enough' soluble P (or K) the crop responded largely to fertilizer P (or K). On other soils with 'too little' soluble P, crops nevertheless obtained enough for full yields and did not respond to fertilizer.

These discrepancies had the beneficial effect of making soil scientists develop methods with a sound theoretical foundation for measuring soluble P and K. Methods using exchange with the radioisotope of phosphorus (^{32}P) showed how much soil phosphate is 'labile' (i.e. capable of becoming ionised in the soil solution and taken up by plant roots). These methods were too tedious for wide practical application but they provided 'yardstick' data against which soluble P, measured by other methods, could be tested. It has been found in most countries that solubility of *phosphorus* in $0.5M$ sodium bicarbonate solution is better than any other simple method and that the *potassium* that is exchangeable with a solution of $1.0M$ ammonium nitrate represents available potassium. Except where otherwise stated only data obtained for soluble P and K by these two methods are given here. Limits that separate poor and rich categories of soils are discussed later.

WHY ARE SOIL ANALYSES NOT WELL RELATED TO CROP RESPONSES TO FERTILIZERS?

Analyses made with these methods for soluble P and K in soils have been extensively compared with responses to P and K fertilizers in field experiments on the land from which the soil samples were taken, these methods were better than others. Nevertheless, in series of annual experiments it has been rare to account for half the variance, usually the proportion has been much less than half. However, in glasshouse pot experiments using soils from many sites of field experiments, soil analyses for P and K by these methods, and the responses of crops to P or K fertilizers, have been closely related. In many pot experiments almost all of the variance in crop response has been accounted for by soil analyses. This means that chemical analyses *can* measure the P (or K) reserves that can dissolve in soil moisture and be useful to crops; the causes of lack of correlation between soil analyses and field experiment results must therefore be sought elsewhere.

Variability in soluble nutrients in soils

Good sampling is essential to obtain satisfactory analytical data. A kilogramme of soil taken for analysis represents only one part in two or three million parts of a hectare of cultivated soil. If a field can be seen to vary in texture, or it is known that parts were differently manured or cropped in the past, these areas must be separately sampled. Repeated sampling of test areas of soil have shown that soluble nutrients sometimes change during the season.

Moisture and temperature alter the solubility of P and K reserves in soils; more important is a *seasonal cycle* in soluble P, and particularly in soluble K, that has been demonstrated in several countries. Large amounts of K are removed by most crops and soluble-K measured in repeated analyses diminishes as the season advances; a large crop may remove half or more of the exchangeable soil K in one season. During winter when crops do not grow, more K (and P) is released from insoluble reserves so that analyses of fresh samples taken in spring may be near to the values of a year before.

Estimating 'available' nutrients by taking a *single* sample at any time of year to represent the soil of a field, cannot be justified. If analyses are used to predict fertilizer responses, 'standard' sampling times must be chosen, preferably coinciding with maximum or minimum values in the cycle of seasonal change in nutrient concentrations. Sampling should be done either immediately after harvest in autumn, or before planting in spring.

Variability in crop uptake

Repeated samplings and analyses of the 'standardised' soils from the Rothamsted long-term and classical experiments show that reliable modern methods can give reproducible amounts of soluble P and K that do not vary when proper precautions are taken to get representative samples at correct times in the year. In the *laboratory analysis* soluble nutrients are removed by the reagent from a weighed amount of fine soil. But *in the field* the amounts of phosphorus and potassium that plants can use depend on weather and soil conditions and on the volume of soil explored by roots. For this reason alone close relationships between soil analysis data and crop performance in field experiments, cannot be expected. (In glasshouse experiments plant roots usually explore the whole soil in small pots so relationships are closer.)

Moisture supply and *temperature* can alter both the chemical solubility of nutrients in soils and also the use that plants make of them. Thus more P is taken up when soils are wetter and warmer than when they are drier and colder. In interpreting soil analysis it is usually assumed that plant roots explore a constant amount of soil. But the actual depth of soil used by a given crop, and therefore the total quantities of nutrients 'available', vary not only from field to field but also from year to year in one field. Root penetration depends on soil compaction caused by cultivations, traffic, and weather, these vary from year to year. The amount of contact of roots with soil particles affects uptake of nutrient ions that do not move far in moisture. Phosphorus hardly moves at all, potassium is a little more mobile and nitrate moves easily (Chapter 1). Soil structure therefore alters uptake; large and impenetrable aggregates which cause coarse rooting diminish uptake of phosphate, but may not affect nitrate uptake.

Variability in responses to fertilizers in the field

It is generally accepted that differences in climate, season, crop and soil management, and the way the tested fertilizers were added to the soil all alter the responses of crops to fertilizers. Differences in weather, and cultivations also change the *value* of soil P and soil K to the crop. Therefore in practice the value to a crop of fixed amounts of both fertilizer P and K *and* soil P and K vary from year to year.

The long-term experiments at Rothamsted have measured the responses to fresh dressings of P and K fertilizers in successive seasons and have given good examples of annual variations in fertilizer responses in standardised conditions. Sampling the soils each year showed that soluble P and K did *not* vary. An example of 3 year's results of tests of fresh P fertilizer in the Agdell experiment was given in Chapter 13 (p. 136). Table 34 shows the results of two years of testing fresh K fertilizer on soils of the Exhaustion Land experiment. Of the three crops, only one (sugar beet) behaved in the same way in the two years (needing no fresh K-fertilizer). The soil K was enough for barley in 1957, but proved too little in 1958; kale behaved in the opposite way needing much K-fertilizer in 1957, but none in 1958. No doubt the difference between crops was from differences in their root systems and in the times when they needed to take up most nutrients.

TABLE 34

EFFECTS OF K-FERTILIZER ON CROPS GROWN ON EXHAUSTION LAND SOIL
AT ROTHAMSTED WITH III PPM EXCHANGEABLE K

	Response/hectare		Fertilizer needed for maximum response kg K/ha	
	1957	1958	1957	1958
Sugar beet (tonnes)	-2.5	-5.0	0	0
Kale (tonnes)	7.5	0.0	120 or >	0
Barley (kilogrammes)	-125	250	0	30

The seasonal changes in the values of fixed amounts of soluble soil P and soil K to crops, which led to this variability in responses, show how much annual variability there must be in the responses to fertilizers measured in annual experiments in farmers' fields, where there is no opportunity of repeating the tests in different seasons. Annual fertilizer experiments with P and K are inadequate sources for advice on manuring individual fields, and may often be misleading. When a soil has little soluble P or K, and response to fertilizer is likely (but not certain, as happens on deficient soils), the actual responses and the optimum fertilizer dressings derived from them, will vary greatly from year to year in ways that *cannot either be explained or forecast*. In developed agricultural systems where much fertilizer is used, a much sounder alternative to advice based on annual field experiments, is to plan manuring so that soil P and K are maintained at levels sufficient for the crops to be grown. (This is discussed at the end of this Chapter.)

Groups of experiments

When large series of field experiments have been done relationships between average responses to fertilizers and average soil analyses are often excellent; such relationships are improved when the soils are grouped by analytical data so that the groups contain equal numbers (ref. 51). When the individual responses to P or K from experiments in such a series are plotted against soluble P and K in the soils, the result is a very scattered array of points which often show no clear trends. Fig. 13 is an example from 50 experiments on sugar beet; there is no clear relationship between analysis and response. When these experiments were arranged in equal sized groups and plotted in Fig. 14, a smooth curve resulted which gave a convincing *average* relationship between response and soil analysis. In these experiments the standard errors of the fertilizer effects were averaged. The straight line marked '2 × S.E.' (i.e. twice the standard error) in Fig. 13 is roughly the size of a response that would be needed to be sure that it was unlikely to arise by chance in these experiments; (any response less than this amount (350 kg sugar/ha) may arise by chance more often than once in 20 tests). An increase in yield of sugar of this magnitude corresponds roughly to the difference between 10 and 20 ppm of sodium bicarbonate-soluble P. When such large differences may occur in individual field experiments, satisfactory relationships between soil analyses and responses to fertilizers cannot be expected.

To summarise this discussion: Most earlier work relating soil analyses to crop responses in series of field experiments ascribed poor correlation to weaknesses in the soil analysis methods, tending to accept the responses measured in the field experiments as correct. There *were* faults in soil analysis methods—as when acids used on calcareous soils dissolved useless P—but it was wrong to blame soil analyses and not the field experiments also. In fact, the crops in the field were growing with a quantity of soil phosphate whose solubility could be measured exactly, but its *value* to crops varied from year to year for reasons that were not recognised and recorded and may not have been measureable. The varying value of the soil P was tested by comparing it with the effects of fertilizer P which also varied not in amount applied, but in actual value to crop, from season to season.

The practical value of soil analysis

In the past too much has been expected of soil analyses for diagnosing deficient soils, forecasting responses and recommending optimum fertilizer dressings. Most of the critical testing has been made on the sensitive part of the curve relating analysis and response, and the conditions have been similar to those of Table 34, responses varying greatly from season to season with soluble P and K remaining constant. Many farmers lack confidence in soil analysis for purposes other than liming. Probably this is because there is no generally accepted numerical definition of 'deficiency' in soil P (or K) and because they cannot assess the risk involved in accepting advice based on analyses. The amounts of soluble phosphorus or potassium above which any given crop is unlikely to give large responses to fertilizer were not defined until recently; this *must* be done to generate confidence in advice on fertilizing based on soil analyses.

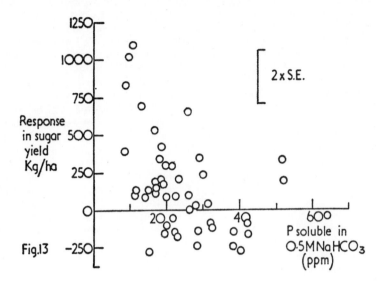

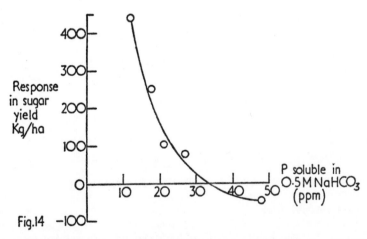

Fig. 13 Relationships between responses of sugar beet to phosphate fertilizer and the amounts of soluble P in the soils on which the experiments were made; individual results are shown.

Fig. 14 Relationships between responses of sugar beet to P-fertilizer and soluble P in the soils after averaging groups of experiments.

Poor versus rich soils. Casual analyses that are not supported by other information on local conditions can only be used to distinguish 'rich' and 'poor' soils when making preliminary improvements in the farming of an area, or when reclaiming land for agriculture. Soil analyses were very useful for these purposes in Britain during the 1939–45 War, but the opportunities for working with such poor soils diminish as agriculture is intensified. Analysing soils to select *only* a *proportion* of fields for manuring is of most value when: (i) the nutrient under test increases yields on only a small proportion of the soils used, (ii) the total quantity of fertilizer available is restricted, and (iii) fertilizers are expensive relative to crop prices. Soil tests are particularly profitable for advising on the manuring of crops that give small average returns from fertilizer and normally receive none.

Average soils. A.D.A.S. (ref. 43) have shown how their advice to farmers on the use of P and K fertilizers is varied by soil analyses and L. J. Hooper (ref. 46) has stated the basis that would be used in England and Wales from 1971. Groups of soil analyses have been established that cover most conditions in Britain and likely fertilizer responses on soils in each group have been published. (Limits with modern analytical methods are given in the next section.) When planning to fertilize soils of 'average' fertility used for growing maximum yields, advice must be given conservatively to avoid the large losses of yield that occur when crops have too little nutrients. Correlations between *average* analyses and *average* responses (such as Fig. 14) can be made more precise by taking account of soil characteristics, weather, and farming systems; but even after these improvements, because the individual responses always 'spread' as in Fig. 13, averages cannot give a *safe* basis for advice for individual fields. On sites where fertilizer prescribed by analysis proved unnecessary the cost of the fertilizer would not be recovered immediately but, if given, its residual effects would have benefited later crops. On the other hand serious losses occur when no fertilizer, or too little for maximum yield, is advised since the value of the crop lost is much greater than the cost of the fertilizer saved. These losses can only be avoided if limits to separate rich and poor soils are set much higher than the average limits suggested by Fig. 14.

ANALYTICAL METHODS FOR SOLUBLE PHOSPHORUS AND POTASSIUM IN SOILS AND THEIR INTERPRETATION

Soluble phosphorus

British work. The N.A.A.S. Conference on soil phosphorus held in 1962 (ref. 52) decided that S. R. Olsen's method (ref. 53) developed in U.S.A. which uses 0.5 molar sodium bicarbonate solution was best in England and Wales. This is amply confirmed by all the work done in Britain to test this method. Best evidence came from 172 manurial experiments on potatoes done between 1955 and 1961 by N.A.A.S. (refs. 35 and 52). The sodium bicarbonate method was better than any of the others, next best being the older N.A.A.S. method using Morgan's solution. The sodium bicarbonate method was not only most effective for the calcareous soils for which it was designed in U.S.A., but was also as good as any other method for non-calcareous sands and sandy loams, and was best for organic soils in Wales. On 99 silt and sand soils (excluding certain soils where all methods did badly) the bicarbonate method accounted for 60 per cent of the variance in the phosphorus responses, compared with 40 per cent for Morgan's solution; for *all* the sand and silt soils in this series of experiments these percentages fell to 23 and 12 per cent respectively. No method was of any use on fen peats. Other methods tested were found to be generally unsatisfactory although they have been widely used in this and other countries.

Other countries. F.A.O. (ref. 54) listed four successful methods of measuring soluble phosphorus. The first described is Olsen's bicarbonate method which has been much tested in many countries. The Commonwealth Bureau of Soil Science listed 103 papers which have tested the method since it was introduced in 1954. Fifty-two of these papers made quantitative comparisons, in 23 papers Olsen's method was clearly best, in 19 papers it was first but with another equally good; in 5 comparisons each the method was second and third of those tested. In no set of comparisons was it completely unsatisfactory.

The sodium bicarbonate method has been so satisfactory in so many different countries that it could be adopted as a national and international standard.

TABLE 35

PHOSPHORUS INDICES AND SOLUBLE P IN SOILS USED BY A.D.A.S. FROM 1971

(After L. J. Hooper (ref. 46) and Ministry of Agriculture, Fisheries and Food (ref. 43)).

Index	Soluble P in soil, ppm		Conditions
	Bicarbonate method	Acetic acid -acetate method	
0	0–9	0–2.0	Arable crops may fail if P fertilizer is not applied
1 ⎫			Glasshouse crops may fail without P fertilizer
1 ⎬	9.1–14	2.1–5.0	
2 ⎭			⎱ Standard rates of P fertilizer for ⎰ cereals, peas, beans and grassland
2 ⎫	14.1–24	5.1–10	
2 ⎭			⎱ Standard rates of P fertilizer for ⎰ root crops, fruit and vegetables
3	25–44	11–20	
4	45–74	21–40	Standard manuring for hops
5 ⎫			Standard manuring for glasshouse crops
5 ⎭	>74	41–70	Phosphate can be reduced for many field crops

Limiting values in Britain. L. J. Hooper (ref. 46) gave values (in Table 35) for soluble P which define the phosphorus indices used by A.D.A.S. in advising on phosphate manuring, together with values by the former N.A.A.S. method based on Morgan's solution (ref. 43). With larger indices than 5, no P fertilizer is needed for glasshouse crops; with still more soluble P crop yields may be diminished by excess phosphate. Examples of standard rates of annual manuring given by L. J. Hooper are:

Spring barley	38 kg/ha P_2O_5	Index 1 and 2
Maincrop potatoes	125–250 kg/ha P_2O_5	Index 2 and 3
	(the actual amount depending on soil type)	
Brussels sprouts	125 kg/ha P_2O_5	Index 2 and 3
Apples	38 kg/ha P_2O_5	with Index 1
	19 kg/ha P_2O_5	with Index 2
Grazed grassland	31 kg/ha P_2O_5	Index 1 and 2

Recent sugar beet experiments showed how successful measurements of sodium bicarbonate-soluble P are for establishing groups of soils to receive differential manuring (ref. 56). The groupings of soil P, responses of crops to phosphate fertilizer, and dressings recommended for each group were:

Bicarbonate-soluble soil P (ppm)	Response to 125 kg P_2O_5/ha kg/ha of sugar	Fertilizer dressing recommended kg P_2O_5/ha
< 10	1080	188
11–15	330	126
16–25	190	63
26–45	60	32
> 46	–125	0

Limiting values in other countries. Figures published in other countries confirm British experience. F. T. Bingham (ref. 57) surveyed the soil tests for soluble P used in U.S.A. and gave the target values in Table 36 for bicarbonate-soluble P.

Indian workers have said cereals respond to P fertilizers on soils with less than 6.5 ppm of soluble P, and responses are possible with 6.5–16 ppm. Much testing has been done in India; for potatoes less than 20 ppm is regarded as deficient; for wheat and rice less than 8 ppm is regarded as 'low', 9–20 as 'medium' and above 20 ppm as 'high'. In Australia 10 ppm of soluble P is the figure that separates soils that are 'deficient' from those classed as 'satisfactory'.

The limits set to separate soils with too little soluble P (where crops are likely to give large responses to fertilizer P) and those with enough P for most crops (where only maintenance doses are needed) are surprisingly similar in different countries although the range is from cold temperate to tropical climates, from podzolised soils in the north to laterites in the tropics and sub-tropics, and from acid to alkaline soils.

TABLE 36

TARGETS FOR SODIUM-BICARBONATE SOLUBLE P

Crop characteristic	Crops	Deficient	Questionable	Adequate
			ppm of soluble P	
Needing				
Little P	Grass, cereals soybeans, maize	4	5 – 7	8+
Moderate P	Lucerne, cotton sweet corn, tomatoes	7	8 – 13	14+
High P	Sugar beet, potatoes celery, onions	11	12 – 20	21+

Soil potassium

British work. A N.A.A.S. Conference on soil potassium and magnesium held in 1963 (ref. 58) decided that several methods which extracted the exchangeable fraction of soil K were roughly equal. Most methods generally used for soluble K extract a fraction of soil potassium that is similar to, or directly related to exchangeable potassium, and

TABLE 37

POTASSIUM INDICES, SOLUBLE K IN SOILS AND THE NEED FOR K FERTILIZERS

(refs. 46 and 58)

Index	Exchangeable soil K (ppm)	Conditions
0	0–60	Arable crops may fail without K fertilizer
1	61–120	Glasshouse crops may fail without K fertilizer
1		Standard manuring for cereals, peas, beans, root crops and grassland
2	120–240	Standard rates for vegetables
3	246–420	Standard rates for fruit and hops
3		
4	426–600	Standard rates for glasshouse crops
5	606–840	Less K-fertilizer can be given for many field crops and fruit

it is not surprising that the data are similar. No method for measuring the reserve of non-exchangeable but potentially-useful potassium is generally accepted. Good methods of measuring exchangeable potassium usually account for no more than 30 per cent of the variance in response to K-fertilizers by *sensitive crops* in series of annual field experiments; crops that respond less to potassium, give poorer relationships. Soil type determines potassium supply, and knowing soil types has often been more successful than knowing soil analyses in interpreting tests of K fertilizers.

L. J. Hooper (ref. 46) recorded that exchangeable potassium measured by shaking with N ammonium nitrate solution would be used by A.D.A.S.; his recommendations are in Table 37.

Other countries. Experiences in relating exchangeable soil K to fertilizer responses fit well with British experience, and the summary in Table 37 may be helpful in many areas.

MAGNESIUM

Deficiencies of magnesium (already discussed in Chapter 6) are becoming more common in Britain and other countries, particularly where crops for sale are grown often on light soils and no organic manures are applied. Analysing soils to forecast where Mg-deficiencies are likely (and therefore where appropriate fertilizers should be used) is useful; most workers use exchangeable magnesium determined in the solution used to measure exchangeable-K. Because other cations interfere with magnesium uptake, methods that take account of the ratios of all cations in the soil solution are often helpful.

Scotland. In a survey in North Scotland (ref. 58) under one-fifth of the soils had less than 31 ppm of soluble Mg which was taken as the limiting value, yield response to magnesium being unlikely with larger amounts. About half of the soils had less than 60 ppm of soluble Mg.

England and Wales. A N.A.A.S. Conference (ref. 58) divided soils used for *arable crops* into 3 classes.

	Exchangeable Mg ppm	
'Low'	0 – 30	Quick-acting Mg fertilizer is needed
'Medium'	30 – 60	Magnesium limestone can be used when lime is also needed
'High'	over 60	Soils contain enough Mg

For *glasshouse* and *fruit crops* the ratio of soluble Mg : K in the soil has to be taken into account. If the ratio of exchangeable K : Mg was more than 2 : 1, some interference of Mg uptake was possible. L. J. Hooper (ref. 46) gave these indices for soluble Mg:

	Soluble Mg ppm in soil
Index	
0	0 – 25
1	26 – 50
2	51 – 100
3	101 – 175
4	176 – 250
5	255 – 350

Vegetable crops are not likely to be deficient with soluble Mg above Index 1 except when potassium is excessive. Magnesium dressings (60 kg Mg/ha) are recommended for fruit with Mg index = 0 or 1; no Mg-fertilizer is advised with Mg index above 2.

BALANCING SOLUBLE P AND K TO FARMING SYSTEM

Field experiments are subject to large errors and in most series variability in responses to P and K cannot be fully explained. By contrast modern methods can measure soluble P and K accurately in most soils. The future value of soil analyses will be as a permanent means of fertility control. Repeated tests should be made on the same soil, and long-term changes in the results used to vary phosphorus, potassium and magnesium manuring (as is now done for liming) so that desired levels of these nutrients are maintained in the soil. Proper precautions must be taken to avoid sampling errors, the samples must be taken at the same time of year and analysed by the same methods. The aim should be to maintain soluble P and K at amounts that are practically sufficient for an average crop so there is no risk of poor crops under adverse conditions when fertilizers do not act well. When using soil analyses for long-term control the relationship between fertilizer dressings and soluble P and K in soil must be known for the cropping systems used. This can be obtained from long-term experiments where the effects of fertilizer and cropping systems on reserves of soluble P and K have been measured. There are few of these experiments in Britain, or indeed in the world. Those started at experimental stations in the last century are the most reliable, but more work with modern large yielding crops is needed.

Fig. 15 shows the amounts of soluble P in soils of long-term experiments at Rothamsted and Saxmundham plotted against the amounts of P applied as fertilizers. From such graphs maintenance dressings of P (and K) may be read off. Figs. 16 and 17 show how soluble P and K depends on the balance between P and K removed in crops and added as fertilizer at Saxmundham and suggests how to maintain desired amounts of soluble P and K in soil.

Phosphate

Fig. 15 compares P soluble in $0.5M$ sodium bicarbonate solution in

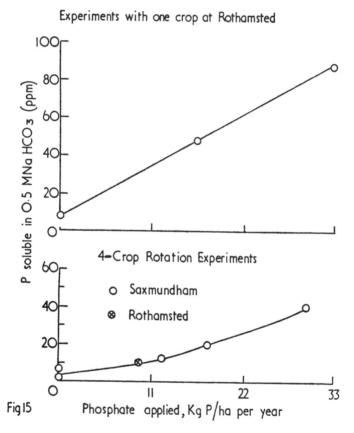

Fig 15

Fig. 15 Relationships between soluble phosphorus in soils of long-term experiments at Rothamsted and Saxmundham, and the amounts of P-fertilizer applied annually.

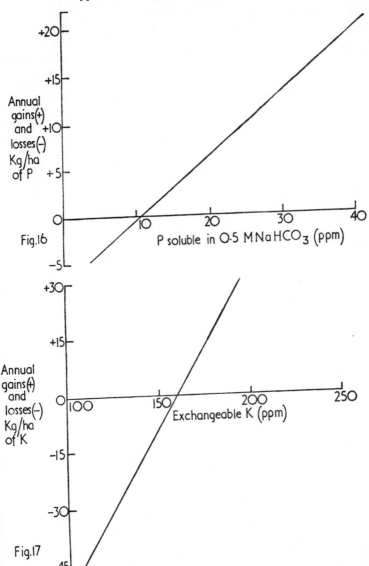

Fig. 16   Relationships between gains (from fertilizers) and losses (in crops) of phosphorus and the amounts of soluble P in the soils of a long-term experiment at Saxmundham.

Fig. 17   Relationships between gains (from fertilizers) and losses (in crops) of potassium and the amounts of exchangeable K in the soils of a long-term experiment at Saxmundham.

soils of long-term experiments at Rothamsted (over 100 years) an
Saxmundham (over 60 years) with the annual amounts of P applied a
superphosphate. In the Rothamsted experiments on single crops th
annual dressing is 33 kg P/ha. Broadbalk wheat experiment also test
an intermediate rate and the line is drawn for this experiment. A
Rothamsted cereal crops receiving no P fertilizer have removed 6-
kg P/ha/year, the grass only 2.2 kg P. Grass receiving P has remove
about 16 kg P/ha; cereals receiving P fertilizer have taken about 12 k
P/ha. The annual surplus of fertilizer P for cereals at Rothamsted ha
therefore been about 20 kg P/ha, and for grass about 17 kg P (this make
no allowance for any P supplied by the soil). In all the four-crop rotatio
experiments wheat, barley, a root crop and a legume were grown. I
Saxmundham RI experiment the annual surplus of fertilizer-P over I
removed in the crop was 6.6 kg P/ha, in the Agdell experiment a
Rothamsted 4.4 kg P/ha. The smooth curve relating soluble soil P an
fertilizer P on these two different soils is very striking.

In the Rothamsted experiments where cereals were grown continuously for over 100 years applying about 5 kg P/ha/year would have beer
sufficient to maintain the soils with about 20 ppm of soluble P while
grain yields ranged from about 2250 to 3000 kg/ha. (This amount i
near to the *extra* P contained in crops that received P fertilizer each yea
as compared with those receiving none.)

In the Park Grass experiment about 7 kg P/ha/year would have maintained soluble P at about 20 ppm. This amount is less than the extra P
removed on P-manured plots (about 11 kg P/ha on plots yielding about
4300 kg hay/ha and 15 kg of P/ha from plots yielding about 6000
kg/ha annually). If gains of P had balanced losses the equilibrium value
for soluble P would have been 22 ppm in Broadbalk wheat experiment
soils and 20 ppm in the Park Grass soils.

G. E. G. Mattingly has used the results from long-term experiments
at Rothamsted to estimate how barley grain yields were increased by
residues of fertilizer phosphate in soils measured by soil analysis. The
increases from an extra part per million of soil P soluble in sodium
bicarbonate solution ranged from 28 kg of grain/ha on very poor soils to
14 kg of grain on soils better supplied with phosphate.

Potassium

Fig. 18 shows the relationships between exchangeable K in the soils of
the Rothamsted Classical Experiments on cereals and grass. To maintain
200 ppm of exchangeable K needed annual dressings of about 33 kg/ha
of K on the cereal experiments. Other work suggests that an equilibrium
value of 150 ppm may be enough for continuous cereal growing. This
would be achieved by about 22 kg K/ha/year for these crops yielding
only from 2250 to 3000 kg/ha of grain. The amounts of soluble K in
soils of the grass experiment are affected greatly by the N applied and
therefore the yield removed:

	K (kg/ha) needed annually as fertilizer to maintain exchangeable K at		
	100 ppm	150 ppm	200 ppm
Plots yielding about 5000 kg/ha of hay (containing 145 kg K)	22	55	90
about 7000 kg/ha of hay (containing 190 kg K)	55	110	170

The 4400 kg/ha of hay removed each year where N and P, but no K
fertilizer is given contains only 25 kg/ha of K. This seems to be as much
as the Rothamsted soil can supply annually after a century of growing
grass that has received 95 kg N/ha plus adequate P, but no K. The hay
harvested from plots receiving 225 kg K/ha as fertilizer contains 2.8–3.0
per cent K in the dry material, much more than is necessary for yield or
desirable for animal health. Hay from herbage receiving no K fertilizer
has only 0.5–0.6 per cent K in dry matter—too little for maximum
yield. The figures suggest that if 55 kg/ha of K is applied each year to
maintain exchangeable soil K at 100 ppm the hay would contain about
1 per cent of K.

If soils have satisfactory reserves of K, fertilizers may supply as much
as the crops remove, so that soluble K in soil does not change. On Broadbalk the amount of exchangeable potassium corresponding with no gain

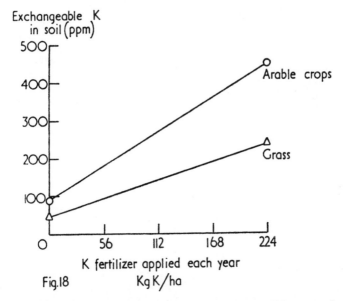

Fig. 18

Fig. 18 Relationships between amounts of exchangeable potassium in
soils of the long-term Rothamsted experiments and the amounts
of K-fertilizer applied annually.

or loss was about 140 ppm of K, at Saxmundham it was 150 ppm. These seem reasonable targets for soil potassium in English soils used for ordinary agricultural crops; the amounts of fertilizer-K needed to maintain them will depend directly on crop yield, if crop yield potentials are increased by other improvements, maintenance dressings of potassium will also have to be increased to prevent soil K diminishing. The amounts needed depend also on K released from the soil; over very long periods this amounts to only 22 kg/ha annually at Rothamsted, but at Saxmundham 45 kg K/ha has been released each year for the last 70 years.

Summary

The work on long-term experiments summarised here shows how soil analyses vary according to both soil and cropping system. Reasonable targets at which soil P and soil K might be maintained are:

	Parts per million of	
	Bicarbonate-soluble P	Exchangeable K
Permanent grass swards (no legumes)	10	100
Cereals, legumes, sugar beet	15	150
Potatoes and out-door vegetables	20	200

In modern farming yields must not be limited by nutrient deficiencies. To achieve this, soluble P and K ought to be maintained above limits where *large* responses to fertilizer are expected, because fresh fertilizer often does not give as large yields as good reserves in soil. In favourable years the P and K freshly applied will increase yields little or not at all, but the dressings will replace nutrients crops remove and augment reserves in the soil. In unfavourable seasons the fresh fertilizer will act as an insurance against the risk that supplies in the soil are not enough, particularly when young crops have small roots.

CHAPTER FIFTEEN

Using Plant Analyses to help with advice on Fertilizing

When plant analyses are used to guide fertilizer recommendations, the crop itself is used to extract nutrients from soil. Nutrient concentrations in parts of the plant are measured; usually growing leaves or stems are used. When plants are *very* deficient applying fertilizer makes them grow, the nutrients may then be 'diluted' and concentrations diminish. With larger doses of nutrient applied as fertilizer both yields and concentrations in the plants increase. Giving more fertilizer than is needed for maximum yield results in even larger concentrations in the plant, these are not useful and are termed 'luxury uptake'; if yield is *depressed* by increasing fertilizer further the nutrient is regarded as toxic. As with soil analyses, plant analyses can only be related to responses to fertilizers when plant materials from suitable field experiments are available. These factors have to be settled: (1) The best leaf or stem, and best stage of development, to represent nutrient status; (2) the nutrient concentrations and ratios associated with best growth and yield; (3) the concentrations of nutrients in leaves that are associated with deficiencies or toxicities. It is possible, if no other factors interfere, to assess the nutritional status of a crop from leaf analyses; manuring is adjusted after comparing measured values with established 'optimum' leaf concentrations. The advantage of leaf analysis, correctly used, is that it indicates how much nutrients the crop was getting from the soil at the time the sample was taken; by contrast soil analysis can only show what the plant *might* get (the actual amount being largely influenced by season).

Most investigations of the value of plant analyses to help fertilizing have been done with perennials and tree crops; for these soil analyses are difficult to interpret as a large mass of soil and subsoil is used by tree roots. Often the composition of the leaves is not clearly related to amounts of soluble nutrients in soil or subsoil. Depth of useful soil and its structure may be more important than nutrient content, but the depth of soil explored by tree roots, and the intensity of root growth, is never known precisely. In addition trees contain large reserves of plant nutrients in trunks, branches and roots.

Methods of using leaf analysis have been developed in Britain for advising on manuring of tree and soft fruit crops and there have been some tentative applications to agricultural crops. In other countries 'tissue tests' are used to control the nutrition of many tree crops, and crops such as sugar cane or pineapples that are sold for processing. This Chapter gives some examples of successful uses and discusses the future importance of plant analysis in planning nitrogen fertilizing. Other examples of leaf analyses used to control the nutrition of tropical crops are in Chapter 22, and the value of leaf and soil analysis for assessing the nutritional status of apple orchards is discussed in Chapter 21.

SEEKING MAXIMUM YIELDS

Plant analyses help to show how large yields may be attained. When the supply of one nutrient determines crop yield, concentrations in leaves can often be used to show how manuring programmes should be developed. Often, however, larger yields require correct supplies of two or more nutrients and maximum yield may only be achieved with one critical ratio. Examples of these effects are in the next paragraphs.

Potatoes need large amounts of potassium. Fig. 19 shows the relationships between yields and %K in the leaves of potatoes grown recently at Rothamsted with a range of manuring treatments involving both K-fertilizer and FYM. Yield was directly proportional to K concentration and because the 'curve' rises so steeply, even larger dressings of fertilizers and FYM seem to be needed for maximum yield and these are being tested.

Raspberries. Work at Long Ashton Research Station (ref. 59) has shown nutrient concentrations in leaves can vary greatly when yields are moderate, but that for maximum yield, percentages of both N and P are critical. Fig. 20 shows lines of equal yields (expressed as per cent of maximum yield) plotted against %N and %P in the dry matter of the leaves. Half the maximum yield could be obtained with a very wide range of N and P concentrations; 75 per cent of maximum yield was achieved with a much narrower range of %N:%P; maximum yield was attained at only one ratio because the NP interaction was so important for this crop.

USE OF PLANT ANALYSES IN ADVISING ON FERTILIZING

Strawberries in Britain

C. Bould (ref. 60) published the tentative standards for classifying the nutrient status of strawberries which are given in Table 38. He stated that these values 'should be fairly reliable (in the absence of non-nutritional limiting factors or extreme nitrogen deficiency) for predicting probable response to P and K fertilizer'. The data in Table 38 suggest that strawberries need little magnesium and, in fact, magnesium deficiency is not often seen in the field.

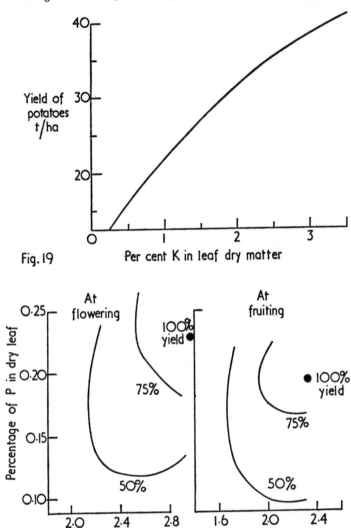

Fig. 19

Fig. 20

Fig. 19   Relationships between the yields of potato tubers and the concentrations of potassium in the leaves of the crops.

Fig. 20   Relationships between the concentrations of N and P in leaves of raspberries (**Rubus idaeus**) and the yields of the crops. The solid points show concentrations needed for maximum yield (100%); isoquants show concentrations that produce 50% and 75% of these maximum yields. (Ref. 59)

Nitrogen manuring based on measurements of nitrate in plants

Fertilizer dressings can be adjusted to the needs of a crop by analysing stems or leaves to see if the plants have enough nitrogen; if tissue contains much free nitrate crops are assumed to have enough N for current growth. Measurements of nitrate-concentrations in single samples of farmers' crops help little with advice on nitrogen-manuring since concentrations of nitrate change rapidly during the season; rate of change in nitrate is more important than amount. Some recent experiments have shown how successive measurements of nitrate can show when top-dressings are needed. If analyses of this kind are to be useful, farmers must act quickly when advice is received since delay in top-dressing diminishes yields; aircraft may have to be used for spreading fertilizer over tall crops.

TABLE 38

TENTATIVE STANDARDS FOR CLASSIFYING THE NUTRIENT STATUS OF
STRAWBERRIES AT FRUITING STAGE BASED ON CONCENTRATIONS
OF NUTRIENTS IN LAMINAE OF RECENTLY MATURED LEAVES
OF 1-YEAR PLANTS

	'Deficient'	'Marginal'	'Sufficient'
	per cent of element in dry matter		
N	<2.0	2.0 –2.5	2.6 –3.0
P	<0.20	0.20–0.24	0.25–0.30
K	<1.0	1.0–1.4	>1.5
Mg	<0.10	0.10–0.14	>0.15

Experiments made by R. J. B. Williams at Rothamsted have shown that measurements of nitrate in plant tissue can be used to forecast the need for extra nitrogen fertilizer. The method has been particularly useful in seasons when large spring rainfall leached a large proportion of the nitrogen normally given when sowing. In years when nitrate in stems of wheat or barley diminished nearly to zero before the end of May there were large responses to extra N-fertilizer. When nitrate in the petioles of sugar beet leaves diminished rapidly in early July, these crops also responded to more N. By contrast in seasons when there was little leaching N-fertilizer given at sowing was efficient. 125 kg N/ha applied in the seedbed maintained much nitrate-N in cereal stems until the end of June, and in sugar beet leaf petioles until the end of July. Top-dressings did not increase the yields of these crops.

Plant analysis applied in this way may become very useful for improving the efficiency of nitrogen applied to farm crops. If crops' needs are not tested in this way, and top-dressings are given at standard rates every year, there will be much waste of N and there is always a risk that unnecessary dressings may diminish yields.

Potassium for grass

It is difficult to plan the potassium fertilizing of grass. Crops that receive 300 kg N/ha usually yield 10–14 t/ha of dry matter when cut several times in the season; this yield removes altogether 200–400 kg K/ha which comes from soil, from K recirculated by excreta, and from fertilizer. Grass rarely shows K-deficiency symptoms, and it does not respond to K-fertilizer until exchangeable potassium in the soil is small. Enough K must be available to the grass to secure the full response to N (see Chapter 20), but if too much is given it is taken up as 'luxury consumption', without a commensurate increase in yield. (The large %K in herbage may increase the risk of hypomagnesaemia.) Workers at the Grassland Research Institute (ref. 61) have found that more than 90 per cent of the maximum yield of grass harvested at grazing or silage stages could be obtained when herbage contained no more than 2 per cent K in dry matter (so a 10 t/ha crop would need 200 kg K/ha). Because grass takes up so much K, exchangeable-K in the soil where it grows may diminish from 'satisfactory' to 'low' amounts in one year. Soil analyses are not good guides to K-manuring because changes occur so quickly, and because grass can extract much K from non-exchangeable reserves. Methods of using herbage analyses to regulate K-manuring are discussed in Chapter 20.

Use of plant analyses in fertilizing tropical crops

Leaf analysis has been developed for planning the manuring of many high-value perennial tropical crops. Two examples are given here, others are in Chapter 22.

Rubber. Workers in Malaya (ref. 62) and other rubber producing countries have established nutrient concentrations in the shaded leaves in the accessible part of the canopy which are used to guide fertilizing. One set of figures that express the minimum concentrations in leaves needed for adequate nutrition are given below, they are used as a basis for advice:

Per cent in leaf dry matter

N	3.30	Mg	0.26
P	0.24	Ca	0.60
K	1.40	(Mn	>50 ppm)

Sugar cane. In Guyana (ref. 63) sampling procedures and analytical techniques have been standardised; examples of optimal concentrations in leaves, and of recommendations of fertilizer appropriate to various nutrient concentrations in leaves are given in Table 39.

Similar work on the use of tissue analysis for advising on manuring has been done in many sugar-producing countries. The leaves to be sampled must be carefully chosen and taken in a standard way; parti-

TABLE 39

ANALYSES OF 18-WEEK OLD SUGAR CANE LEAVES AND CORRESPONDING
FERTILIZER RECOMMENDATIONS

	N per cent	P per cent	K per cent
	(in leaf dry matter)		
Optima			
Plant cane	2.25	0.20	1.20 or more
Ratoon cane	2.20	0.20–0.22	1.20 or more
Plant cane			
Level, per cent in leaf	>2.0	>0.18	>1.2
Recommendation	None	None	None
Level, per cent in leaf	1.8–2.0	0.15–0.18	1.0–1.2
Recommendation/ hectare	25 kg N	28 kg P	67 kg K
Level, per cent in leaf	<1.8	<0.15	<1.0
Recommendation/ hectare	50 kg N	42 kg P	134 kg K

cular care is needed in interpreting percentages of nitrogen as values decrease with age of leaf and with drought, and they are sensitive to soil conditions and temperature. The very elaborate system of 'crop-logging' which has been developed in Hawaii to control nutrition and growth of sugar cane is described in Chapter 22.

CHAPTER SIXTEEN

Plant Nutrient Cycles

When agriculture is intensified the larger yields produced need extra plant nutrients. In early improvements fertilizers are used simply for their immediate effects; but, to make a new system stable, account must be taken of residual effects of fertilizers and crops, and of the cycle of nutrients between crops, soil, air and rain. For a full knowledge of nutrient cycles, amounts in crop, soil, rain, animal excreta, leaf fall and crop residues must be known. Most important is how the yield is used, for this determines whether nutrients used to produce the crop remain in the system, or are lost. Nutrient balance sheets may be made for a country or region, a whole farm, or a single field or plantation. Calculations for a whole region aid planning. Balance sheets for a forest or perennial plantation help to understand nutrition where fertilizer experiments cannot be done, or take too long. For a whole farm they take account of sources of nutrients other than fertilizers, and of the nutrients lost in produce sold. Calculations of nutrient balance for single fields supplement the results of field experiments and soil and plant analyses in advisory work and reveal weaknesses in manuring systems.

Large amounts of fertilizer-N are usually needed to make large yields, but the 'natural' nitrogen cycle through the soil is very important in understanding the effects of previous land use and cropping. Intensifying agriculture usually involves using fertilizers that increase the stock of soil P and cycles involving this nutrient have little practical value. Calculating the K removed and added to soils is important because crops take up large amounts and can easily exhaust reserves. Except in calcareous soils, calcium balance is important but is usually indicated by measuring reaction; changes in pH reflect losses and gains of Ca. Nutrient balance sheets are particularly useful for assessing the need to supply nutrients which are not generally applied as fertilizers because they are provided by traditional organic manures and by soil and rain. Sulphur, magnesium, sodium and micronutrients are examples.

COMPONENTS OF NUTRIENT CYCLES

Gains

In calculating nutrient balance for an individual field or plantation the important gains in nutrients are from: (1) organic manures and crop residues added, (2) fertilizers, (3) nutrients released from soil—important for both N and K, and (4) natural fixation of N by legumes.

The nutrients applied in *fertilizers* should be known precisely, those in *organic manures* can be estimated accurately enough. It is more difficult to estimate nutrient returns in crop residues, and more difficult still when grazing animals use the crop. Much of the nitrogen in excreta may be lost to the air or by leaching. All nutrients in excreta from livestock on grassland fall in patches which make the soil locally irregular.

Weatherable minerals in soil continually release cations; estimates of the K released/year in *long-term* British experiments range from 20–100 kg/ha, varying with kind of crop and type of clay. An average annual release from soils ranging from loam to clay under grass in Britain is 45 kg of K/ha, but some clays release twice as much.

Legumes add much nitrogen to many agricultural systems. Symbiotic fixation in England ranges from 50–150 kg N/ha fixed by annual arable legumes, or clovers growing with grasses in dry areas, to 200–400 kg N/ha fixed by tap-rooted clovers and lucerne. Clover growing in New Zealand pastures has fixed 600 kg N/ha in a year. In addition non-symbiotic organisms fix more nitrogen; usually in agriculture the amounts are uncertain and probably small except when rice is grown on flooded land (Chapter 22) where blue-green algae may contribute much.

Losses

The crop and the way it is used determine the nutrients removed. In a forest little is lost until the timber is removed. In artificial plantations where only a little of the annual growth is harvested losses are small (as with latex from rubber trees), but they may be very large if fruit is removed (oil palm). Amounts of nutrients removed in yields of annual arable crops may be relatively small or very large, depending on crop. Cereal grain removes only moderate amounts of N, P and K; the straw contains some nutrients and the way it is used determines whether they are lost. Root crops (e.g. sugar beet and potatoes) often remove five times as much nutrients as cereals grown on the same land. Ploughing in sugar beet tops conserves nutrients, using the tops for cattle fed elsewhere removes N, P and K. Large yields of grass deplete nutrient reserves; 12 t/ha of dry grass often contain 300 kg N, 50 kg P, 250 kg K/ha. These nutrients are completely removed if the grass is used for hay, silage or feeding green, but most are returned if livestock graze the herbage.

Leaching from plants. All nutrients are leached from growing plant by rain and when washed into soil may be used again. This nutrient cycle is important in forests and in perennial plantations. Probably the most important example with agricultural crops is the loss of cations

(K and Ca) from ripening cereals. A good cereal crop will contain 150 kg K/ha altogether when the grain is forming; during the few weeks when the crop is ripening, half or more of this will be returned to the soil in leaf fall and by leaching. Consequently harvested cereals remove little K *if* the straw is ploughed in, but much more is removed if the crop is cut green for silage.

Leaching from soil. Only small amounts of nutrients are leached from soils under forests or unfertilized plantations but there are large losses in most intensive farming systems in humid climates, particularly with annual crops. Leaching losses depend on the amount of water passing through the soil and this can be assessed for each season of a year by calculating the excess of rainfall over evapotranspiration.

The losses of N are serious when much rain falls in spring. We have found 30 mm of drainage water passing through soil in 24 hours in April/May after spring manuring has been done removed 20–30 kg/ha of N. Even with modern rates of manuring P is rarely leached except from the coarsest sandy soils, but residues from regular large dressings may move into subsoils. Potassium is more mobile and, when annual dressings are much larger than crops take up, some K may be leached into subsoil and part may be lost. Losses of calcium have been discussed in Chapter 7.

Although leaching is very important it is difficult to assess *how much* nutrients per hectare are lost in this way. Qualitative estimates of losses of nitrate are however worth making for land that is underdrained; it is relatively easy for an adviser to sample drainage water and measure nitrate concentrations, if much is leached top-dressings of N will be justified.

Erosion causes serious losses of plant nutrients in many countries; most can be prevented by better cultivations, drainage and control of surface water. In countries with temperate humid climates, such as Britain, soil erosion tends to be ignored because it is rarely spectacular. It does however occur on much of our arable, and also on steeply-sloping grassland.

Nitrogen in soil

Nutrient reserves depend much on the nature of the soil and how it has been used; changes are the nett result of many processes.

Inorganic nitrogen in soils is ephemeral; growing crops take it up quickly; any not used by plants is likely to be leached or denitrified. The large and long-lasting reserves of N in soils are all combined with organic matter. A part of the organic matter in most soils is very ancient, but part is derived from recently-added crop residues or organic manures. A small part of the old organic matter is decomposed to release a few kg of N each year. Turn-over of the more recent organic matter is much quicker and soils containing residues of leguminous crops, grass-land, or organic manures may release 100 kg N/ha/year, or even more.

The total supply of organic matter in a soil cannot be altered quickly. Reserves of organically-combined N increase where organic manures are regularly used and when land is sown to grass; they diminish when grass-land is ploughed or when arable land is cultivated more often. Fertilizers have little effect, the extra crop residues grown by large manuring seem to compensate for the increased oxidation caused by more intensive cultivation.

H. L. Richardson (ref. 65) found at Rothamsted, when arable land was laid down to grass, % total N in soil changed from 0.12% (in old arable land) to 0.20% in soil under 40-year old grass and to 0.25% N under very old grassland. When old grassland is ploughed organic matter diminishes rapidly, particularly in the first years. For example when one field at Rothamsted was ploughed from grass that was over a century old and had over 7% of organic matter, 12 years of arable cropping removed one-third of the stock of organic matter; even a single ploughing with immediate reseeding allowed some loss. All changes in soil organic matter that occur when land use is changed are accompanied by changes in nitrogen reserves. In the 12 years after ploughing the old grassland about one-third of the total N in the soil was mineralised, equivalent to 2000 kg N/ha in the top 22 cm layer. Crops grown after this land was ploughed from grass did not respond to N-fertilizer for many years, whereas there were large responses to N in the parallel experiment made on old arable land (ref. 44).

To decide the fertilizer needed in ley and arable cropping systems, the N released by leys of different types must be known. In an experiment at Jealott's Hill Research Station clover-glass leys that had lasted for 1, 2 and 3 years were ploughed and kale was grown, followed by wheat. The 3-year ley had fixed 450 kg N/ha that became available for following arable crops, as well as the N needed for its own growth. The total N available from shorter leys was less: 290 kg N/ha from a 2-year ley, 200 kg N/ha from a 1-year ley and the arable soil without ley provided only 100 kg N/ha.

For any system of cropping, manuring, and cultivation, there is an equilibrium value for % N in soil that depends on climate and soil type. When land is used differently, slow changes begin and % total N tends towards the equilibrium value for the new conditions. These are general statements applicable to farming everywhere. In the famous Morrow Plots in Illinois (U.S.A.), begun in 1876, three rotations were tested: (1) continuous maize; (2) maize and oats, with catch crops in the oats; and (3) maize, oats, red clover. Where fertilizer, lime and FYM have been used the maize-oats-clover rotation has maintained % N in the soil near its original value whereas under continuous maize without manuring, it has been roughly halved.

Phosphorus and potassium in soil

In contrast to the complexities of nitrogen relationships in soil, P and K behave simply. If more P and K are supplied than crops take up the surpluses accumulate in soil to increase the stock of potentially soluble P and K (but when very large dressings of K are used regularly, some may be leached). We believe the whole of these residues of P and K dressings may ultimately become available to crops. The effects of long continued manuring on soluble P and K in experiments where all the produce was removed have been shown already in Figs. 15, 16, 17, and 18.

EXAMPLES OF NUTRIENT CYCLES

Forests

Established forests contain large stores of nutrients in the trees and other vegetation; the system conserves nutrients efficiently. Because the soil is not disturbed, and roots usually fill the upper layers, only very small proportions of the meagre stock of mobile nutrient ions (Ca^{2+}, Mg^{2+}, NO_3^-, SO_4^{2-}) are lost by leaching each year. Phosphate and potassium ions are also conserved by 'fixation' processes in the soil. For example, the amounts of nitrogen and calcium in the components of a 35-year old Douglas-fir ecosystem were investigated by D. W. Cole and S. P. Gessel (ref. 66). Only 4 kg/ha of Ca was lost from the system annually by leaching from the soil, an amount that is no doubt made good by calcium released from weathering minerals and supplied in rain. Nitrogen was conserved efficiently and the annual loss by leaching was estimated as less than 1 kg/ha of N. The amounts of N in rain and fixed biologically were unknown, but were no doubt sufficient to replace the amounts taken up each year by the trees.

Annual arable crops

There is a great contrast between nutrient cycles in natural vegetation and the nutrients involved in growing an arable crop. Fig. 21 shows the nutrients which a good crop of sugar beet (40 t/ha) takes up in one year (the N is nearly as much as a forest accumulates in thirty years' growth). This crop removed from the soil a quarter of the soluble P and over half of the exchangeable K. The fertilizer dressing shown is a normal recommendation for beet, it supplies as much P as the crop takes up, but much less N and K. Amounts of N, P and K in the crop are divided roughly equally between roots and tops. The roots are taken to a factory for processing and the N, P and K they contain is lost to the farm. The use made of tops determines whether the nutrients they contain are returned to the field (by ploughing in) or whether they also are removed when the tops are taken for stock to eat indoors or on other fields.

Nutrient cycle balance sheets are of little use in planning to use N and P fertilizers on sequences of annual arable crops. Enough N must be applied to produce the yield required, allowances being made for supplies from soil and crop residues, for season and losses by leaching. Dressings of P recommended when agriculture is being intensified always supply more than crops remove and reserves accumulate (later the rate of manuring is best controlled by soil analysis). But modern K-manuring does not *necessarily* result in extra K accumulating in soil; a balance of soil potassium must be calculated so that 'optimum' recommendations

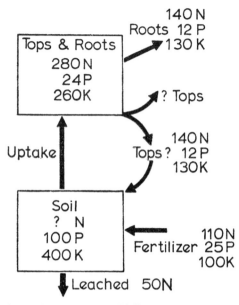

Fig 21. Kg / ha of N, P and K

Fig. 21   The nutrient cycle in a sugar beet field.

will supply enough K to maintain reserves in the soil after allowing for any K naturally released by the clay. Many arable crops, potatoes and wheat are examples, respond to fresh dressings of K-fertilizer more than grasses do. When grass and arable crops are grown alternately, the potassium applied must replace that removed in the (unresponsive) grass so that following crops which are more responsive do not suffer. Soils with good reserves of K often give larger yields than poorer soils dressed liberally with fresh K-fertilizers. Replenishing soil K should not be left until reserves are exhausted as the soil may not then give maximum yields, however much fresh K fertilizer is applied.

Potassium cycles

The components of a potassium cycle vary greatly. Herbage and fodder crops, potatoes and sugar beet *remove* much K. Cereal grain removes relatively little K and the straw contains about as much as the grain; but cereals at earing often contain twice as much K as at the green crop may have taken up as much as potatoes do. Grazing animals return much K, often 200 kg K/ha in a year; 40 t/ha of rich FYM may supply 300–400 kg K/ha. These examples emphasise the need to use enough K; but it is easy to use too much and the surplus may be taken up (as 'luxury') without extra yield, or it may be leached from light soils.

There is always a danger that too little K may be used when crop sequences include many crops that have a reputation for needing little K-fertilizer because they are unresponsive. Table 40 shows calculations for a Rothamsted experiment.

TABLE 40

GAINS AND LOSSES OF PLANT NUTRIENTS IN A FOUR-COURSE ROTATION

	N	P	K
		kg/ha	
Removed in 4 crops	437	58	348
Added by fertilizer* with 20 per cent N 4.4 per cent P 8.3 per cent K	359	81	151
10 per cent N 4.4 per cent P 8.3 per cent K	364	157	303
Added by farmyard manure (25 t/ha)	56	34	168

(*The amounts of compound fertilizer being adjusted to supply roughly the same amount of N)

Kale, barley, ryegrass and wheat were grown. All of these are considered to need much N but relatively little P and K and 'high-nitrogen' compound fertilizers are commonly used. Using the standard fertilizer (with %N—%P—%K = 20–4.4–8.3) supplied enough N and rather more P than the crops used. But the crops removed nearly 200 kg K/ha more than was supplied. To balance the nutrients removed a compound fertilizer with nearly equal percentages of N and K was required. Alternatively a large dressing of FYM to the kale, or feeding off the grass and kale on the field, would restore the potassium balance.

Grassland

In climates that suit grass-clover associations (as in New Zealand), large yields of 20 t/ha of dry matter or more are possible without using N-fertilizer. In other regions, including most of North-Western Europe, nitrogen fertilizers must be used to obtain maximum yields from grassland. Grass that has enough water and 300–400 kg N/ha/year can produce 12-14 t/ha of dry matter in one season in England; productive grasses grown with much N in the tropics can yield two or three times as much. Very large yields of grass need much fertilizer to produce them and maintain the sward. A 12 t/ha crop of grass may contain 300 kg of N and K and 60 kg of P and these are 'at risk' when the grass is used. If grass is cut and removed continuously, fertilizers must be used to replace the P and K if reserves of soluble P and K in soils are to be maintained. If the grass is grazed continuously the situation is more complicated. Most of the P, Ca and Mg, and nearly all the K in th grass eaten, are returned in excreta. But because they are returned i patches the nutrients are distributed irregularly making it difficult t plan the P and K manuring needed to maintain the sward.

Although no simple scientific solution of these problems has bee obtained, experience suggests that once a good stock of P and K ha accumulated, only a little extra is needed as fertilizer to maintain graze grassland. Nitrogen is more difficult. Not only is the return irregular bu ammonia may be lost when urea in urine decomposes and some of th large amounts of nitrate formed may be lost by leaching. Althoug nitrogen accumulates in grazed swards, and some of it may b used more than once in a season, much N is lost by volatilisatio and leaching besides the amount retained by the grazing animals to mak protein. Many experiments have shown that soil under grazed gras accumulates more N than soil under cut grass; yield is increased becaus the N that is retained is used again. In one example from New Zealan where clover contributed very much nitrogen, 130 kg/ha of extr nitrogen was available to the system in a grazing as compared with cutting regime. An English example by workers at North Wyke Experi mental Station (ref. 67) showed that as the total nitrogen available t the sward increased, yield was directly proportional to N supplied i cutting systems; in grazing systems the effect of the N available progres sively increased as the dressings increased.

Glasshouse crops

Intensive glasshouse cropping involves more nutrients in circulatio than any of the systems discussed above, the soils receiving and losin large amounts. N.A.A.S. workers (ref. 68) calculated the nutrient balanc in Table 41 from a glasshouse experiment on tomatoes where additio and losses were measured for 3 seasons. Much more N, K and M was used than was needed. The glasshouses were flooded in winte and 80–90% of the water applied drained away. Each 110,000 litres/h (equivalent, roughly, to 10 mm of water) removed 7–10 kg of N, 6–7 c K, 33–38 kg of Ca and 4–6 kg of Mg/ha. In the growing season onl nitrogen losses were serious (about 70 kg/ha). Table 41 suggests that th

TABLE 41

DIFFERENCE BETWEEN NUTRIENTS ADDED AND LOST IN GLASSHOUSE
CROPPING FROM SPRING 1957 TO AUTUMN 1959

	Added	Lost kg/ha	Difference
Nitrogen	1652	569	+1083
Potassium	3392	322	+3070
Calcium	1720	1911	−191
Magnesium	1263	201	+1062

manuring used wastes much N and allows unnecessarily large amoun of K and Mg to accumulate. Other estimates confirm the large losse aused by leaching in glasshouse cropping. In one example glasshouse omatoes removed 620 kg K/ha in fruit while 250 kg K/ha was lost in ummer drainage and 460 kg K/ha through winter flooding. To use ertilizers more efficiently in glasshouse cropping, the balance of nutrients n crop, fertilizer and drainage must be calculated.

Tree crops

Some tree crops need very large amounts of nutrients because much s removed in the yield; other tree crops may need much fertilizer to :stablish them but quite small fertilizer dressings are sufficient to main-ain the mature crop. Two contrasted examples are discussed here:

Rubber is an extreme example as the harvest of latex contains very ittle nutrients. The following amounts were found to be needed to ·eplace losses in latex and to grow the trees (ref. 69):

	N	P	K kg/ha	Mg	Ca
For annual tree growth	78	11	34	17	22
For 1120 kg dry rubber	7	1.3	4.7	1.1	0.04

The nutrients needed for tree growth were similar to those used by ;ome temperate crops, but most were immobilised in trunk and branches. n early years the trees took up nutrients rapidly and had to be fertilized o grow well. Less nutrients were needed later in the plantation's life and ·ecycling in dead litter and by rainwash from the trees provided much »f the nutrients needed each year.

Oil palm. The fruit of oil palm, which is harvested, contains more 1utrients per hectare than many temperate crops as the figures in Fable 42 for a 20-year old plantation (148 palms/ha) in Nigeria show. Before harvest the N, P and K in fruit were more than half of the total umounts on the site. Much K is removed in fruit and K-deficiency ;oon develops in oil palm in Nigeria.

TABLE 42

NUTRIENT CONTENTS OF OIL PALM IN NIGERIA

	N	P	K kg/ha	Ca	Mg
Tree	390	55	250	220	230
Roots	70	5	90	14	30
Fruit	430	90	500	76	65
Total	890	150	840	310	325

Very different fertilizer treatments are needed for these two tropical :ree crops. Manuring rubber with similar fertilizers to those used for :ereals in Europe (much N, moderate P and K) appears reasonable. Oil palm appears to need fertilizers rich in N and K, such as are used for potatoes in Europe.

THE EFFECTS OF ANIMAL FARMING SYSTEMS ON THE CONSERVATION OF PLANT NUTRIENTS

Balance for the U.K.

The following estimates (ref. 64) show that grass grown in U.K. contains much more nutrients than arable crops contain:

	N	P	K
		thousands of tonnes	
Arable crops	430	60	300
Grassland	1000	115	850

If the whole of the produce from grassland is fed to housed stock their excreta may be returned to the land to grow more grass; if it is disposed of in some other way, the nutrients must be replaced as fertilizers.

Table 43 shows total nutrients in crops grown in 1968, and the amounts available for recycling because the crops containing them were used by animals or were wastes returned to the land. The figures suggest that in Britain half of the N, two-thirds of the P and most of the K could be returned to the land to be used again. If this was done uniformly, relatively little P and K fertilizer should be needed, but much nitrogen would have to be used. In mixed animal-arable farming, once soil fertility has been built up by increasing the stock of nutrients in the soil by fertilizers, whether much or little P and K fertilizers is needed to maintain the system depends on how well nutrients in stock excreta and crop wastes are returned to the land that produced the crops. There are serious practical problems in returning wastes from many modern animal farming systems which were discussed in Chapter 2. All the nutrients in wastes from crops and livestock should be returned to land and distributed uniformly. Our success in doing this (or lack of success) will determine how much fertilizer P and K we need in future.

TABLE 43

ESTIMATES OF TOTAL NUTRIENTS IN U.K. CROPS AND AMOUNTS AVAILABLE FOR RE-USE THROUGH ANIMAL EXCRETA IN 1968

	N	P	K
		thousands of tonnes	
Total nutrients in crops	1430	175	1150
Available for recycling			
Arable crops	260	40	230
Grassland	500	85	750
TOTAL	760	125	980
Difference, assuming no waste	670	50	170
Fertilizers used	748	205	367

Balance for an English farm

The need for long-term fertilizer policies is now recognised on many large farms. A plan for the 1970s, made for Bridgets Experimental Husbandry Farm (ref. 70), takes account of the nutrient cycle between soils and crops; it is a good example of how all the information available from soil and crop analysis and from field experiments must be synthesised to plan the manuring of different farming systems. Part of the farm is in continuous cereals, another part is arable, but slurry is given to some crops and FYM to potatoes; outlying grass fields are mostly cut for conservation, and grass near the dairy unit is intensively grazed. These systems have very different effects on the nutrient reserves in soils. For example, soil potassium has been depleted where whole barley is used for silage, and has increased under grazing by dairy stock. Phosphate is needed to maintain yield, but single dressings annually or treble dressing every three years have been equally good in experiments.

This account of work at Bridgets Farm suggests that the results of short-term experiments on parts of the system on the farm could be misleading; as least as good fertilizer practice is likely to result from balancing the nutrients added and removed in a rotation. The nutrients removed in crops used in various ways are shown in Table 44 (from ref. 70). Much slurry is used; 112,300 litres of winter slurry supply 130 kg N, 30 kg of P and 109 kg of K; it is used for potatoes at 400,000 litres/hectare and supplies about twice as much nutrients as the FYM that is used for other potatoes. The nitrogen used varies greatly. Grazed grass get 300–370/kg N/ha, conserved grass 225 kg N for two cuts. Intensive cereals receive 125–150 kg/ha, other wheat and barley 100 kg N/ha and 75 kg N/ha respectively.

TABLE 44

CROP YIELDS AND THE P AND K THEY CONTAINED AT BRIDGETS EXPERIMENTAL HUSBANDRY FARM

	Yield/hectare	Nutrients removed	
		P	K
		kg/ha	
Grass			
grazed	11,230 litres of milk	10	20
conserved	7.5 tonnes dry matter	24	208
Wheat grain (+ straw)	4600 kg grain	6	18
Barley			
grain (+ straw)	4250 kg grain	5	17
whole silage	8 tonnes dry matter	10	104
Potatoes	30 tonnes	23	149
Peas	3100 kg	12	26

Annual losses of P and K in the separate farming systems used on different parts of this large farm were calculated from these figures.

In planning fertilizing it was assumed that 36 kg K/ha comes each year from soil and that 80% or more of the K applied is taken up, but that only 20-25% of the P applied is used (half of the fields are very low in soluble P). The maximum amounts of P and K needed annually were 46 kg P/ha and 83 kg K/ha in a four-year rotation of grazed grass, grass silage and wheat followed by barley used for silage. Least potassium was needed in rotations where continuous cereals were taken or grass was grazed for 5 out of 7 years.

These phosphate dressings needed were planned for direct effect and were therefore much larger than crops removed. As their residues accumulate soluble P in the soils will increase and less P will be needed as fertilizer. Experiments with phosphate suggests that a bulk dressing given to a new ley will last for 4 years; such infrequent applications save spreading costs. Distributing the potassium through the rotation is not simple. Excess on grazed grass can lead to hypomagnesaemia, too little on grass or barley cut for silage (both remove over 100 kg K/ha) results in serious deficiency. Because cereals *require* much more potassium than is taken off in grain and straw (much K being returned to the soil before harvest) dressings given are larger than Table 44 suggests are necessary; 30 kg K/ha is proposed for the cereal area to allow for this.

Balance on Dutch farms

Similar methods are used in advisory work in the Netherlands. The Dutch Ministry of Agriculture published the nutrients in manure produced by various classes of stock and calculated the amounts needed by crops grown on clay and sand soils. When large numbers of stock are kept, large amounts of plant nutrients are brought on the farm and most will be present in manure. Table 45 shows the P and K in manure produced by different classes of stock in a year. The amounts of P and K estimated as needed by grass and other crops on soils having satisfactory amounts of soluble nutrients are shown in Table 46. As stocking increases the P and K in manure increases too and eventually may be more than is needed by the crops. Using fertilizers to supply more P and K under such conditions would be at best wasteful and might harm health and growth of cattle.

An example is given of a mixed farm of 12 hectares of sandy soil carrying 15 milk cows, 3 two-year and 5 one-year old young stock, 7 calves and 200 fattening pigs. The nutrients in manure (allowing for 10% loss in storage), and the needs of the crops grown (2 ha of grazed grass, 7 ha of hay (cut once), and 1 ha each of barley, rye, sugar beet and turnips) were calculated:

	P	K
	kg per year	
Available from manure	612	1454
Needed by crops	251	1278

TABLE 45

AMOUNTS OF P AND K IN MANURE PRODUCED BY LIVESTOCK

(Estimates from the Netherlands)

	Production (kg/year)			
	P	K	P₂O₅	K₂O
Type of stock				
1 Cow	7	37	17	45
10 Calves	6	18	13	22
1 Average fattening pig	3	4	6	5
1 Sow (housed)	5	8	12	10
1 Sow (grazed)	1	2	3	2
100 Broilers	4	7	10	8
100 Laying hens (free range)	17	13	38	16
100 Laying hens (batteries)	33	27	76	32

On such a farm with a large surplus of P and a small surplus of K, fertilizer P and K will not be needed. Indeed using the manure produced will build up much P in the soil and increase the K reserves too which may damage the health of stock; therefore it is necessary to sell some of the manure.

TABLE 46

ESTIMATES FROM NETHERLANDS OF NUTRIENTS NEEDED BY COMMON FARM CROPS

	Nutrients needed kg/ha		
	P	K	
		clay or peat soils	sandy soils
Grassland (grazing)	11	17	50
Grassland (hay—one cut)	20	83	116
Grassland (each additional cut)	13	50	66
Barley	30	17	83
Other cereals	17	17	83
Potatoes	39	190	133
Potatoes (industrial)	39	108	62
Fodder beet	30	116	208
Sugar beet	30	66	133
Turnips	13	66	66

CHAPTER SEVENTEEN

Making Recommendations and Selecting Fertilizers

This chapter shows how information about soils, crops and the effects of fertilizers measured in field experiments are co-ordinated to make a practical fertilizer recommendation. It draws on other chapters and summarises some of the discussions of the effects of manures and fertilizers given earlier.

MAKING A FERTILIZER RECOMMENDATION

Developing areas

Recommendations to use fertilizers may be made because deficiency symptoms have been seen on crops, or because simple experiments have shown that yields are likely to be increased by applying one or more nutrients. Usually there will be no local information on shapes of response curves and therefore no means of deciding *how much* nutrient should be applied. All that can be done is to use amounts which produced responses in local experiments, or which had been satisfactory elsewhere. The process of translating the recommendation into actual fertilizers is the same as in more intensive agricultural systems, and where there is much more information about the effects of fertilizers.

Intensive agriculture

All the available scientific information relevant to the particular crop and field should be integrated in making a fertilizer recommendation.

Response curves appropriate to *crop, climate* and *soil type* (Chapters 11, 12, 13) are the basis; allowance must be made for interactions between nutrients, and between other factors (e.g. irrigation) and nutrition. The amounts of nutrients considered likely to give maximum yield and profit will apply to crop and region; they must be adjusted to fit the particular farm and field.

Previous cropping and manuring are responsible for large variations in the nitrogen needed. Allowance should be made for extra N remaining in soil from previous crops, and also for *rainfall* in the *preceding winter* because this alters the amount of N that is actually available to the next crop.

Residues of manures and fertilizers that supplied phosphorus and potassium for long periods are detected by *soil analyses* which should be made every few years to check whether reserves of P and K are accumulating or diminishing. Dressings of P and K fertilizers should be altered to allow for these trends. Changes in soluble P in soil will be the main reason for varying dressings of *phosphate*.

Potassium dressings should be varied to allow for changes in exchangeable K in soils, but also for the balance between losses of K in crops and gains through manures and fertilizers (Chapter 16). Some farming systems that remove much K and supply little can exhaust soil potassium quickly. A check should always be made that the manuring regime used balances losses of potassium; allowance must be made for the amount of K that soil itself can supply without becoming exhausted.

USING ORGANIC MANURES AND FERTILIZERS TOGETHER

In Britain about a third of our area of root crops, a fifth of the area of grassland, and 7 per cent of the cereal area gets a dressing of 35–45 t/ha of farmyard manure each year. Other land receive slurries; poultry manure, sewage sludge, and other organic wastes are used on smaller areas. All these manures supply much plant nutrients and lessen the amount of inorganic fertilizer needed, optimum dressings must be adjusted accordingly. The compositions of organic manures have been discussed in Chapter 2. The section below briefly recapitulates the adjustments that should be made.

Farmyard manure (FYM)

An average dressing of 25 t/ha of farmyard manure supplies about 125 kg of N, 75 kg of P_2O_5 and 125 kg of K_2O. These quantities are not completely available to crops in the year of application, in particular the nitrogen is very slow acting. Only one-third of the N is likely to be useful to a first-year crop. About two-thirds of the phosphate may be effective and most of the potassium will be available. *The actual value of 25 tonnes of farmyard manure on a hectare of root crops is equivalent to about 40 kg of N, 50 kg of P_2O_5 and 100 kg of K_2O.* When FYM is used the normal fertilizer dressing should be diminished by these amounts. Farmyard manure is generally used on root crops that also receive a compound fertilizer (generally of *High-Potash NPK* type with 1 : 1 : 1½ or 1 : 1 : 2 plant food ratios). A rough, but satisfactory, way of allowing for the use of farmyard manure on root crops is to use the same type of compound as when manure is not given, but to apply only one-half to two-thirds of the normal quantity. Where cereals receive FYM, no P or K fertilizer will be needed, and the normal nitrogen manuring should be cut down by 40 kg of nitrogen per hectare.

Quality of farmyard manure depends on its content of plant foods and these depend on the way the manure was made and how it was looked after. Most of the plant foods in manure come from the feed given to the stock that make it and only a small proportion from the straw. If much straw relative to excreta is used the manure will be poorer. Fattening stock retain less of the plant foods in the crops and feeds they eat than do young stock or dairy cattle; other things being equal, manure from fattening beasts will be richer.

Care must be taken of farmyard manure to prevent loss of its plant foods. *Nitrogen* is lost as ammonia gas when heaps of manure are turned and also when FYM is carted into a field and allowed to lie in heaps, or spread over the land for any length of time before it is ploughed in. Both N and K are lost in the black liquor which often oozes away from a manure heap. Heaps should not be turned, they should be kept as compact as possible and should be protected from rain to lessen losses by leaching. When FYM is applied to the land it should be spread and ploughed in immediately. Even a day's delay in ploughing can result in much loss of ammonia.

Liquid manures

Disposing of slurry and other liquids from intensive animal farming is now a serious problem in many countries; it has already been discussed in detail (Chapter 2). The compositions of farm effluents depend on their origin. The kind of stock producing the excreta and their food, how urine and dung are handled, and dilution with water used for washing, etc., all have effects. The wide variability in compositions of liquid manures are shown by the recent analyses of samples collected in the United Kingdom given in Table 8. Farmers should consider the factors that cause variations in slurry compositions (Table 8) and estimate the amounts of plant nutrients that may be supplied. If no better basis can be devised these figures may be useful:

	kg per 1000 litres		
	N	P_2O_5	K_2O
From dairy cows	4	0.5	6
From pigs	6	1	4
Average of slurries of many types given by H. T. Davies (ref. 10)	7	2	7

(To convert these amounts to pounds per 1000 gallons, multiply them by 10.)

The amounts of nutrients applied as slurry should be deducted from the optimum fertilizer dressing. In some experiments the nutrients in liquid manures have been as effective as the same amounts supplied as fertilizers, in other experiments less effective. A fair valuation is to expect the nitrogen and phosphate to be two-thirds as effective as N and P in fertilizers and the K to be fully effective.

Poultry manure

Manures from poultry vary in composition according to the type of birds, the way they are kept and their food. Mixing the manure with litter or sand lowers its concentration, and long exposure, or bad storage, both cause a serious loss of nitrogen. Poultry manure is rich in N and P but poor in K; the nitrogen is not more effective than two-thirds as much inorganic nitrogen supplied as fertilizer; the phosphate is insoluble in water and acts slowly. Fresh poultry manure may contain from 0.9 to 1.5% N, about 1% P_2O_5 and from 0.4 to 0.6% K_2O. A tonne of such manure will contain 9 to 15 kg of total N, 10 kg of total P_2O_5 and 4 to 6 kg of total K_2O. Although these amounts are not all immediately available a dressing of several tonnes per hectare of poultry manure will supply enough nitrogen and phosphate for most arable crops, but extra potassium will also be needed for potatoes and sugar beet and other root crops. When smaller amounts of poultry manure are used the usual dressing of fertilizers should be reduced by 10 kg of N, 10 kg of P_2O_5 and 5 kg of K_2O for each tonne of manure applied.

Sewage sludge

Sludges vary in condition from sticky materials containing half their weight of water to well-dried powders, easy to handle and spread. All sludges contain much N and P but little K. Used alone sludge is suitable for leafy crops such as cabbages and kale, but if it is given to potatoes, sugar beet or vegetables the normal quantity of potassium fertilizer must be given as well. For practical purposes 25 tonnes per hectare of raw sludge will supply plant foods equivalent to about 250 kg of N, 125 kg of P_2O_5 and 25 kg of K_2O. These values may be used as a basis for adjusting fertilizer dressings by assuming that only a half of the N and P may be available to a first-year crop.

Other organic materials

Most other organic manures are unimportant on the average farm but they may be used by horticulturists. *Brewery wastes* are sometimes available; they contain about 2% each of N, P_2O_5 and K_2O. *Seaweed* is used near the coast; a 25 tonne per hectare dressing may supply 50 kg of N, 25 kg of P_2O_5 and 50 kg of K_2O, besides some organic matter. *Sawdust* is of variable composition and has little value as manure; wood ashes may contain up to 5% of K_2O. *Peat* decomposes slowly in soil and releases some nitrogen but very little P and K. *Town refuses* are poor in plant foods and often contain only about 0.5% N, 0.3% P_2O_5 and 0.2% K_2O. If used at moderate rates, their contribution to plant food supplies can be ignored. Some analyses of town refuses and other wastes are in Table 6.

Composts

Composts can be made by rotting down straw and other farm wastes. Although these manures are often important in market gardening they are rarely used on ordinary agricultural crops. Dry materials like straw may be rotted down with water, or better by pumping liquid manure and drainings from cattle sheds and yards over the heaps. Nitrogen fertilizer must be added as well to assist micro-organisms to break down the straw, 60 kg of ammonium nitrate for each tonne of straw is needed and 50 kg of ground limestone should be spread as well to keep the heap sweet. About 4000 litres of water will also be needed. Composts are usually much poorer in plant foods than an equivalent bulk of FYM; those made from farm wastes contain less available N and are poorer in P and K than FYM. Unless composts have been enriched with fertilizers there is no need to alter the normal fertilizer dressings when moderate dressings of composts are used as well.

Straw ploughed in directly affects plant food supplies in two ways: it locks up some available nitrogen and it supplies a little potassium. When straw has been ploughed in at ordinary rates the only adjustment needed to the normal manuring of the following crop is to apply an extra 10 kg of N for each tonne of straw that has been ploughed in.

CONVERTING RECOMMENDED AMOUNTS OF NUTRIENTS INTO A FERTILIZER DRESSING

When all adjustments have been made the final recommendation will be in terms like 'x kg N, y kg P_2O_5 and z kg K_2O per hectare'. These have to be translated into practical fertilizers before the dressings can be bought and applied.

Following a recommendation

Recommendations on using fertilizers are usually given to farmers in one of four ways:

(a) As weights of straight fertilizers, often kg/ha of ammonium sulphate, superphosphate and muriate of potash.

(b) In terms of compound fertilizers of stated analysis: For example a recommendation might be: '1000 kg/ha of a compound fertilizer containing 7% N, 7% P_2O_5, 10.5% K_2O or alternatively 600 kg/ha of a fertilizer containing 12% N, 12% P_2O_5, and 18% K_2O.' (These analytical specifications are often abbreviated by leaving out the signs for percentages of plant food. 7% N, 7% P_2O_5, 10.5% K_2O and 12% N, 12% P_2O_5, 18% K_2O then become 7–7–10½ and 12–12–18 respectively.)

(c) As kg/ha of N, P_2O_5 and K_2O, (or for present practical purposes in Britain as UNITS (hundredths of 1 cwt), or pounds per acre of these plant foods).

(d) In terms of a group of compound fertilizers for which the plant food ratio is stated, together with the number of kilogrammes of N (or P_2O_5) needed per hectare (or the number of UNITS or pounds of nutrients per acre).

In applying these recommendations there is no difficulty in following the first and second methods, since the appropriate fertilizers are bought and used. The other methods are a little more difficult to apply.

Recommendations in terms of straight fertilizers can be put into practice with a compound fertilizer by using Table 11 to work out the kilogrammes of plant food required and then proceeding as described below.

Recommendations in terms of kg of N, P₂O₅ and K₂O can be converted to straight fertilizers by using Table 11. Usually, however, they are converted into an equivalent dressing of a compound fertilizer in the following way:

i. Work out the plant food ratio of the desired compound;
ii. use the ratio to pick a suitable fertilizer, and then
iii. use the recommended number of kilogrammes of N (or P_2O_5) to work out the dressing of compound fertilizer per hectare. An example will make this clear:

Recommendation for potatoes: 'Use 100 kg N, 100 kg of P_2O_5 and 150 kg of K_2O per hectare, broadcast over the furrows before planting.'

In terms of straight fertilizers Table 11 shows that this recommendation implies dressings per hectare of

475 kg of ammonium sulphate (21% N)
500 kg of ordinary superphosphate (20% P_2O_5)
250 kg of muriate of potash (60% K_2O)

The *plant food ratio* of the appropriate compound fertilizer is 100 : 100 : 150 or $1 : 1 : 1\frac{1}{2}$. Table 24 (supplemented when necessary by other current fertilizers in manufacturers' lists) shows that several fertilizers would suit, since they have this plant food ratio:

Fertilizer A is a liquid and contains 7% N, 7% P_2O_5, 10% K_2O
Fertilizer B contains 12% N, 12% P_2O_5, 18% K_2O
Fertilizer C contains 15% N, 15% P_2O_5, 23% K_2O

The choice between these alternative fertilizers may be decided by calculating the values of the nutrients they contain; one may be cheaper than the others; perhaps one may be bought easily while the others are difficult to obtain. In any case the amount to be applied is decided by working out the number of kg required to supply 100 kg of N. Fertilizer B contains 12% N so 100 kg of N are supplied by $\frac{100 \times 100}{12} = 833$ kg of compound. Fertilizer C contains 15% N and 100 kg supplies 15 kg of N, so to give 100 kg of N $\frac{100 \times 100}{15} = 667$ kg (approximately) of the compound should be applied per hectare.

With all recommendations worked out in this way, if the plant food ratio of a compound fertilizer has been correctly chosen, it is only necessary to work out the weight per hectare required for one plant food; the amounts of the other plant foods will be correct automatically.

Other units of weight

These examples are given in metric units, but the calculations are made in the same way and just as easily in the other units discussed in Chapter 3. If the British Imperial system is used, with a (long) ton of 2240 pounds (lb) and a hundredweight (cwt) of 112 lb, all calculations are simplified by using the conventional UNIT of 1.12 lb; a 1 cwt bag of fertilizer contains the number of UNITS of N, P_2O_5 and K_2O shown by the analysis usually printed on the bag. Superphosphate with 20% P_2O_5 has 20 UNITS of P_2O_5 in the 1 cwt bag. American measures are equally simple; the 'short' ton has 2000 lb and the 'hundredweight' bag contains 100 pounds so the analyses give the numbers of pounds of N, P_2O_5 and K_2O in the bag. An advisory recommendation given in UNITS (Imperial) or pounds (U.S.A.) is quickly converted into 'hundredweight' bags (or tons) in either system.

CHAPTER EIGHTEEN

Timing and Placing of Fertilizer Dressings

Fertilizers must be applied so that they can be reached by the roots of crops just when they are needed if they are to be used efficiently. Inorganic nitrogen fertilizers are soluble in water, they move in soil moisture and it is usually (but not always) satisfactory to spread them on the top of the soil for rain or irrigation water to wash in. Phosphate and potassium fertilizers combine with soil and do not move easily, so they must be put below the surface where they can be reached by roots when the crop begins to grow. P and K must always be incorporated in the soil if they are to act quickly, dressings broadcast after planting an arable crop are likely to be inactive. Often the most effective way of applying P and K fertilizers is to place them in a band close to the seed when the crop is planted; sometimes, but not so often, there are gains from applying nitrogen fertilizers in this way. Special methods of applying fertilizer close to the plant are given the name *'fertilizer placement'*. With some crops better yields are obtained by placement, with others the ordinary method of broadcasting the fertilizer before sowing gives just as good yields. The value of placement depends greatly on the plant food reserves in soils. Putting fertilizer close to seed or roots is usually of most benefit on the poorest soils.

NITROGEN FOR CEREALS

As nitrogen fertilizer is liable to be washed out from the surface soil by heavy rain, particularly in winter, dressings must be timed correctly to ensure that the crop has nitrogen when it needs it, and yet to avoid waste.

Winter cereals

There is not a good case for giving nitrogen when sowing cereals in the autumn except on poor soils. In British experiments autumn dressings have generally given poorer yields than equivalent amounts used in spring. There is not enough experimental evidence, however, to condemn the common practice of giving *some* nitrogen in autumn. On the more difficult classes of heavy land it may be impossible to use a fertilizer distributor early enough in spring, and autumn-applied nitrogen may supply the crop for a time. Where autumn nitrogen *is* justified the amount applied at sowing time should not exceed one-quarter to one-third of the total nitrogen dressing which the crop requires. Levington Research Station workers (ref. 71) showed that the effect of autumn nitrogen in terms of a spring dressing depended on the rain falling between November and March. Their results suggest that efficiencies of autumn-applied N vary in this way:

November to March rain mm	Response to autumn-N as percentage of response to spring-N
125	75
250	50
375	25

Timing of spring nitrogen for cereals

In most series of experiments on spring cereals which have tested seedbed dressings applied at sowing time against top-dressings in late April-early May, and in tests of early (March) against late (May) top-dressing for winter cereals, results have varied greatly with season. Spring rainfall is the critical factor, and its distribution is at least as important as the total amount. When heavy rain (20 mm or more) falls in short periods (a day or less) in April and May before crops have made much growth the soil is severely leached and much nitrate is lost. Top-dressings applied *after* such rainfall are usually much more effective in increasing grain yields than are early dressings. If the weather is dry in April and May, and there is little rain to wash in top-dressings, the N is more effective when applied early. In other seasons, with well-distributed light rain, early and late dressings are equally effective. As spring weather cannot be forecast it is best to plan to insure against both excessively wet and excessively dry periods in April-May by giving half the N when sowing spring cereals (or in March for winter cereals), and the other half as a May top-dressing. This gives an opportunity to vary the amount of the top-dressing according to the needs of the crop, judged by its colour or, preferably, by measurements of nitrate in plant stems.

In experiments made in Illinois (U.S.A.) the recovery of N applied in December/January depended greatly on the precipitation in the following five months. When rainfall was 500 mm in this period uptake of N was 25 per cent less than in a dry winter with only 250 mm of rain. Losses were most when much N was used in nitrate forms and in very wet winters. In other experiments in Illinois 1 kg N applied in spring was as effective in increasing wheat yields as 1.5 kg N applied in autumn.

Lodging: Risks from lodging sometimes influence decisions on timing of nitrogen for cereals. Trials in Eastern England on medium and heavy soils showed that when 60 kilogrammes N per hectare were needed in the season, best yields of winter wheat were obtained from a single May top-dressing. When 100–120 kg N per hectare were needed, dividing the dressing and giving part in autumn and part in May, or part in March and part in May was best. Applying large dressings wholly in March gave crops that looked well in June, but they were liable to be damaged by lodging; there was little lodging with May top-dressings.

Future policy. Very large amounts of nitrogen fertilizer are used on cereals wherever they are grown. In arable areas of England the (unsubsidised) cost of the N needed for maximum yields may be a quarter of total production costs. It is very important that waste of part of these large sums should be avoided, but at present it is rare to recover in cereal grain plus straw more than half of the fertilizer-N applied to the crop. Dressings must be adjusted to fit farming systems as these determine the amounts of N that soils can supply (Chapter 13). They must also be applied at times that minimise loss by leaching. Systems will be developed for applying a portion of the N early in the season and the remainder when the crop is growing well, using both crop appearance and plant analyses to decide amount and time for the second dressing.

TIMING OF PHOSPHORUS AND POTASSIUM FERTILIZERS

Arable land

Most of the arable areas of Britain have for many years received P and K from regular dressings of fertilizers and FYM. In most of the soils the residues of P and K accumulated are sufficient for the immediate needs of many crops. Results of recent experiments summarised in Table 31 show that only potatoes frequently give roughly equal responses to N, P and K fertilizers. Practically all potato fields should receive dressings of these three nutrients when the crops are being planted and correct timing and placement are important. On average land immediate returns from P and K fertilizers used for cereals are likely to be small; responses of sugar beet and other root and fodder crops to P and K are usually less than to N and correct placement is not important. These changes in the fertility of English soils have lessened the amounts of P and K fertilizers needed by cereals and most other crops except potatoes; they have also made it unnecessary for *every* crop to be fertilized. The P and K fertilizers used on soils with average amounts of soluble P and K serve two purposes: They *maintain* soil fertility by replacing the nutrients removed by crops: they also *insure* against the risk of an unsuspected deficiency in land not recently analysed, and also provide some extra P and K that may be useful to stimulate root growth in an unfavourable season.

Fertilizers applied for 'maintenance' and 'insurance' purposes need not be applied in spring when all labour is needed for cultivating and sowing. P and K fertilizers for sugar beet, kale and forage crops (and sodium fertilizers for beet) can be applied in winter; often the dressings are satisfactory if ploughed in. Under many conditions maintenance dressings, particularly of P, need not be applied every year. If potatoes or sugar beet are in the rotation, the dressings given to these crops can be planned so that they leave enough for following cereals or fodder crops. These possibilities have been tested in experiments. At Rothamsted we found that large triennial dressings of phosphate (about 400 kg/ha P_2O_5) had large residual effects and gave practically the same yields of barley, potatoes and swedes as applying one-third as much P annually. At Boxworth Experimental Husbandry Farm on the calcareous Hanslope Series soil 280 kg/ha P_2O_5 applied once in 5 years produced as good yields of cereals as 56 kg/ha P_2O_5 applied every year. The experiment also compared basic slag and super; slag was slightly less effective than super when used annually, but was just as good as super when used once in 5 years. In similar experiments at Bridgets Experimental Husbandry Farm (Hampshire) triennial dressings of P and K fertilizers maintained yields just as well as annual dressings supplying one-third as much P and K. Considerable flexibility in timing of P and K fertilizers is possible without lessening yields. For many crops P and K can be applied in winter, (or less frequently than annually) and much or all of the N being applied *after* the crops are sown; liquid fertilizers may be used to supply the nitrogen if they suit labour management.

PLACEMENT METHODS FOR CEREALS

Placement of fertilizer for cereals is usually by 'combine-drilling', seed and fertilizer being dispensed by separate mechanisms but sown together by the same coulter.

Nitrogen for spring cereals

In Rothamsted experiments, combine-drilling 280 kg/ha of ammonium sulphate gave about 250 kg/ha more barley grain than broadcasting the same amount of ammonium sulphate at sowing time. In tests made in other areas on soils poor in N, large responses occurred and combine-drilling ammonium salts gave larger yields than broadcasting. Combine-drilling has no advantage with nitrates because they check early growth more and because they are more mobile in the soil. Combine-drilling of urea is dangerous and liable to damage germination; dressings placed in separate bands $2\frac{1}{2}$ cm to the side of the seed are safe and give good yields. Fertilizers containing much or all of their nitrogen as nitrate or urea are unlikely to give better yields when combine-drilled than when broadcast, as they harm germination more than ammonium salts do. These results apply to both solid and liquid fertilizers containing these sources of N.

Phosphate and potassium for cereals

On deficient soils all the P and K needed by cereals should be applied with the seed by combine-drill since, on such soils, both superphosphate and muriate of potash are *twice as effective* when combine-drilled as when broadcast. A given yield is achieved by using only half as much P or K applied by combine-drill as is needed if the plant foods are

roadcast before sowing. *Top-dressings* of P and K given after cereals are
own generally have little effect on the crop to which they are applied.

On the richer soils where cereals have been grown for long, and have
een regularly fertilized, reserves of soluble P and K have accumulated
nd there is little or no response to fresh dressings whether drilled or
roadcast. Recommended dressings are intended to maintain fertility by
eplacing the nutrients removed in yields; they can be applied when it is
lost convenient—in winter, or to a preceding root crop.

POTATOES

Potatoes are sensitive to shortages of P and K in soils and they
espond more often to P and K fertilizers than other agricultural crops
rown in Britain. The whole of the dressing of N, P and K recommended
hould be applied immediately before, or at planting. Later top-dressings
ave no place in potato growing. Fertilizers given to potatoes should
e applied so that they are close, but not too close to the seed. Potatoes
lanted in contact with heavy doses of fertilizer may be damaged when
hey sprout. Growth will then be checked and yields lessened.

When potatoes are planted in the furrows of ridged land spreading
he normal complete fertilizer dressing over the furrow sides before
lanting is safe and efficient. After fertilizing and planting potatoes in
his way the dressing is concentrated around the seed when the ridges
re split. Dressings broadcast on the flat before drawing out the furrows
re less effective. A good guide to the value of placement methods for
otatoes grown in Britain is still a series of experiments completed 20
ears ago. In these trials 1900 kg/ha of 7–7–10.5 fertilizer broadcast
efore drawing the furrows were needed to give the same yield as was
btained from only 1250 kg/ha of this fertilizer broadcast over the
urrows before planting.

When mechanical planters are used on flat land broadcast fertilizer
annot be concentrated near to the seed. In experiments with a mechani-
al planter compound fertilizer broadcast early and cultivated deeply
nto the soil gave much the same results as dressings broadcast on the
seedbed just before planting, but placing the dressing in one band 5 cm
to the side of the seed and 2.5 cm deeper gave 2 t/ha more yield. As
in the experiments described above 1900 kg/ha of broadcast 7–7–10.5
compound fertilizer gave about the same yields as 1250 kg/ha of the
same fertilizer placed beside the seed. Similar returns were obtained on
most ordinary soils in England, and the extra cost of having a fertilizer
placement attachment on a potato planter may be written off after planting only a few hectares. In these experiments broadcasting 1900 kg/ha
of compound fertilizer gave an average profit of £77/hectare but the
same level of profit was obtained by *placement of only half as much*
fertilizer.

Some potato planters place fertilizer in contact with the seed in the
planting shoe. Under dry conditions this practice may be harmful when
optimum fertilizer dressings are applied. *Contact placement* is safe only

if small dressings of fertilizers are used; it is dangerous and can lead to loss of yield when large dressings are used, particularly on light soils on badly prepared seedbeds and in dry years; the remedy is to use planter which places fertilizers *at the side of the seed.*

Two recent reviews (refs. 72 and 73) have discussed experiments done by N.A.A.S. which showed that the response curves for placed and broadcast fertilizers were differently shaped and that *each nutrient* placed was more efficient than when broadcast. About 50 per cent more fertilizer was needed for maximum response when dressings were broadcast on flat land before planting. Optimal dressings were:

	In furrows of ridged land	Broadcast on flat seedbed
	kg/ha	
N	150	225
P_2O_5	200	265
K_2O	190	>280

In spite of the gains from placement that are possible on most potato growing soils, about half of the growers visited in Surveys in England applied their fertilizers on flat land before planting. Presumably this was because suitable machines were not available, or because growers found that labour management was better when fertilizer was broadcast although more had to be applied.

ROW CROPS

Safe methods of placement

Many farmers who have used combine-drills for crops such as sugar beet, swedes and peas have found that even small dressings of compound fertilizer may damage germination when they are drilled with the seed. The extent of damage depends on the nature and amount of fertilizer used, the texture and moistness of the soil at sowing, and on the weather that follows. Dressings of N or K fertilizers placed either *with,* or *directly below* the seed are very likely to damage germination and these methods of application cannot be recommended. Placing any kind of fertilizer in a band 5 cm to the side of the seed, and about 2.5 cm deeper, is safe and effective for crops grown in wide rows. This method has been tested on most important row crops and some British findings are described in the following paragraphs.

Peas and beans

These crops do not often give a worthwhile return from using *broadcast* fertilizer on *average* land, but higher yields are obtained by placing P–K compound fertilizer 5 cm to the side of the seed and 7.5 cm below the soil surface. Trials show that for both green and threshed peas, and for winter and spring beans, placing 350 kg/ha of fertilizer gave larger yields than broadcasting twice as much. Broadcasting fertilizer for threshed peas and both kinds of beans lost money at current prices in the 1950s, though it gave a small cash return on green peas. Placing the fertilizer was profitable on all crops, and for green peas it yielded a high return of about £50/ha. Such gains from placement more than pay for the fertilizer and over a few seasons justify buying a special drill. In these trials the benefits from placement were due to both the P and K components of the mixture. With crops that give larger responses to fertilizers applied by special methods, the way the fertilizer is to be used must always be specified in making advisory recommendations, since both profit and optimum dressings depend on this factor.

Vegetables

In trials on market-garden crops (ref. 74) granulated 7–7–10½ compound fertilizer applied in a band 5 cm to the side of the seed and 2.5 cm deeper was superior to broadcasting the fertilizer and working it into the seedbed. Most of the vegetables grown gave higher yields from 280 kg/ha of fertilizer placed beside the seed than from 560 kg/ha broadcast; placing the heavier dressing increased yields still further. Placement also had the important advantage over broadcasting that some vegetables grown with placed fertilizer were ready for market earlier. In a few other experiments on carrots, broadcasting fertilizer was better than placing it.

These trials were made on ordinary arable land carrying a rotation that included vegetables; none were on intensively manured horticultural holdings. Placement for vegetables is likely to be most useful when they are taken as part of an ordinary arable rotation of farm crops, or where intensive horticulture is being introduced on ordinary soils and before plant food reserves are built up by heavy manuring and fertilizing. The high cost of using both bulky and concentrated organic manures for vegetables is justified by growers on the grounds that the crops grow more quickly and uniformly and that they mature rapidly. Placing fertilizer dressings in correct positions at the side of the seed produces full yields with the least amount of fertilizer and gives crops of uniform quality. Placement methods may be valuable where sufficient supplies of the traditional manures cannot be obtained.

Swedes, turnips, mangolds, sugar beet and kale

The traditional way of growing *swedes* and *turnips* in Scotland is to sow them on ridges; for crops planted in this way fertilizer can be broadcast before ridging; after ridging the dressing is concentrated in the centre of the ridge and acts efficiently. For swedes grown on the flat, broadcasting fertilizer is less efficient and placing the dressing at the side of the seed gave higher yields in English trials.

For sugar beet, mangolds and kale placing a complete compound fertilizer in a safe position at the side of the seed gave about the same yields as broadcast dressings worked into the seedbed. Nothing was gained from placement methods *except* that treated crops got a better start and were ready for singling earlier. This advantage of early growth is also obtained by drilling a small quantity of water-soluble phosphate either with, or directly under the seed as a 'starter', the rest of the phosphate and all the N and K needed being broadcast separately before sowing. Placement for these crops has the practical advantage that seed and fertilizer are sown in one operation; with good organisation, labour and tractor time can be saved by doing the two jobs together. Another advantage is that one run of the tractor over the seedbed (to spread fertilizer) is avoided.

HERBAGE CROPS

Sowing leys

In Scottish experiments, using a combine-drill to sow phosphate fertilizers with grass and clover seeds gave better establishment of temporary grassland than broadcasting the same quantity of phosphate before sowing.

Lucerne

In English trials placing PK fertilizer at the side of the seed did not give better early growth or yield of lucerne than broadcasting the dressing. When 250 kg/ha of ordinary superphosphate was drilled 5 cm *directly below* the seed to act as a 'starter', the lucerne was established very quickly and made much better growth during the first year than was obtained with much larger dressings of broadcast superphosphate. Similar results showing the benefits from starter doses of phosphate placed *under* lucerne seed have been obtained in U.S.A.

Established crops

Established herbage crops always receive broadcast fertilizer; this method has been compared in trials with placing the fertilizer below the surface of the sward. For established crops of sainfoin and lucerne, and for permanent grass, PK compound fertilizer placed in bands 7.5 cm below the soil surface gave smaller yields of hay than the same fertilizer broadcast over the sward, and nothing was gained from placement.

SPECIAL METHODS OF BROADCASTING FERTILIZERS

The efficiency of fertilizers can often be improved by broadcasting them at particular stages in the cultivation of arable land. The benefits of placement are obtained with potatoes simply by broadcasting fertilizer over the furrows in which the crop will be planted. For other crops, methods which can be put into practice without special equipment are often better than the usual practice of broadcasting fertilizer on the surface of a prepared seedbed.

'Early' broadcasting

In English trials on peas and beans PK compound fertilizer broadcast 'early', and either ploughed or cultivated about 10–15 cm deep into the seedbed, often gave larger yields than later dressings broadcast on the seedbed and worked in shallowly. In dry seasons fertilizer mixed with the surface soil was useless and both placed fertilizer and 'early' broadcast fertilizer gave larger yields because they were incorporated deeply. In wet seasons early dressings ploughed-in were not more effective than late dressings worked into the seedbed. Farmers who do not have special placement drills should broadcast fertilizer for beans and peas before cultivating to prepare the seedbed; if possible the dressings should be ploughed in.

Ploughing in fertilizers

No general recommendation can be made to plough down all or part of the P and K normally given to arable crops. The method may, however, be worth a trial for deep-rooting crops grown on light soils. On poor soils where immediate responses are expected part of the PK may be ploughed-in, the remainder should however be held back and broadcast on the seedbed in spring to start the crop.

'Split' applications

Trials have been made on peas, beans, sugar beet, kale and carrots of broadcasting half of the normal dressing of compound fertilizer and placing half beside the seed. The extra yields obtained by splitting the dressing in this way, as compared with placing the whole dressing, were not sufficient to cover the cost of the extra operation of broadcasting part of the fertilizer.

Mid-season top-dressings

Some farmers hold back part of the normal dressing of nitrogen for crops with a long growing season and apply it as a top-dressing during the summer, fearing that some of the nitrogen fertilizer applied in spring may be lost by leaching before the crop can take it up. As with cereals, which were discussed above, rainfall determines whether top-dressings are justified for other crops. In very wet springs when much of the nitrate given at sowing is leached before the crop can use it, top-dressings will increase yields of crops such as sugar beet and kale. In seasons when spring rainfall is light and evenly distributed top-dressings will have no advantage; in dry summers they will be at a disadvantage.

Potatoes. Dressings of nitrogen applied wholly to the seedbed gave larger yields of potatoes than equivalent amounts applied partly to the seedbed and partly as top-dressings just before the last earthing-up. There is no case for top-dressing potatoes with nitrogen fertilizer if adequate amounts are used when planting. It appears to be difficult for top-dressings to be washed *into* the ridges and down to the roots of the crop.

Sugar beet. Some farmers apply a mid-season dressing of nitrogen fertilizer (often consisting partly or wholly of nitrate) to sugar beet to stimulate the crop to make rapid growth after the set-back caused by singling. In recent trials in years with normal rain in spring nothing was gained from mid-season top-dressings for beet when sufficient nitrogen had already been applied before sowing; split dressings, part being

applied on the seedbed and part as top-dressing, gave the same yield
as applying the whole amount to the seedbed. In other years much
rain falling in short periods leached nitrate from the soil and top-dress
ings increased yields.

Applying fertilizers as sprays

Plant foods applied as sprays are taken up through the leaves. Some
times soil conditions prevent crops taking up micro-nutrients through
the roots, and sprays are more effective. Trace-element deficiencies are
often treated by a spray correctly timed to carry the crop through to
harvest.

As sprayers for controlling weeds are now generally used on arable
farms, many farmers have been interested in the possibilities of spraying
on the growing crop part or all of the normal fertilizer dressing. Although
the major plant foods *can* be applied this way to produce a full crop
there are a number of snags, apart from the corrosion of the equipment
by strong solutions of fertilizer salts. Only small quantities of the major
plant foods can be absorbed through the leaf at any one time, and
spraying must be repeated a number of times for the crop to receive
sufficient. Sometimes sprays have advantages; thus urea sprayed on the
leaves has proved an effective way of getting nitrogen into apple trees
in grassed-down orchards where crops often suffer from nitrogen
deficiency. In wheat experiments ammonium nitrate solution sprayed
on the crop in May and June was as effective as the solid fertilizer
applied to the soil at the same time. Top-dressings of liquid fertilizers
can be combined with weed-control sprays. Solutions containing
ammonium nitrate, or urea (or both) have been tested in experiments
together with weed-killer spray. Both solutions caused some 'scorch'
on the crops, but they usually recovered fairly quickly.

GENERAL GUIDANCE ON METHODS OF APPLYING FERTILIZERS

Advice on how to apply recommended fertilizer dressings is given in
Chapter 19. Placement used for crops like peas may provide a profit
from the outlay on fertilizer under conditions where broadcast manuring
would not be justified. For potatoes smaller quantities of placed than of
broadcast fertilizers are needed to give full yields. For crops like sugar
beet, placed fertilizer may give a good start, but may not produce
higher yields than broadcast fertilizer. For such crops a small quantity
of water-soluble phosphate drilled either with, or directly under the
seed as a 'starter', often gives rapid establishment, but the sodium and
potassium fertilizers and any extra phosphate needed, must be broadcast separately before sowing. On many soils a small quantity of phosphate drilled in this way as a 'starter' will satisfy the needs of crops
completely in the early stages; later they take supplies from soil
reserves.

Methods of application must always be considered at the same time as fertilizer dressings are being chosen; correct placing and timing, whether achieved by a special machine or a particular way of broadcasting, is necessary if the best return is to be obtained from fertilizers. The benefits from placement are likely to be greatest on soils that are deficient in 'available' plant foods, and the technique often brings yields up to a high level quickly. Placement may not be so useful on soils that have been adequately manured and fertilized for many years, but on such land it can be a suitable way of applying the small dressings needed as insurance against unsuspected soil deficiencies, attacks by pests and diseases, or unfavourable growing conditions. Special methods of application may be needed to make best use of individual fertilizers: For example, nitrogen is likely to be lost when urea is broadcast on the soil surface, but this material is very efficient when drilled (as solid or as a solution) in a band beneath the soil surface.

Anhydrous ammonia. When special equipment has to be used to apply fertilizer, the costs of the work *per hectare* (and therefore the *applied cost* of the fertilizer) depend on the area of land treated. The greater the number of hectares treated, the smaller is the proportion of overhead costs of machinery that each hectare has to bear. This is true for all placement methods and a farmer having only a small area cannot afford special machines unless the crops are very valuable. An interesting example of how applied cost of N depends on area treated was given by G. W. Lugg (ref. 74a) for anhydrous ammonia. Ammonia must be injected into soil to avoid loss and the machines needed are heavy and expensive. Lugg compared costs of buying and applying solid ammonium nitrate with corresponding costs of ammonia applied in the field. The figures below show that the advantage of ammonia (in being cheaper to buy per unit of nitrogen) is only realised when large areas are treated. (Ammonium nitrate was assumed to cost £32/tonne or 9.4 np/kg of N; the two prices for ammonia—£54 and £42/tonne correspond to 6.6 np and 5.1 np/kg of N respectively.)

Ammonia used on farm per year tonnes	Price of ammonia, per tonne	
	£54	£42
	Differences in cost of N applied to the field (Costs of N in solid minus costs of N in ammonia, expressed as percentages of cost of solid N)	
20	2.6	18.1
30	10.7	26.2
50	17.1	32.7

When ammonia was bought at the larger price and only 20 tonnes/year were applied, there was little saving as compared with using solid N-fertilizer; considerable savings were possible when 50 tonnes of ammonia per year were used.

The examples of gains from special methods of application given here have mostly been taken from British experiments. The results are, however, applicable in many countries. Wherever experiments testing similar methods of application on the same crops have been made in similar climates and on similar soils in Britain and U.S.A., the results have been similar. The main factors governing fertilizer efficiency apply all over the world. The rewards from using special methods of application are likely to be greater where agriculture is being developed because fertilizers are more expensive, and soils are poorer, than in the developed agricultural areas of temperate regions.

DIRECT DRILLING AND MINIMUM CULTIVATION SYSTEMS

Direct drilling

In recent years techniques have been developed for sowing the seed of crops such as cereals, oil-seed rape and fodder crops directly into undisturbed land. Herbicides (usually Paraquat) are sprayed immediately before drilling to kill all weeds and give the new crop a clean start; usually special drills are used which cut grooves for the seed in the consolidated soil. Few experiments have been made on fertilizing crops sown in this way. The little evidence that exists shows that direct-drilled crops need as much or more nitrogen than crops sown by conventional drills on prepared seedbeds; dressings of N applied on the soil surface appear to be satisfactory. Applying phosphorus and potassium fertilizers is more difficult since the P and K in these materials are fixed in the upper zone of the soil and do not penetrate quickly to the crop roots. There is a risk that crops sown on soils deficient in P or K (Index O, pp. 151, 153) may fail unless soluble P and K fertilizers are placed near their roots. Direct drilling should not be practised on such soils. On better land (Index 2 or above for soluble P and K) cereals and brassicae crops will get enough P and K from the soil and slow penetration of fertilizers applied on the surface *may* suffice to feed later crops, but there is no evidence to prove this. Until more experimental work is done it is best to advise that land used for direct-drilled crops should be ploughed or cultivated once in every 4 years or so, so that dressings of P and K fertilizers, and lime where needed, can be mixed with the topsoil. Drills are now being developed to place fertilizers in the soil below and to the side of seeds sown directly into the soil; some of these apply fertilizers in liquid form. If they are successful they may provide a solution to the problem of fertilizing direct-drilled crops.

Minimum cultivations

Much energy is used in ploughing land. Tractor fuel is saved by new systems of cultivation where ploughing is avoided and surface soil is stirred by cultivators and harrows, the seed being sown in a prepared seedbed. Crops grown in this way should receive exactly the same fertilizer dressings as crops sown in more conventional systems.

CHAPTER NINETEEN

Fertilizing Arable Crops in Britain

This Chapter offers advice on manuring most of the arable crops grown in Britain. The *background* of recent experimental work on fertilizers for cereals, sugar beet and potatoes is briefly reviewed and other factors that affect decisions on amounts and times and methods of application are discussed. This is followed by general recommendations on manuring the crops.

The quantities recommended are in the conventional measures of amounts of N, P_2O_5 and K_2O; these are easily converted to dressings of straight and compound fertilizers by using Tables 11 and 24. Where compound fertilizers are recommended the plant food ratios suggested may not fit the recommendation exactly. Since N gives larger responses than P or K for all crops except legumes, it is always most important to use the right amount of nitrogen; *small* variations from ideal in plant food ratios are of little importance to yields provided the weight of compound fertilizer provides the correct nitrogen dressing.

The dressings recommended are for average yields now grown in Britain. Where yields are consistently larger, more fertilizer than is recommended must be applied to build the bigger crop.

The straight and compound fertilizers recommended are only some suggestions of products that will be suitable. Table 24 gives examples of the important *types* of current materials; as several hundred compound fertilizers are now made in Britain it would be difficult to list all of them. Other proprietary materials with either the *same plant food ratio,* or a *very similar ratio,* will usually be just as effective as the fertilizers recommended here. Of the compound fertilizers recommended, a few are not listed in Table 24; they occur in manufacturers' current lists.

Amounts of plant nutrients in tabulated recommendations are stated in terms of kilogrammes per hectare, and UNITS (of 1.12 lb) per acre. (In some tables pounds per acre are also given to help readers in other countries.) Where amounts of N, P_2O_5 and K_2O are given in the text only to illustrate an argument, often these are only in metric units.

Very detailed advice on manuring crops is now available from other sources for England and Wales (ref. 43) and for Ireland (ref. 47). There-

fore this Chapter is more concerned with principles governing the use
of fertilizers than with attempts to recommend for all the combinations of the soil and climate that are possible. The Responses to fertilizers
and optimum dressings in British experiments are similar to those
recorded in other countries of North-West Europe, and from some parts
of North America. Broadly this advice applies to the crops named
when they are grown in cool temperate areas with farming systems
similar to those in Britain. It also assumes that rainfall is well distributed through the year with total average rain exceeding total average
evaporation by at least 125 mm, but not by more than about 600 mm.

CEREALS

RESULTS OF EXPERIMENTS

Nitrogen

Modern stiff-strawed varieties of cereals make it possible to use
sufficient nitrogen to get full yields and maximum profit with little risk
of loss by lodging if care is taken to adjust dressings to crop rotation.
The newer varieties of wheat and barley give larger returns from
nitrogen fertilizer than older varieties and they must be fully fertilized
to realise their potential yields.

Lodging is caused partly by root disease that weakens stems, but
mainly by too much nitrogen supplied either by fertilizer, or from soil
in late spring and early summer, causing excessive growth of leaf and
stem. Wind and rain determine the severity of lodging in a crop
that is liable to be damaged. Cereals grown on old arable land without
organic manure rarely lodge even when much N-fertilizer is used. The
trouble usually occurs when soils contain large residues of N; sometimes
these are from N-fertilizer given the year before, but mostly they are
residues of organic manures, a preceding legume crop, or ploughed grassland that either contained legumes or had been grazed (or both). Reserves
of organic-N of this kind decompose and release extra nitrate; the most
N becomes available after a warm dry autumn and winter, and also
during warm wet weather in late spring and early summer. Because
weather in the season ahead cannot be forecast, nitrogen release in
summer is unpredictable in amount and time. But most risk from lodging
can be removed by adjusting N dressings to allow for previous cropping
and then making a second adjustment to allow for weather in the past
autumn and winter so far as it affected leaching. If, in spite of these
adjustments, lodging often occurs (as may happen after ploughing old
grass) one of the chemicals that shorten straw (such as CCC (2-chloro-
ethyltrimethyl-ammonium chloride)) should be applied as a spray in
spring.

In fertilizer experiments wheat has needed from none to 190 kg N/
hectare for maximum yield, barley from none to 150 kg. The largest
amounts have been needed for crops on well-leached old arable soils,
following short leys of grasses alone that were cut, and where root

disease or bad soil conditions diminished uptake by the crop and made the fertilizers less efficient. Least N has been needed where soil had large reserves of organic nitrogen left by grazed grassland containing clovers, by arable legumes, by organic manures, and after dry winters. In experiments which have been continued for several seasons response has varied from year to year. Table 32 shows an example of how previous cropping and history affected the N needed for maximum yield at Rothamsted; for each set of previous cropping conditions the amount of N needed for maximum yield varied within the range shown. Uncertainty in the N needed was least in continuous arable cropping and most where grass and legumes were grown and FYM was used for potatoes.

With present knowledge and facilities nitrogen manuring cannot be made exact but much of the uncertainty can be removed by taking account of the factors discussed here. In addition to variations in N needed caused by local conditions, there are regional differences caused by the effect of climate on farming system. In high rainfall areas in Western and North-Western England, in Wales and over much of Scotland, except the South-East, less nitrogen should be given than is recommended for cereals in other areas. Predominance of grassland and of animal farming in these areas ensures a good supply of N from soil.

Phosphorus and potassium

Seventy-seven experiments on spring barley done by N.A.A.S. from 1964 to 1966 on typical cereal-growing soils showed these results:

Dressing		Increase in grain yield	
kg/ha	UNITS/acre	kg/ha	cwt/acre
75	60 P_2O_5	+112	+0.9
75	60 K_2O	−25	−0.2

Levington Research Station was responsible for 63 other experiments on spring barley which tested P but not K. The response to 50 kg P_2O_5/ha (40 UNITS P_2O_5/acre) was about 150 kg of grain per hectare (1.2 cwt/acre). There was no relationship between soluble P in the soils and the fertilizer responses.

Where cereals are grown on land traditionally used for arable crops sizeable responses to P and K fertilizers are only obtained on soils containing small amounts of soluble P and K. The dressings of P and K fertilizers recommended must therefore be chosen to maintain soluble nutrients in the soils rather than for immediate effect. A barley crop yielding 4400 kg/ha contains about 44 kg of P_2O_5 and 65 kg of K_2O; the grain usually contains most of the P but less than half the K.

RECOMMENDATIONS FOR CEREALS

Nitrogen

Dressings of nitrogen should be varied to allow for local conditions of soil and previous cropping that have been discussed in Chapters 12,

TABLE 47

DRESSINGS OF NITROGEN FERTILIZER NEEDED BY WINTER WHEAT AND SPRING BARLEY ALLOWING FOR THE EFFECTS OF PREVIOUS CROPPING, LAND USE AND MANURING

| Previous cropping | Old arable land | | | | Permanent grass or leys that were grazed and ploughed in the last 10 years | | | |
| | Much | | Little | | Much | | Little | |
Leaching in previous winter	Wheat	Barley	Wheat	Barley	Wheat	Barley	Wheat	Barley
	kg of N/ha							
Cereals / Cut grass	190	150	150	125	95	75	65	40
Roots	100	75	65	40	65	40	25	0
Clover / Beans / Peas	75	50	50	25	50	25	25	0
or Roots grown with FYM / 3 years of lucerne / Grazed grass-clover leys (2 years or more)	50	25	25	0	25	0	0	0
	Units of N/acre							
Cereals / Cut grass	150	120	120	100	75	60	50	30
Roots	80	60	50	30	50	30	20	0
Clover / Beans / Peas	60	40	40	20	40	20	20	0
or Roots grown with FYM / 3 years of lucerne / Grazed grass-clover leys (2 years or more)	40	20	20	0	20	0	0	0

13 and 16. The factors that have the largest effect on N reserves in soil, and hence on the fertilizer-N needed to supplement them to produce maximum yield, are: (1) Previous cropping (both the cropping of the previous 1, 2 or 3 years and also the longer-term history of the land). (2) Weather of the previous winter; this alters the size of the reserves that are left from previous crops. (3) Kind of soil and climate. Other factors being equal heavy soils tend to supply more N than light soils; reserves of organic matter (containing more N) are larger in wet than in dry areas (probably because more grass is grown where rainfall is large).

General recommendations for winter wheat and spring barley that allow for previous cropping and land use and for winter rainfall are in Table 47. These are suitable for most soils in central, eastern and southern England. Specific recommendations for a few local soil types, and for other areas, are given by A.D.A.S. (ref. 43).

Spring wheat crops should receive the same nitrogen dressings as spring barley.

Oats. Fewer experiments on oats have been done than on wheat and barley. Dressings that are two-thirds as large as Table 47 shows for spring barley should be satisfactory top-dressings for winter oats or as seedbed dresssings for spring oats.

Ley farming. As Table 47 shows, great care in choosing N for cereals is needed when leys are grown in the rotation, and when organic manures have been used, or where cereals follow grass on which cattle grazed. When cereals are grown after ploughing up very good grass, nitrogen fertilizer may not increase the yield of grain, and on such land farmers must be cautious with this plant food. If light dressings of N are applied, a few unfertilized strips should be left across the field; the benefits from the fertilizer, estimated by comparing the strips and the rest of the field, should be used as a basis for adjusting dressings in later years. On the other hand very poor pasture, or a poor grass ley, when ploughed may release little nitrogen, and more fertilizer nitrogen may be necessary. (Poor grass is also likely to be deficient in phosphate, and it may be short of potassium too, extra dressings may be needed for the first few years after ploughing up.) Grass that is treated with much N-fertilizer and cut continuously, and grassland that is often grazed, must be regarded as completely different crops. Grass swards without legumes which are cut accumulate little N; during the time the grass roots are rotting cereals may need even more N than is used in continuous cereal cropping.

Phosphorus and potassium

Maintenance manuring recommended to replace P and K removed in average grain crops is 40 kg each of P_2O_5 and K_2O per hectare annually; (equivalent to 30 UNITS P_2O_5 and K_2O/acre). Proportionately more may be given every 2 or 3 years. Larger amounts are needed on stockless farms where most produce is sold, less where grain is fed to animals on the farm and manure or slurry is returned. Cereals grown in rotation with well-manured roots usually receive enough P and K from residues of dressings given to the roots. By contrast, cereals grown in rotation with mown leys (which remove much potassium), need larger dressings.

Deficient soils need much larger dressings to secure maximum yields of cereals. They should be detected by soil analysis. Table 48 shows how P and K manuring should be adjusted by soil analysis and how to allow for previous cropping.

TABLE 48

DRESSINGS OF PHOSPHORUS AND POTASSIUM NEEDED BY WINTER AND SPRING VARIETIES OF WHEAT AND BARLEY AND BY OATS, ALLOWING FOR SOLUBLE P AND K IN SOILS AND FOR PREVIOUS CROPPING

	Cereals after cereals		Cereals after root crops	After FYM applied for current cereal crop	After cut grass ploughed for cereals
	straw retained	straw sold			
Soluble P (ppm)			kg/ha of P_2O_5		
to 5	125	125	65	65	125
5–10	65	65	40	0	65
10–15	40	40	0	0	40
> 15	0	0	0	0	0
Soluble K (ppm)			kg/ha of K_2O		
to 100	100	125	65	65	125
100–150	40	65	0	0	65
150–200	0	25	0	0	25
> 200	0	0	0	0	0
Soluble P (ppm)			UNITS/acre of P_2O_5		
to 5	100	100	50	50	100
5–10	50	50	30	0	50
10–15	30	30	0	0	30
> 15	0	0	0	0	0
Soluble K (ppm)			UNITS/acre of K_2O		
to 100	80	100	50	50	100
100–150	30	50	0	0	50
150–200	0	20	0	0	20
> 200	0	0	0	0	0

METHODS AND TIMES OF APPLYING FERTILIZERS
FOR CEREALS

Phosphorus and potassium

If immediate responses are expected on poor soils all P and K for cereals should be combine-drilled; broadcast dressings are much less effective; top-dressings given after the crop is sown will not increase yield. Phosphate drilled with the seed should be a water-soluble form. Compound fertilizers containing nitrogen are quite suitable for drilling with the seed of spring cereals (provided they contain only a small proportion of urea or, preferably, none at all). If used to supply no more than 75 kg N/ha (60 UNITS N/acre) compounds containing most of their N in ammonium form are not likely to damage germination of crops grown on medium and heavy soils.

Nitrogen

Inorganic nitrogen fertilizers may be leached from the soil by heavy rain and they must be put on at the right time to avoid this loss. This is particularly important on light soils and not quite so important on medium and heavy land. Skilfully chosen top-dressings may increase yield and efficiency of N.

Winter cereals. There is no evidence of benefits from giving a little N to winter cereals in autumn; nor is there enough evidence to condemn the practice provided no more than 25 kg N/ha (20 UNITS/acre) is applied. Farmers who consider their crops need nitrogen in autumn should apply a little, either broadcast or drilled with the seed, together with the P and K needed, by using a *High-Phosphate, High-Potash* (i.e. *low-nitrogen*) NPK compound fertilizer with plant food ratio of $1-2\frac{1}{2}-2\frac{1}{2}$. Nitrates should not be used for autumn dressings. Spring top-dressings should provide at least three-quarters of all the nitrogen given to the crop. Where large total dressings of N are given on medium and heavy soils, and no nitrogen has been applied in autumn, the fertilizer should be split, half being given in March and half in early May. Where the crop has had autumn nitrogen the spring top-dressing should be delayed until late April or early May to avoid risk of lodging. Where losses of N by leaching may be serious, on light soils or in wet areas, give half the total dressing early in March and half in May.

Spring cereals. Our experiments on arable land in the Eastern Counties have shown consistent gains of 125–250 kg/ha (1–2 cwt/acre) of extra grain from combine-drilling as compared with broadcasting a compound fertilizer suitable for spring barley or wheat. Combine-drilling is often slower than drilling and spreading fertilizer in separate operations and many farmers consider the small gains in yield that are possible are not enough to compensate for the later sowing that using a combine drill may cause. On land where nutrient reserves have been built up by past treatment, combine-drilling is unlikely to have any advantage at all.

There is little risk of damage to germination from drilling 70 kg/ha (60 UNITS/acre) of nitrogen with the seed on light, medium and heavy soils; 110 kg (90 UNITS) combine-drilled has delayed germination on light soils and when seedbeds were dry and contained little fine soil, but gave less trouble on heavy land in good condition. Combine-drilling of dressings supplying up to 70 kg/ha (60 UNITS/acre) of N is generally recommended as being safe on medium and heavy land if the nitrogen is supplied as an ammonium salt. But trials with nitrates and urea have shown that they can cause a severe check to germination if combine drilled at this rate. If compound fertilizers containing some nitrogen as nitrate or urea are combine drilled for cereals the amount contributed by these sources should not exceed 30 kg N/ha (25 UNITS/acre). Much of the nitrogen in British compounds is still as ammonium salts, but increasing amounts of nitrates and urea are being used in concentrated fertilizers; extra care is needed in using them. Farmers should ask their agents and salesmen what form of nitrogen was used to make the concentrated solid compounds, or liquid fertilizers they buy.

Farmers who do not use combine drills have several options:

(1) A compound fertilizer may be applied at or before sowing to supply all the PK needed, and part or all of the N.

(2) PK may be applied earlier. On good land maintenance dressings may be ploughed in, or twice the normal dose can be applied every second year. Then all the N can be applied a little time before or after drilling to speed up the actual sowing.

(3) When P and K are applied separately, and the soil is rich in N, the small amount of fertilizer N needed is best applied after the crop is established in late April-early May. If much N is needed (100 kg N/ha (80 UNITS N/acre) or more) much may be gained in some seasons from splitting the dressing and giving half at sowing and half in late April/early May; this splits the risks of N-fertilizer being inefficient because of loss by leaching, or because dry weather after application prevents it being taken up.

Top-dressings are valuable when:

(1) There is evidence that much of an earlier dressing of N has been lost by leaching.
(2) The early dressing was not sufficient.
(3) Larger percentages of protein in grain are needed to improve baking quality of wheat or feeding quality of barley.

Analyses of cereals during growth are now being made to guide the timing of top-dressings and improve the efficiency of N-fertilizers (Chapter 15). Extra N as top-dressing will almost certainly increase yields in England if nitrate-nitrogen in cereal stems falls below 50 ppm before the end of May.

Using compound fertilizers

Except where straight nitrogen fertilizers are sufficient, compound fertilizers are commonly used for cereals. The kinds of fertilizers shown below are useful, either combine-drilled or broadcast in autumn, for autumn-sown wheat, oats and barley.

Type of compound fertilizer	Plant food ratio N : P₂O₅ : K₂O			Suggested analyses			Suggested rates		
				N %	P₂O₅ %	K₂O %	kg/ha	lb/ acre	cwt/ acre
Low-Nitrogen NPK (or High-Phosphate, High-Potash NPK) compound	1	3	3	5	15	15	250	220	2
	1	2½	2½	6	15	15	250	220	2
				9	25	25	160	140	1¼
Phosphate-potash compound	0	1	1	0	20	20	190	170	1½
	0	1	2	0	14	28	250	220	2

Spring cereals. NPK compounds applied at sowing can be used to supply maintenance requirements of PK on ordinary soils and part or all of the N. Table 49 gives examples. When more N is needed than is conveniently or safely provided by a compound fertilizer the remainder may be applied as a later top-dressing. Table 49 shows a few suitable compounds, their plant food ratios, analyses and the dressings that will supply about 40 kg/ha each of P_2O_5 and K_2O. (The N supplied by these amounts of compounds is also stated.)

SUGAR BEET AND MANGOLDS

Very many field experiments on the manuring of sugar beet have been made; their results provide up-to-date advice. It is more difficult to make precise recommendations for mangolds, partly because there have been fewer trials, but also because mangolds have no standard value, as they are usually fed on the farm where they are grown. Mangolds respond to NPK fertilizers and to sodium in much the same way as sugar beet, and there is little reason for manuring the two crops differently. Where the indirect returns obtained from using mangolds on the farm make a hectare of the crop less valuable than a hectare of beet, smaller dressings than are suggested here may be used. A. P. Draycott has published a full review of sugar-beet nutrition (ref. 107).

RESULTS OF EXPERIMENTS

Nitrogen
The results of two series of experiments testing nitrogen fertilizers on beet were published in 1970. Because the recommendations made differ, some details of them are given here. Staff of Broom's Barn Experimental Station and the British Sugar Corporation did 170 experiments between 1957 and 1966 (ref. 39). On most sites the fertilizer increased yields sharply and linearly up to an optimum beyond which yield changed little or decreased only slightly; each 1 kg/ha of N increased yield by 25 kg/ha of sugar. On about a fifth of the fields, beet needed no fertilizer

TABLE 49

COMPOUND FERTILIZERS SUITABLE FOR CEREALS IN BRITAIN

	Plant food ratio N:P₂O₅:K₂O (approximate)	Suggested analyses N %	P₂O₄ %	K₂O %	Rates* to supply maintenance dressings of PK					
					Dressing			N supplied		
					kg/ha	lb/acre	cwt/acre	kg/ha	lb/acre	UNITS/acre
At sowing										
High-Nitrogen NPK compounds	1½ 1 1	20	14	14	300	280	2½	60	56	50
	2 1 1	20	10	10	400	340	3	80	68	60
	2¾ 1 1	25	9	9	450	390	3½	112	98	88
General-Purpose NPK compounds	1 1 1	10	10	10	400	340	3	40	34	30
		15	15	15	260	220	2	39	33	30
In winter or early spring, before sowing										
PK compound	0 1 1	0	20	20	200	170	1½	—	—	—
Potassic basic slag	0 1 1	0	12	10	400	340	3	—	—	—

*Amounts of fertilizers given in metric and Imperial units are only approximately equivalent.

nitrogen. In 7 of the 10 years, the average optimum was 75–100 kg N/ha (60–80 UNITS N/acre); the other three years were dry and less was needed. After crops other than cereals, beet needed less N; on some soils derived from Chalky Boulder Clay beet responded more than average. Apart from these examples, differences in responses to N between individual sites could not be explained, and 125 kg N/ha (100 UNITS/acre) were generally recommended, with more to beet on Chalky Boulder Clays and on light sands poor in organic matter. Less is needed where crop residues may supply much N. The 125 kg/ha of N is more than the average need for maximum yields, but losses from giving too much N are only the cost of the unnecessary fertilizer whereas much more money is lost when yields are smaller because of giving too little N (Chapter 12).

In 43 other experiments on sugar beet done by the Staff of Levington Research Station (ref. 75) optimum rates of N were:

		kg N/ha	UNITS N/acre
East Anglia	Chalky Boulder Clays	151	121
	Glacial sands and gravels	104	83
Lincolnshire	Wold chalk	182	145
	Cliff and heath	172	137

Responses to N were not clearly related to sowing and harvesting dates, soil texture, previous cropping, winter or summer rainfall, or summer sunshine. Probably drought limited responses on glacial sands and gravels.

The effects of previous cropping were shown well by other experiments. The N needed by beet at Broom's Barn Experimental Station depended on previous cropping:

	kg N/ha	UNITS N/acre
After cereals	150	120
After beet	100	80
After lucerne	50	40

Much N given to the previous crops diminished the need for fresh N. Beet grown after potatoes given 190 kg N/ha or after ploughed in trefoil, needed only 63 kg N/ha; after winter wheat or a ryegrass ley, 125 kg were needed, this was also the amount needed by beet after barley or potatoes that had 65 kg N/ha.

Beet on fen peat soils now seems to need little fertilizers. In recent trials responses to all fertilizers were small and rarely profitable on deep light peats in the area round the Wash. In experiments on fen peat soils, done between 1934 and 1949 about 125 kg P_2O_5/ha and more than 150 kg K_2O/ha were needed. Probably this change is due to the residues of P and K accumulated from the heavily fertilized crops commonly grown since 1949 on these soils. As fertilizing should maintain the present amounts of soluble nutrients, 63 kg/ha (50 UNITS/acre) P_2O_5

and 125 kg/ha (100 UNITS/acre) K_2O were recommended for beet on the peats. Little nitrogen was required, but a small insurance dressing of less than 50 kg N/ha (40 UNITS N/acre) would be justified.

Phosphorus, potassium and sodium

The large numbers of experiments done from 1934 to 1949 showed that sugar beet gave large responses to P fertilizers only on soils that had small soluble P. Responses to K were less than to N; sodium (supplied as sodium chloride, called 'salt' here) gave larger responses than potassium and, when sodium was applied, the response to potassium was quite small. All these results have been confirmed by the many experiments made since 1955. The important results with salt in the early experiments are still relevant:

Dressing		Extra yield of sugar	
kg/ha	UNITS (or cwt) /acre	kg/ha	cwt/acre
150 kg K_2O	120 UNITS K_2O	439	3.5
628 kg salt	5 cwt salt	628	5.0
150 kg K_2O +628 kg salt	120 UNITS K_2O +5 cwt salt	740	5.9

Liquid fertilizers

In experiments on sugar beet an ordinary granular NPK compound fertilizer and a liquid NPK fertilizer (where N and P were supplied by urea and diammonium phosphate) gave the same yields of sugar when the solid was broadcast and the liquid was sprayed on the seedbed. Liquid fertilizer placed to the side and 15 cm (6 in.) below the seed gave consistently larger yields than liquid placed 5 cm (2 in.) deep, but was no better than broadcast solid fertilizer. Anhydrous ammonia injected while preparing the seedbed, and solid N fertilizer applied to the seedbed, gave the same yields. Injecting the ammonia during early spring, or beside the crop before singling was less effective than injecting into the seedbed. Some of the ammonia applied early was lost; late injecting ultimately provided as much N as seedbed injection, but supplied it too late for maximum yield.

Magnesium for sugar beet

Workers at Broom's Barn Experimental Station (ref. 76) have reviewed 50 experiments on beet which showed increases from none to 10 t/ha (4 tons/acre) of roots from dressings of Mg fertilizers. Leaf symptoms of *previous* beet crops are not a sure guide to Mg deficiency in subsequent crops, analysis of the growing crop is, but is made too late for Mg dressings applied to the soil to be effective. It was recommended that soil samples taken during winter before growing beet should be analysed to determine whether Mg fertilizer was needed. No responses occurred on any field with more than 35 ppm of exchangeable Mg;

where Mg increased yield by 5% or more the soil usually had less than 20 ppm of Mg.

Exchangeable Mg in soil ppm	Magnesium dressing recommended			
	as Mg		as kieserite	
	kg/ha	lb/acre	kg/ha	cwt/acre
Less than 25	100	90	630	5
25 to 50	50	45	315	2½
Over 50	0	0	0	0

Sugar beet in England, especially in East Anglia, often shows symptoms of magnesium deficiency, which may be confused with those of Virus Yellows disease. The area reported to show Mg-deficiency increased from 3,500 acres (1,400 ha) in 1946–49, to 33,200 acres (13,000 ha) in 1958–61 and to 54,400 acres (22,000 ha) in 1962–65. The deficiency in beet occurs early in the season, usually on soils with little exchangeable Mg and in poor growing conditions; symptoms disappear with warmer weather and rain.

In experiments on fields on light sandy soils where magnesium deficiency was expected, 630 kg/ha (5 cwt/acre) of magnesium sulphate (kieserite) increased average yields by 460 kg/ha (3.7 cwt/acre) of sugar; the greatest response was 1,630 kg/ha (13 cwt/acre). Mg fertilizer was profitable at 11 of the 17 centres where experiments were made. (The average responses to potassium and sodium dressings were not altered by magnesium in these experiments and were 340 kg/ha (2.7 cwt/acre) and 430 kg/ha (3.4 cwt/acre) of sugar).

The possibility of Mg being deficient may not be known when sowing, so it would be convenient to correct a deficiency by spraying the crop when it developed, but foliar sprays have given smaller increases than applying Mg before sowing. Hence soil dressings seem the only reliable method of applying magnesium, and probably 380 kg/ha (3 cwt/acre) of kieserite is enough.

N.A.A.S workers found that in at least 60 per cent of fields investigated Mg-deficiency symptoms in East Anglian crops were accentuated by bad soil conditions, produced by compaction, or unstable soil structure.

RECOMMENDATIONS FOR SUGAR BEET

When salt is not used beet and mangolds should receive the following dressings per hectare (per acre).

Soils with more than 15 ppm soluble P	P-deficient soils (less than 15 ppm soluble P)
125 kg (100 UNITS) N	125 kg N (100 UNITS)
63 kg (50 UNITS) P₂O₅	125 kg P₂O₅ (100 UNITS)
188 kg (150 UNITS) K₂O	188 kg K₂O (150 UNITS)

There are very few conditions in which sodium manuring is not justified for sugar beet and mangolds and related crops. The extra response to K where Na has been applied is barely enough to pay the cost of the dressing; but Na gives a profit whether or not K is applied. Sugar beet tops contain much potassium and if they are removed for feeding there is a substantial drain on soil K reserves. More K is removed in the beet sold to the factory. Some K should be given to sugar beet, even when salt is used, so that the reserves in the soil are not lessened by growing the crop.

The general recommendation which includes salt is per hectare (per acre):

> 125 kg (100 Units) of N
> 63 kg (50 Units) of P₂O₄
> 100 kg (80 Units) of K₂O
> 380 kg (3 cwt) of salt (NaCl)

As alternatives to applying straight agricultural salt, sodium nitrate (which also supplies nitrogen) or compound fertilizers containing salt may be used; kainit can be used to supply both sodium and potassium.

Adjustments to standard recommendations

Farmyard manure applied at about 25 t/ha (10 tons/acre) lessens the fertilizer needed by about one-third. Crops grown with FYM, but without salt, should receive per hectare (per acre) about 90 kg (70 Units) of N, 45 kg (35 Units) of P₂O₅ and 125 kg (100 Units) of K₂O. When 380 kg/ha (3 cwt/acre) of salt is used as well as FYM, the fertilizer dressings should be: 90 kg/ha (70 Units/acre) of N, 45 kg/ha (35 Units/acre) of P₂O₅ and 60 kg/ha (50 Units/acre) of K₂O.

Soil type. There is no *general* reason to vary either nitrogen or potassium manuring according to the area where the beet is grown. Exceptions are: (i) *Fen peats;* where little or no nitrogen is needed but where P and K may be justified. (ii) *Chalky Boulder Clay* soils in general need more than average nitrogen and less than the normal K dressing, 125 kg (100 Units) of K₂O (without salt) should be enough; at least 150 kg/ha (120 Units/acre) of N will be needed on these soils.

Soil analysis should be used to confirm that the standard dose of phosphate (60 kg/ha) (50 Units/acre) of P₂O₅ will be satisfactory; on soils with 15 ppm or less of soluble P this dressing should be doubled.

Previous cropping has not had consistent effects on the responses of beet to N, except where it is grown immediately after ploughing grassland or a leguminous or grazed ley. Then the nitrogen dressing may be diminished to 60 kg N/ha (50 Units of N/acre). Where beet follows long runs of cereal crops it is likely that more N will be needed and 150 kg N/ha (120 Units/acre) should be given. Where beet is grown in arable rotations where much fertilizer is used, there is little reason to depart from the 125 kg N/ha standard recommendation. Beet is a deep rooting, long-season crop, and seems able to recover residues from previous dressings of N that have been leached into the subsoil.

Winter rainfall is correlated with the effect of nitrogen on beet yields. 25 mm (1 inch) above (or below) the mean rainfall per month in winter increases (or decreases) the gain from nitrogen by about 300 kg/ha (2½ cwt/acre) of sugar. After an abnormally *wet* winter nitrogen manuring may be *increased* by 25 kg/ha (20 UNITS/acre) of N; after a very *dry* winter the dressing may be *decreased* by the same amount.

Top-dressings of N after singling are considered necessary by some farmers. There is not enough evidence to recommend this practice generally, but on light soils, and particularly in wet springs, it will give better yields than applying all the nitrogen before sowing. Where no sodium salt has been applied before sowing, sodium nitrate is a suitable top-dressing. Liquid fertilizers may also be used.

Leaf analysis can be used to determine whether later top-dressings of N are justified, either because the spring dressing was insufficient, or because it has been leached out by intense rain. If leaf petioles contain less than 300 ppm of nitrate-nitrogen at any time before the beginning of July in British conditions an extra dressing of 60 kg N/ha (50 UNITS /acre) is likely to be justified.

Using straight fertilizers for beet and mangolds

Savings in costs of fertilizers can often be made by adjusting the manuring of beet and mangolds to suit the fertility of particular fields; labour can be saved by applying straight fertilizers before or after sowing instead of giving manufactured NPK compounds when the crop is being drilled. 'Straights' must be used when no suitable compound fertilizer can be found to apply plant foods in the ratio of the adjusted recommendation. Mixtures of ammonium sulphate, superphosphate and muriate of potash may be applied just before sowing, but these are not likely to result in much saving in cost. It is often more convenient (and cheaper) to apply fertilizers at times that avoid the busy period when the beet is sown.

Basic slag is suitable for beet grown on most soils, but it may not be as quick-acting as water-soluble phosphate on soils with reserves of limestone or chalk. Slag should be applied in winter and worked into the land well before the beet is sown. 'Maintenance dressings' of phosphate on soils with 15 ppm or more of soluble P can be given as basic slag. PK fertilizers can be ploughed-in in winter without loss of yield except on poor soils.

Nitrate of soda applied at 380 kg/ha (3 cwt/acre) as a top-dressing makes salt unnecessary and supplies 60 kg (48 UNITS) of nitrogen. The rest of the plant foods needed 65 kg N, 63 kg P_2O_5, 100 kg K_2O per hectare (52 UNITS N, 50 UNITS P_2O_5, 80 UNITS K_2O per acre) can be given as a '*High-Potash*' NPK compound fertilizer (with a plant food ratio of $1:1:1\frac{1}{2}$) harrowed into the seedbed before sowing; 500 kg/ha (4 cwt/acre) of a fertilizer containing 12% N, 12% P_2O_5, 18% K_2O would be suitable.

Kainit at 630 kg/ha (5 cwt/acre) supplies about 110 kg (90 UNITS) of K_2O, together with sufficient salt for a good crop.

Winter dressings of phosphate, potassium and salt should be followed by a straight nitrogen fertilizer supplying 125 kg/ha (100 UNITS/acre) of N; this can be broadcast before or after sowing; on very poor soils it may pay to work an additional dressing of 380 kg/ha (3 cwt/acre) of an NPK compound fertilizer (with a 1 : 1 : 1 ratio) into the seedbed to start the crop.

How and when to fertilize sugar beet and mangolds

The usual way of applying fertilizers for beet and mangolds is to harrow the N, P and K into the seedbed before sowing. Alternatives using straight fertilizers applied at other times are discussed above. Combine-drilling is dangerous with beet and mangolds, but it is quite safe to drill fertilizers in a band 5 cm (2 in.) to the side of the seed by using a placement drill. In England sideband placement does not give higher yields than broadcasting the fertilizers, but it makes beet grow more quickly early in the season.

All salt and kainit intended for beet and mangolds should be applied at least a few weeks before drilling; if desired the dressings may be ploughed in. Nitrogen fertilizers should not be ploughed in during autumn or winter as much or all of their value will be lost by leaching before the crop is sown.

Manuring sugar beet seed crops

Good yields of red beet and sugar beet seed have been grown with about 250 kg (200 UNITS) N, 125 kg (100 UNITS) P_2O_5 and up to 250 kg (200 UNITS) K_2O per hectare (acre). Crops grown where sown should have all the P and K, and a third of the N, applied during their first year; transplanted crops should have almost all their fertilizer applied to the bed before planting; 380 kg/ha (3 cwt/acre) of salt may be beneficial.

Compound fertilizers for sugar beet

The special needs of sugar beet grown on most British soils are for compound fertilizers low in phosphorus and either rich in potassium (if no sodium is applied) or low in potassium as well if sodium is given. Most requirements are met by two classes of compounds; fertilizers containing magnesium, and others with boron, are also made for beet. Some suitable materials are listed in Table 50.

POTATOES

Potatoes give large responses to all three nutrients, N, P and K, and they receive more fertilizer than other arable crops in Britain. Inter-action effects are very important in building up yields and the fertilizers needed must be planned by considering both the likely effects of N, P and K and their interactions. About a third of British crops get FYM at average rates of about 35 t/ha (14 tons/acre) and these dressings are important sources of extra nutrients, and particularly of potassium.

TABLE 50

COMPOUND FERTILIZERS SUITABLE FOR SUGAR BEET

Type of compound	Plant food ratio N : P_2O_5 : K_2O			Suggested analyses N %	P_2O_5 %	K_2O %		Amount to supply 125 kg N/ha (100 UNITS/acre) kg/ha	cwt/acre
For use without sodium									
Low-phosphate NPK compound	2	1	3	12	6	18		1040	8.3
				14	6	20		890	7.1
				17	8	24		740	5.9
For use with sodium (assuming 380 kg NaCl/hectare (3 cwt/acre) is given)									
Low-phosphate NPK compound	1½	1	1½	18	12	18		690	5.5
				(if tops are removed)					
High-nitrogen NPK compound	1½	1	1	20	14	14		620	4.9
	2	1	1	20	10	10		620	4.9
				(if tops are ploughed in)					
With magnesium				N %	P_2O_5 %	K_2O %	Mg %		
High-nitrogen NPKMg compound	2	1	1	16	8	8	4	780	6.2
With boron				N %	P_2O_5 %	K_2O %	B %		
High-nitrogen NPKB compound	2	1	1	20	10	10	0.3	620	4.9
High-potash NPKB compound	1	1	1½	13	13	20	0.3	960	7.6

Most potato crops justify N, P and K fertilizers and compound fertilizers are very convenient and are used on most fields.

RESULTS OF EXPERIMENTS ON POTATOES

Two large series of field experiments done in 1955 to 1962 form the basis of current advice on potato manuring.

N.A.A.S. experiments

Largest responses (ref. 35) to P were in soils with impeded drainage, largest effects of K were, as might be expected, on light soils; responses to both P and K were larger in the West and North than in other parts of Britain. Responses to N varied more from year to year than for any other reason. Apart from seasonal differences, N responses were larger on light than on heavy soils and greater after cereals than after other root crops or leys. The largest dressings of N tested (150 kg/ha) were only justified after cereals on light soils; they diminished yields seriously when used on heavy soils where roots or leys preceded the potatoes. Interaction effects found in these experiments have already been used as an illustration in Chapter 12.

Soil analyses for soluble P and K showed how much fertilizer can be saved by giving less on the richer soils and extra yields can be obtained by giving more than average on the poorest land. Average analytical data and crop responses are in Table 51.

TABLE 51

RELATIONSHIPS BETWEEN SOLUBLE P AND K IN SOILS AND RESPONSES OF POTATOES TO P AND K FERTILIZERS

	Increases in yield of potatoes (tonnes/hectare)	
	Silty loams, clay loams and clays	Loamy sands and sandy loams
Soluble P ppm (sodium bicarbonate method)	Response to 125 kg P_2O_5/ha	
< 10	4.8	6.3
10–20	4.0	4.9
20–30	3.3	4.6
30–40	2.6	1.9
> 40	1.7	1.0
Soluble K ppm (Morgan's method*)	Response to 190 kg K_2O/ha	
< 50	5.1	7.7
50–100	0.6	2.7
100–200	0.0	1.7
> 200	0.2	1.1

(*Extraction with acetic acid-ammonium acetate solution (Morgan's method modified by N.A.A.S.))

Levington experiments

These experiments (ref. 38), done in about the same period as the N.A.A.S. experiments, were sited in many parts of Great Britain. The overall response to 134 kg/ha of nitrogen was 3.5 t/ha—about the same as the 4.2 t/ha from 150 kg N/ha in the N.A.A.S. work. Site-to-Site variations in response were very large. Variation in response to N was largely attributed to previous cropping and winter rainfall. Much more N was available where grass was in the rotation and where stock were kept; the value of these reserves depended on winter rainfall; Table 52 shows the effects of these conditions on optimum dressings. Most N was needed for potatoes grown in arable systems, and particularly after cereal cropping. There was no effect of soil texture on nitrogen response.

The mean response to 134 kg P_2O_5/ha was 2.7 t/ha. Although there was a trend for the largest responses to occur on the most deficient sites, the experiments gave no certain guide as to how phosphate dressings should be varied to allow for varying soil fertility.

The mean response to potassium was 4.7 t/ha from 180 kg K_2O/ha. The range was from a depression of $2\frac{1}{2}$ t/ha to an increase of 25 t/ha. The variation was largely accounted for by the variations in citric acid soluble K, which are shown in Table 52.

TABLE 52

RESULTS FROM POTATO EXPERIMENTS SHOWING HOW OPTIMUM NITROGEN RATES DEPEND ON WINTER RAINFALL AND PREVIOUS CROPPING AND HOW OPTIMUM POTASSIUM RATES DEPEND ON ACID-SOLUBLE SOIL K

History of land	October-March rainfall (mm)	Optimum rate
	Results with nitrogen	kg/ha of N
Grassland and mixed cropping	$<$300	38
	$>$300	90
Arable cropping	$<$300	136
	$>$300	176
	Results with potassium	kg/ha of K_2O
Citric acid soluble soil K (ppm)		
to 58	–	269
58–166	–	216
$>$166	–	55

RECOMMENDATIONS FOR POTATOES

The two series of experiments reviewed above show how impossible it is to make general recommendations for potato manuring that have any value. For long the average recommendation was 125 kg/ha each of N and P_2O_5 and 188 kg/ha K_2O. These figures *can* be taken as a base but maximum yields will not be obtained, nor will the fertilizers used be efficient (except on the few chance sites where these figures *are* appropriate), unless considerable adjustments for local conditions are made.

The *N.A.A.S. experiments* (ref. 35) give these average recommendations for optimum dressings, differentiated by soil texture:

	All loamy sands and sandy loams		All silty loams clay loams and clays	
	kg/ha	Units/ acre	kg/ha	Units/ acre
N	125	100	100	80
P_2O_5	113	90	175	140
K_2O	163	130	113	90

Levington experiments (ref. 38) led to these recommendations *for nitrogen*:

	Wet winter		Dry winter	
	kg/ha of N	Units/ acre of N	kg/ha of N	Units/ acre of N
After grass and in mixed rotations	90	71	56	44
In arable rotations and after cereals	168	134	134	107

At least 134 kg P_2O_5/ha (107 Units/acre) was recommended, except on light soils rich in P. The K recommended varied from 56 kg K_2O/ha (44 Units K_2O/acre) to 270 kg K_2O/ha (215 Units K_2O/acre) depending on the amount of acid-soluble K in the soils.

Adjustments to standard recommendations

Farmyard manure gives better returns from potatoes than from most other farm crops. Manure lessens the response to plant foods supplied by ordinary fertilizers and when FYM is given the normal fertilizer dressing should be diminished by about one-third. (The contributions of FYM to nutrient supplies were stated for sugar beet earlier; potatoes benefit similarly.) Manure should be ploughed-in in winter for potatoes.

Farming system, previous cropping and *winter rainfall* all affect the nitrogen needed.

Soil type. Fertilizer recommendations for potatoes should allow for type of soil as described above and in more detail by N.A.A.S. (ref. 43). On fen peats potatoes usually need much less than the standard dressing of nitrogen, these soils are often deficient in phosphate.

Soil analysis results, together with experience of the land and its previous cropping, should be used to decide whether phosphate may be diminished or increased. Soil type and soluble soil K together should be used to decide potassium manuring.

How and when to fertilize potatoes

Potatoes should *always* receive all of their fertilizer dressing at planting time. Top-dressings given after planting are likely to be wasted. The fertilizer must be applied so that it is close to the seed after planting but *not* in contact with it. Placement for potatoes has been discussed in detail in Chapter 18; the main points are repeated here:

Where potatoes are planted in the furrows of ridged land, fertilizer should be spread over the furrows before planting; splitting the ridges after planting will concentrate the dressing over and around the seed. Granular fertilizers tend to roll down into the bottom of deep furrows so that they come into contact with the seed and check sprouting in a dry season; a little soil pulled down from the side of the ridges to cover the fertilizer before planting the seed will avoid damage. When crops are planted by a machine working from flat land, the planter should have a fertilizer attachment to place bands about 5 cm (2 in.) to the side of the seed potato and about 2.5 cm (1 in.) deeper, dressings broadcast before planting are less effective than placed dressings. *All* recommendations made here for manuring potatoes assume that the dressing is applied by the most efficient method. If broadcast dressings are used before ridging for hand-planted potatoes, or for potatoes planted from the flat by a machine, the dressings recommended here *must* be increased by 50% to achieve the desired yield.

Early potatoes and seed potatoes

The manuring of *early potatoes* should be decided in much the same way as main crop varieties if no farmyard manure is given. 'Earlies' give a very good return from farmyard manure (and from seaweed in coastal districts); if these 'organics' are used less fertilizer, and particularly less potassium, will be needed. Dressings may be rather less than for main-crop potatoes on similar land. The returns from early potatoes depend on how soon they can be harvested, consequently it is important to have sufficient fertilizer for maximum growth, but to apply it so that the soluble salts do not check growth. Using compounds based on potassium sulphate (and not chloride) may diminish risks of scorch in early crops and may also produce more seed size of main crops.

Compound fertilizers for potatoes

The '*High-potash NPK*' groups of compound fertilizers (typical plant food ratio was $1:1:1\frac{1}{2}$) were developed for potatoes when the standard recommendation per hectare was 125 kg N, 125 kg P_2O_5 and 188 kg K_2O (100 UNITS of N and P_2O_5, and 150 UNITS of K_2O per acre). Compounds with plant food ratios of about $1:1:1\frac{1}{2}$ are still widely used for the crop. There is a wide selection that allows for local variations in soil fertility and previous cropping; some are made with potassium sulphate instead of chloride. Some fertilizers sold for potatoes have the exact $1:1:1\frac{1}{2}$ ratio (e.g. 12–12–18), some are richer in K (e.g. 12–12–30), others have proportionately less K (e.g. 15–15–19). The section which reviewed experimental results shows that almost any of the NPK compound fertilizer ratios in Table 24 will be useful for potatoes in particular local conditions. Detailed recommendations for particular fields of potatoes should be derived by studying this and earlier Chapters; where possible an Advisory Officer should also be consulted.

OTHER ARABLE CROPS

The field experiments testing fertilizers which have been done in the last 15 years on cereals, potatoes, sugar beet and grassland have made it possible to manure these crops much more precisely by adjusting the dressings to local soil, farming system and climate. Consequently the relatively simple basis given for fertilizing these crops in 1960 (ref. 77) is now out of date as the preceding parts of this Chapter show. By contrast very little recent experimental work on our less important arable crops has been published and the recommendations made in the remainder of this Chapter are substantially those published in 1960 (ref. 77). All the crops discussed except carrots and peas are grown for animal food (but some of the swedes and turnips grown on farm scale are sold for human use). It is difficult to make precise recommendations because the values of the crops are uncertain and change from season to season according to the prices of other foods; relationships between yield and supplies of N, P and K fertilizers are not well defined for any of the crops. The factors that affect the reserves of N, P and K in soils which have been discussed above for cereals and root crops grown for sale will affect the fertilizers needed in just the same way.

About 320,000 hectares (800,000 acres) of fodder crops were grown in U.K. in 1970. Beans accounted for nearly a quarter of this area. Turnips, swedes and fodder beet occupied about 100,000 hectares; mangolds are rapidly diminishing in importance in Britain, only 10,000 hectares were grown in 1970. Statistics show nearly 120,000 hectares of 'other fodder crops', probably most of this area was under kale and other leafy cruciferous crops.

KALE AND CABBAGE

Although large areas of kale, cabbage and related fodder crops are grown there have not been enough trials on the manuring of these crops to provide precise recommendations on fertilizing them.

In practically all experiments kale and cabbage responded well to nitrogen and *at least* 100 kg N/ha (80 UNITS/acre) is justified. It does not matter whether the N fertilizer is applied at sowing or as a later top-dressing. There is little information on the P and K manuring of these crops grown on average land, but as they contain considerable amounts of both P and K these plant foods should be applied as fertilizers before sowing. A 50 t/ha (20 tons/acre) crop of kale can contain about 200 kg each of N and K_2O and 60 kg of P_2O_5 (160 UNITS/acre of N and K_2O and 50 UNITS of P_2O_5). If kale is grazed in the field most of the P and K will be returned in excreta but if removed for feeding elsewhere soil K reserves are seriously depleted. It is better to take account of how the kale *was* used when manuring the next crop. After cutting and removing a large crop of kale an *extra* 40 kg of P_2O_5 and 80 kg of K_2O per hectare (35 UNITS P_2O_5 and 70 UNITS K_2O per acre) should be given to the following crop except on soils rich in P and K.

General recommendations

Kale, cabbage and similar leafy crops should receive:

per hectare	*per acre*
125 kg of N	100 UNITS of N
125 kg of P_2O_5	100 UNITS of P_2O_5
125 kg of K_2O	100 UNITS of K_2O

Adjustments

Farmyard manure lessens the need for P and K, but as the crops are such 'gross feeders' their nitrogen fertilizer requirement may not be affected when FYM is used. A suitable dressing with FYM is the same N as shown above but only half as much P and K.

Soil analysis should be used to guide the amounts of P and K fertilizers chosen. Where soluble soil P is >15 ppm and exchangeable K is >200 ppm, half the standard dressings will be enough.

Kale grown after ploughing up pastures or leys with a good proportion of vigorous clover will need only half the general recommendation of N.

The way of using all these leafy crops affects the amounts of fertilizer that are justified. If heavy yields are grown frequently for cutting, more P and K must be given than is used for crops fed on the land.

How to apply fertilizers for fodder crops

A fertilizer mixture containing all three plant foods should be harrowed into the seedbed before sowing or it can be applied in a band 5 cm (2 in.) to the side of the seed and a little deeper by a placement drill. Placement gives these crops a good start, but may not give better yields than broadcasting. If large yields of green stuff with high protein contents are required, a top-dressing of 60 kg/ha (50 UNITS/acre) of N should be applied in mid-season. Top-dressings should be applied carefully as they may 'scorch' the plants if fertilizer sticks to the leaves in damp weather.

Straight fertilizers can be used for kale and similar crops. Basic slag is suitable on most classes of soil; on acid soils (with pH values of 6.0 or less) finely ground rock phosphates are satisfactory. Both these phosphate fertilizers cost less than an equivalent dressing of superphosphate.

Summary

A summary of fertilizer recommendations for kale and cabbage is in Table 53.

TABLE 53
FERTILIZER DRESSINGS FOR KALE, CABBAGE, SWEDES AND TURNIPS

| | Amounts of plant food† | | | | | | | | | NPK Compound fertilizer | | | | | |
| | kg/ha | | | lb/acre | | | UNITS/acre | | | Type | Type and analysis | | | Rate† | |
Type	N	P₂O₅	K₂O	N	P₂O₅	K₂O	N	P₂O₅	K₂O		N %	P₂O₅ %	K₂O %	kg/ha	cwt/acre
KALE AND CABBAGE															
Standard conditions															
Without FYM	125	125	125	112	112	112	100	100	100	General purpose	15	15	15	830	6½
With FYM*	125	62	62	112	56	56	100	50	50	High nitrogen	20	10	10	625	5
Other conditions															
Following good-quality ploughed grass { no FYM	62	125	125	56	112	112	50	100	100	Low nitrogen	8	20	16	775	6¼
with FYM*	44	88	88	40	80	80	35	70	70	Low nitrogen	8	20	16	550	4½
Soil with more than 15 ppm soluble P 150 ppm exchangeable K { no FYM	125	62	62	112	56	56	100	50	50	High nitrogen	22	11	11	570	4½
with FYM	88	44	44	80	40	40	70	35	35		22	11	11	400	3¼
SWEDES AND TURNIPS															
Standard conditions															
Dry areas { Without FYM	100	100	100	90	90	90	80	80	80	General purpose	17	17	17	600	4⅔
With FYM*	50	50	50	45	45	45	40	40	40		17	17	17	300	2⅓
Wet areas { Without FYM	50	100	100	45	90	90	40	80	80	Low nitrogen or High PK	10	25	25	500	4
With FYM*	25	50	50	22	45	45	20	40	40		10	25	25	250	2
Soils with more than 15 ppm of soluble P and 150 ppm of exchangeable K															
In dry areas { Without FYM	100	50	50	90	45	45	80	40	40	High nitrogen	20	10	10	500	4
With FYM*	50	0	0	45	0	0	40	0	0	Straight-N fertilizer	33	0	0	150	1¼

SWEDES AND TURNIPS

In the districts where they are normally grown swedes and turnips require a good dressing of phosphate, but only moderate amounts of nitrogen and potassium. Field trials in south-east Scotland showed that 90–110 kg P_2O_5/ha (70 to 90 UNITS of P_2O_5/acre) was sufficient for maximum profit. In the wet areas where swedes are a common crop the leafy tops grown with large nitrogen dressings may be damaged by mildew. Trials in north-east Scotland showed that optimum dressings for swedes and turnips without FYM were: 75 kg/ha of N (60 UNITS/acre) and 150 kg/ha (120 UNITS/acre) each of P_2O_5 and K_2O. When FYM was applied the best dressing was about 60 kg of N and K_2O and 125 kg of P_2O_5 per hectare (50 UNITS of N and K_2O and 100 UNITS of P_2O_5/acre).

General recommendation
Swedes and turnips should receive the following dressings:

per hectare	*per acre*
100 kg of N	80 UNITS of N
100 kg of P_2O_5	80 UNITS of P_2O_5
100 kg of K_2O	80 UNITS of K_2O

(In the few areas where experiments have been done recently, as in south-east and north-east Scotland, and in northern England, more appropriate local recommendations can be obtained from the Advisory Services.)

Adjustments
Where farmyard manure is used the standard dressing should be decreased by one-half.

Soil analysis should be used to adjust the manuring of swedes and turnips. On soils that are 'very low' in P (or K), raising the dressing may be justified; on soils that are 'satisfactory' in these plant foods the generally recommended dressing may be halved.

In very wet areas the nitrogen dressing should be decreased to 50 kg/ha of N (40 UNITS/acre).

Using straight fertilizers
Basic slag is suitable for swedes; where the crop is grown on acid soils with pH values of less than 6.0, rock phosphates may be used to supply part or all the P needed. Both phosphates are cheaper per kg of P_2O_5 than superphosphate.

How to apply fertilizers for swedes
Swedes and turnips have a relatively short growing season and they benefit from fertilizer placed close to the seed, but *not* in contact with it. Soluble phosphate near to the seed gives crops a good start and makes

them grow rapidly, so that they resist attacks by pests such as flea beetle or wireworm. When swedes are sown on ridged land fertilizer should be broadcast on the flat; ridging the field places the fertilizer in the middle of the ridge where it is of most use. Swedes grown on the flat benefit from side-band placement; when a suitable drill is available the complete fertilizer dressing should be applied in a band 5 cm (2 in.) to the side of the seed and 2.5 cm (1 in.) deeper. Swedes do not require any top-dressing.

These fertilizer recommendations for swedes and turnips are summarised in Table 53.

CARROTS

Carrots have responded well in experiments to potassium fertilizers and also to agricultural salt (crude sodium chloride). An example of average results of experiments done in 1941–2 was:

	Tonnes/hectare	Tons/acre
Without potassium or salt	34.9	13.9
With 112 kg K_2O/ha (90 UNITS K_2O/acre)	38.4	15.3
With 375 kg/ha (3 cwt/acre) of salt	38.2	15.2
With potassium plus salt	39.9	15.9

As with sugar beet, salt increased yields even when muriate of potash had been applied; nitrogen and phosphate gave only small increases in yields. Recent experiments by A.D.A.S. (refs. 113 and 114) confirmed older work: 50 kg N/ha was enough for maximum yields even on light soils; phosphate was beneficial and 150 kg K_2O/ha was needed on light soil, 75 kg K_2O/ha on better soils. Salt gave larger increases than potassium and increased yields even when K was given.

Recommendation

For carrots grown without salt the following dressings are suggested:

per hectare	per acre
55 kg N	45 UNITS of N
55 kg P_2O_5	45 UNITS of P_2O_5
110 kg K_2O	90 UNITS of K_2O

A 'High-potash' NPK compound fertilizer with plant food ratio of 1:1:2 would be suitable; 550 kg/ha (4½ cwt/acre) of a 10–10–18 compound fertilizer would supply roughly these amounts of plant food.

Salt applied at 375 kg/ha (3 cwt/acre) lessens the need for potassium; when it is used a *General Purpose* NPK compound fertilizer of 1:1:1 plant food ratio should be used to supply 55 kg/ha (45 UNITS/acre) each of N, P_2O_5 and K_2O. 375 kg/ha (3 cwt/acre) of a 15–15–15 compound fertilizer would be satisfactory.

Where *farmyard manure* is applied only about half as much of the same fertilizer dressing will be needed as is recommended for carrots grown without FYM.

How to apply the dressing

Salt should be worked into the soil well before sowing carrots; the other fertilizers needed should be worked into the seedbed. Only a few placement experiments have been done on carrots, but placing fertilizer in a band at the side of the seed gave smaller yields than broadcasting; this method has no value for the crop.

LEGUMES

Nitrogen

Peas, beans, clover, lucerne and vetches do not require nitrogen fertilizer, as nodules on their roots contain symbiotic bacteria that fix enough nitrogen from the air; usually nothing is gained from supplying this plant food as fertilizer for legumes that are harvested when mature. Sometimes the soil does not contain the organism responsible for forming nodules and fixing nitrogen. The remedy is to innoculate the seed with a culture of bacteria. This is often necessary when lucerne is sown on land where it has not been grown before. Where other legumes are traditionally grown their appropriate organism is usually present in the soil, and any trouble is likely to be due to some cause other than the nitrogen supply. Although some farmers give small doses of nitrogen to peas and lucerne, experiments rarely show that yields are increased although the crop may look greener. Large dressings of N are often recommended for beans that are grown as vegetables and picked green for canning or freezing.

Phosphorus and potassium

The traditional fertilizer for legumes in this country is a good dressing of phosphate. But in most recent experiments in the arable areas potassium fertilizers gave profitable increases in yield much more frequently than phosphate fertilizers; on many soils a dressing of K is essential for legumes to persist and to yield well.

BEANS AND PEAS

Farmyard manure is traditionally used for beans; its main value for the crop is the good dose of potassium that it supplies. On most arable farms all the farmyard manure that is made is given to cash root crops, since they give the best financial returns. If FYM at about 25 t/ha (10 tons/acre) can be spared for beans or peas, these crops should not need any other fertilizer.

Only small dressings of broadcast P and K can be recommended for beans and peas, and worthwhile returns are only to be expected if soil analyses show that there is a deficiency of one or both of these plant nutrients. On soils with less than 15 ppm of soluble P give 50 kg P_2O_5

/ha (40 UNITS P_2O_5/acre); with less than 150 ppm of exchangeable K 100 kg K_2O/ha (80 UNITS K_2O/acre) will be needed. Where broad cast P or K fertilizer is justified for beans and peas on deficient soils, the dressings should be applied early and worked deeply into the soil by the cultivations given to prepare the seedbed. Experimental work described in Chapter 18 showed that placing P and K in bands near to the seed will increase yields of peas and beans on soils where responses to broad cast dressings could not be expected; where large acreages of these crops are grown regularly the outlay on a placement drill will be repaid after a few years. *Combine-drilling* PK fertilizer *cannot* be recommended for peas and beans, since the potassium salt is likely to damage germ ination.

Recommendation
On soils of average fertility *phosphate-potash compound fertilizer* with a plant food ratio of 0 : 1 : 2 should be placed 5 cm (2 in.) to the side of the seed and 7.5 cm (3 in.) below the soil surface; 370 kg/ha (3 cwt/acre) of 0–14–28 will be satisfactory. On rich soils it is unlikely that placed fertilizer will give any worthwhile return.

LUCERNE

Lucerne requires much potassium and only remains vigorous for several years on soils that are well-supplied with this plant food. The amounts of K fertilizers that are justified depend very much on local soil conditions and precise advice on fertilizing the crop can only be obtained through local experimental work and from soil analyses.

Recommendation
The following advice is a general and conservative guide:

Basal manuring must be given before sowing to provide much of the plant food needed during the life of a crop of lucerne, since top-dressings of phosphate applied in later years may not be used so efficiently. Fertilizers supplying 125 kg P_2O_5 and 250 kg K_2O/ha (100 UNITS P_2O_5 and 200 UNITS K_2O/acre) should be ploughed or worked deeply into the seedbed before sowing. This dressing can be of straight super phosphate or basic slag plus muriate of potash; alternatively 900 kg/ha (7 cwt/acre) of 0–14–28 compound fertilizer can be used.

A good starter for lucerne is 250 kg/ha (2 cwt/acre) of ordinary superphospate (50 kg/ha (40 UNITS/acre) of P_2O_5) *placed* directly beneath the lucerne seed and 2.5–5 cm (1–2 in.) deeper. This dressing promotes rapid growth until the crop has a sufficient root system to use the phosphate previously mixed with the soil. This starter dressing can be applied easily by simple alterations to a combine-drill. The ordinary coulter is used to apply the phosphate and an extra hose is mounted so that it drops the seed in the shallow furrow left by the fertilizer coulter. Harrowing and rolling will cover the seed satisfactorily.

Top-dressings of muriate of potash may be necessary in subsequent years if experience has shown that growth is likely to be limited by K deficiency.

CLOVER

Little work has been done on the manuring of clover grown as an arable crop. As with other legumes, a good supply of potassium is important, and many failures of clover to grow or persist are due to a shortage of 'available' soil K. Phosphate is also needed on soils with less than 15 ppm of soluble P. Since clovers are generally undersown in the preceding cereals, provision must be made for them before the cereal is sown; it may be too late to manure the clover after the nurse-crop is harvested. The small dressings of P and K that are sufficient for combine-drilled cereals may not leave enough of these plant foods to maintain undersown clover. On soils deficient in P (and/or K) it will usually pay to *broadcast* dressing of fertilizers and work them into the soil before sowing the cereal. Without precise guidance from trials accurate recommendations cannot be made, but 60 kg P_2O_5 and 120 kg K_2O/ha (50 UNITS P_2O_5 and 100 UNITS K_2O/acre) should be enough for both cereal and clover. These quantities may be applied as straight fertilizers or as 450 kg/ha ($3\frac{1}{2}$ cwt/acre) of 0–14–28 PK compound fertilizer.

Where one-year clover or clover-ryegrass leys are grown in an arable rotation for cutting, the K that the crop removes should generally be replaced, even if the ley itself does not respond to this plant food. A single-year ley that is cut removes so much K that yields of following wheat crops may be lessened unless the amount that the ley takes up is replaced. At least 125 kg K_2O/ha (100 UNITS K_2O/acre) will be needed; the potassium will have most value for following crops if it is applied just before the ley is ploughed up.

NEW CROPS

Two crops (oil-seed rape and maize) have become common in the last few years. Little experimental work on fertilizing either of these has been done and these paragraphs give the best advice available.

Oil-seed rape is grown as a cash crop. Winter rape should have 50 kg N/ha applied in autumn; a further dressing of 100–150 kg N/ha should be applied early in spring, the actual amount depending on previous cropping of the land. Spring-sown rape should receive 50 kg N/ha in the seedbed and a later top-dressing of 50–100 kg N/ha. Both kinds of rape should have about 50 kg/ha each of P_2O_5 and K_2O. A 1–1–1 compound fertilizer (such as 17–17–17 at 300 kg/ha) can be used to supply 50 kg/ha each of N, P_2O_5 and K_2O.

Maize is increasingly grown for silage-making and a small area is grown for grain in south-east England. On ordinary arable soils of average fertility maize grown for either purpose should receive per hectare 100 kg N, 60 kg P_2O_5, 60 kg K_2O; a high-nitrogen fertilizer (such as 21–14–14 at 500 kg/ha) will be satisfactory. For maize grown on soils deficient in P or K 100 kg/ha of P_2O_5 or K_2O should be applied.

CHAPTER TWENTY

Fertilizing Grassland

We have learned much about the effects of fertilizers on grass in the last twenty years, but progress in applying this knowledge has been slower than progress in manuring arable crops. There are several reasons:

(i) Grass has no standard value; its value depends on what it is used for and it may be difficult for a farmer to work out a system of manuring that will be profitable.

(ii) Clovers fix nitrogen from the air. It has not been easy to strike a balance between the use of clover to supply nitrogen to associated grasses and the use of fertilizer nitrogen on a pasture, or to decide when fertilizer must be used and when clover will supply enough nitrogen.

(iii) In the past the income per acre from grassland farming has often been less than from arable farming and there has been less incentive to use fertilizers to achieve large production. Fertilizers used on arable cash crops produce an immediate return in extra produce that can be sold; fertilizers used on grass do not produce a return until the crop is 'processed' by an animal. Systems with quick turnover, such as dairying, offer the clearest incentives.

(iv) Farming systems often determine the amounts of fertilizers that should be used. On farms that are mainly arable a portion of the plant foods taken up by grass finds its way into farmyard manure, which is normally used on arable land. So there is a slow transference of plant foods from grassland to arable land. On such farms more fertilizers may be needed per acre to maintain large production from grass than are necessary in other systems where farmyard manure is used on the grass. One of the merits of ley farming where the whole of the land is in arable cropping from time to time, is that fertility differences caused by transferring plant foods tend to even out and the soils on a farm become more uniform. On heavily-stocked dairy farms with little or no arable land, most or all of the farmyard manure is applied to grass. Not only does the manure contain most of the plant foods in hay or silage fed indoors, but it should also contain the extra plant foods brought on to the farm by purchased straw and feeding stuffs. Under such a system little fertilizer P and K may be needed to maintain grassland productivity. In some modern systems of keeping animals the manure is produced as a slurry. These semi-liquids are much more difficult to store, transport and distribute uniformly over the land than is FYM. The value of slurry and problems of using it are discussed in Chapter 2.

In planning the use of fertilizers, the type of grassland and the use to which it is put must be taken into account. Three main purposes of fertilizers used on grass are distinguished here; they are:

   (i) For leys that are to be sown.
 (ii) For established grass that is cut.
(iii) For established grass that is grazed.

Planning manuring for leys that are to be sown, and also using fertilizer to produce more from a single cut of grass, are both relatively simple matters which involve much the same principles as manuring for arable crops. Much more difficult problems are involved in making the most profitable use of fertilizers over longer periods on established grass.

The Chapter is divided into three parts. The first is a brief review of some recent British information on two problems—using nitrogen on grass, and deciding how much potassium to apply. The second part discusses applications of the information we have and some of the practical problems arising from manuring grass, the third part gives practical advice. The immediate intention is to help British farmers and most of the experimental results used were obtained in Britain. The nature of the problems, the results of experimental work, and the advice offered, are, however, all applicable to countries with a cool temperate climate like that of North-Western Europe and parts of North America, and where rainfall in the year averages at least 100 mm more than evaporation.

1. BACKGROUND OF EXPERIMENTAL RESULTS

NITROGEN

An excellent and comprehensive account of experimental work on most aspects of nitrogen fertilizing of grassland was published in 1970 (ref. 78). It should be consulted when planning to use more nitrogen to increase grass yields. Important new results of A.D.A.S. experiments were published in 1975 (ref. 118).

Response relationships

Responses of cut grass have often been directly proportional to fertilizer nitrogen supplied up to about a total of 330–450 kg N/ha/year; many experiments then show a sharp transition (as in Fig. 11) to a smaller rate of response to each kg of additional N. With dressings larger than about 550–650 kg N/ha in the year there has been little increase in dry matter although the 'crude protein' (i.e. $\%N \times 6.25$) in the crop will continue to increase. Even larger rates may diminish yields.

Response relationships were fully explored recently by D. Reid in Scotland (ref. 79) where 21 amounts of N in the range 0–900 kg N/ha were tested on perennial ryegrass grown alone and with white clover. The nitrogen was applied in 5 equal dressings for 5 cuts. Over 3 years yields of ryegrass increased almost linearly up to the point of inflexion at 340 kg N/ha, but then the extra yield became steadily smaller and with more than 560 kg N/ha there was little further response. The grass plus clover sward also responded almost linearly from 50 to 340 kg N/ha, yielding more than grass alone. With more than 340 kg N/ha, the two swards responded similarly. Maximum dry matter responses in the three years ranged from 34 to 39 kg dry matter per kg of N from grass alone and from 15 to 18 kg from the grass-clover sward. Assuming that 1 kg of dry grass grazed *in situ* by a cow giving 3,600–4,000 litres per lactation is worth about 1.37 np, to pay for N at 7.8 np/kg the response needed is about 5.7 kg of dry matter. Reid calculated the maximum dry matter return was from 450–500 kg N/ha used on either sward. Including clover in the seeds mixture was profitable only when less than 225 kg N/ha was given. The amount of fertilizer-N needed to give the same yield of dry matter as the grass-clover sward gave without N ranged from 140 to 175 kg N/ha during the three years; to get equivalent crude protein yields needed 200 to 330 kg N/ha.

Responses to N in terms of extra animal production

Professor W. Holmes (ref. 80) concluded that for dairy cows milk production per hectare increased almost linearly up to 450 kg N/ha used in a year. Each kg of N yielded 1.05 grazing days and about 15 kg of milk. Up to about 200 kg N/ha in the year, responses by beef cattle were about 1 kg liveweight gain per kg of N applied. Professor Holmes published some targets for animal production from nitrogen used on grass:

Fertilizer applied kg/ha of N	Herbage yields of dry matter kg/ha	Production expected in 180 days of grazing			
		Dairy cows		Beef cattle	
		no./ha	milk kg/ha	no./ha	liveweight gain kg/ha
0	5500	2.2	5880	4.6	820
150	9800	3.9	10500	8.1	1460
300	11700	4.6	12500	9.7	1750
450	13200	5.2	14100	10.9	1970

Averages of the experimental results at present available may make these figures seem ambitious. However they are justified as targets at which grassland farmers should aim; certainly they will be achieved much more generally than at present as standards of fertilizing and grazing and cutting management improve.

Some experiments with both cattle and sheep have shown that individual performance from each grazing animal has not been as good when grazed grass has received much N fertilizer as when leguminous swards without N were grazed. Other experiments have shown that rates of live-weight gain of beef animals, and milk yield per cow and milk quality were the same on clover-grass pasture and on grass receiving much N-fertilizer. However, sheep, and particularly lambs, do seem to do better when they graze pastures containing much clover. Some recent experiments have tested very large rates of fertilizer-N (up to 900 kg N/ha/annually) and there is no evidence that these large amounts, which cannot be economically justified at present in Britain, have any detrimental effect on animal performance and health.

Many other papers have discussed the place of fertilizers in grassland farming, and particularly the value of nitrogen in increasing yield; references 81, 82, 83 and 84 are among those that will be found useful.

USING HERBAGE ANALYSES TO ADVISE ON POTASSIUM MANURING

The amounts of potassium needed by herbage used in different ways, and grown on different kinds of soil remain very uncertain. Potassium concentration in herbage is often suggested as a means of deciding whether the crop has sufficient K. Some suggestions are that the critical %K in herbage, above which yield responses should not be expected, is 1.6% of dry matter for grass and 1.8% of dry matter for clover. In many experiments reported, the herbage never contained as little K as this even where there were small responses to K-fertilizers. Workers at Hurley (ref. 61) have suggested that increasing the %K in herbage dry matter above 2% does not give worthwhile increases in yield and increases the risk of hypomagnesaemia. N.A.A.S. workers (ref. 85) have described a test of this method of assessing K-status of grass under practical advisory conditions. Very little fertilizer-K was needed to maintain over 2% K in dry matter of *grazed* herbage. In the West Midlands 10% of silage herbage and 6% of grazed herbage had less than 1.6% K in dry matter, while 17% of silage and 37% of grazed herbage had more than 3% K. It seems that analyses of herbage for %K can be used for recommending K-fertilizer dressings on grass; but if the method is to be useful, limits for herbage composition still have to be agreed and sampling and analysis will need great care.

2. *PRACTICAL PROBLEMS AND DECISIONS*

THE ROLES OF CLOVER NITROGEN AND FERTILIZER NITROGEN

Vigorous clovers grown alone, or in a grass-clover association, can generally fix 160–190 kg of N per hectare per year, and some British experiments have shown fixation up to 250 kg.

A grassland soil that has not received animal excreta or fertilizer rarely supplies more than about 100 kg of N per hectare in a year. This quantity comes from soil reserves, from rain and from biological fixation that is not connected with legumes; it sets a limit to the possible yield of a sward that has no clover. If no fertilizer N is used grasses grown alone often produce less than 2 tonnes and rarely more than 4 tonnes of dry matter per hectare in a year. Of the 190 kg of N fixed by clover in good pasture, part is used by the clover itself and the grasses obtain part. In Britain the yearly limit of forage production where legumes in *established* mixed herbage are the only source of nitrogen is about 6.0–7.5 tonnes/hectare of dry matter, depending on how the grass is used. One-year grass-clover leys that are cut often produce more on good soils. If more forage per hectare is required there is no alternative to using nitrogen fertilizer, and a grassy sward will yield about 11-13 tonnes/ hectare of dry matter in a season, provided it receives 300–360 kg N/ha.

Nitrogen fertilizers applied to mixed herbage stimulate grasses and depress legumes so the latter contribute less N to the total supply. Fertilizer-N given continuously through the season in *small* doses may simply replace clover-nitrogen without much increase in total yield. When more fertilizer nitrogen than this is given yields rise to about 12½ tonnes/ hectare when 360 kg of N are used in the season—and this is about the maximum annual production from grassland that is possible in many parts of Britain. There is little point in simply using small quantities of nitrogen fertilizer throughout the season if these do nothing more than replace N that clover might otherwise supply. However, when this amount of fertilizer nitrogen is used at specific and carefully chosen times in the season, and clover is allowed to grow at other times, fertilizer-N and clover-N *both* contribute to the total yield, which will be higher than clover alone can give. It pays to sow clover in the sward even when fertilizer nitrogen is to be used heavily but most of the gain from clover may be in the early life of the ley.

TIMING OF NITROGEN FERTILIZER

With grass swards containing no clover, timing of nitrogen is simple; except in spring there will be little growth without fertilizer-N and a dressing of 100 kg N/ha before each cutting, or 75 kg of N before each grazing will be needed for large production. With mixed herbage containing clover the time of fertilizing and of cutting (or grazing) affects both the return from the dressing and the effect of the fertilizer-N on the later performance of the sward. Dressings early in spring give the largest immediate return. When single dressings are given in February-March to increase growth for 'early bite' or for cutting the longer the initial cut or grazing is delayed, the greater is the proportion of fertilizer nitrogen removed in the first harvest, the less is its residual effect and the more is clover suppressed. In fact, the bigger the benefit from a single dose of fertilizer-nitrogen, the more likely is clover to be checked.

A good financial return usually comes from fertilizer-N used in early spring and in late summer to produce extra grass early and late in the season. The direct contribution of clover to yield is usually greater in the first half of the year than later. Fertilizer-N used in mid-season (June or early July) is more effective than clover in maintaining summer growth; by using fertilizer-N it is possible to maintain yields through the period in July and August when grass-clover swards normally produce much less than earlier in the season.

Whether extra grass produced by fertilizer nitrogen is used early or late, the subsequent contribution by clover is likely to be less than without nitrogen. If early spring dressings given for 'early bite' grazing, or for silage making, depress clover seriously, subsequent growth of herbage may not be enough for grazing. All nitrogen dressings should be part of a careful plan for the whole farm designed to produce the right amount of grass exactly when and where it will be wanted. Dressings should be given to clover-grass swards only for specific purposes; spring dressings will produce more for early grazing or extra yields of hay or silage, applied in June-July they will overcome the common fall in production after midsummer, and given in August-September they provide extra grass for late grazing until the end of November.

RESULTS FROM NITROGEN

The returns expected from nitrogen fertilizer used on grass must be estimated before rational plans can be made for manuring. Increases in yield vary with type of sward, time of applying nitrogen, and of using the crop, and with grazing or cutting management.

One kilogramme of nitrogen applied in February-March increases the yield of grass from mixed swards by 10 to 40 kg of dry matter; the actual extra yield obtained depends on the time of using the grass, and is greater the longer the herbage is left. The greatest immediate return from nitrogen on grass swards will usually be when grass is allowed to grow to hay (or late silage) stages before cutting, then the response may be as much as 40 kg of dry matter per kg of nitrogen. Autumn and spring dressings used to produce out-of-season grass usually give only 10 to 15 kg of extra dry matter per kg of N; similar increases may be expected from N used in mid-season on mixed herbage intended for grazing.

Where legumes are not an important part of the sward, production is directly proportional to the total amount of N given in the season up to about 300–360 kg of N/ha. A rough working basis is to assume that extra grass will be produced at the rate of 20 to 25 kg of dry matter per kg of N. If much more than about 350 kg N/ha is used the return will be less than this.

PHOSPHORUS AND POTASSIUM

Much permanent grass, particularly in the west and north, is on soils that are too acid for large yields; usually unimproved land of this kind is also very deficient in phosphate. Until these shortages are made good, satisfactory output from grassland cannot be expected. Phosphate alone, or with lime, is essential for promoting white clover, which, in turn, increases the nitrogen supply to grass and improves the pasture. On light land potassium manuring is often essential for establishing legumes. This is also true on some classes of heavy land, though many clay soils have enough soluble potassium for grasses and clovers. It is not difficult to advise on manuring of grassland at the 'pioneering stages' when improving very poor soils. After making good any shortage of lime, liberal dressings of phosphate and potassium will be needed for several years to build up nutrient reserves in the soil and to improve the sward. It is more difficult to advise satisfactorily on the P and K dressings that should be used for highly-productive swards on good land. The actual returns from fertilizer on a particular farm vary with soil type, management system and composition of the sward. The plant foods in grass are 'at risk'; whether they are of further use to grass or other crops grown on the field depends on the way the herbage is used. Cattle grazing continuously return to the soil in their excreta most of the N, P and K in the grass eaten; almost all of the P and K and some of the N returned can be used again. Cattle that graze for part of the day and are housed for the remainder void a proportion of the plant foods indoors, and these may be lost in washings, or conserved in farmyard manure and liquid manures. When grass is cut continuously and removed from the field *all* the P and K in it are carried away.

P and K for grazed grass

It might appear that when good grassland is grazed continuously only small amounts of P and K fertilizers are needed to maintain the supplies of these nutrients, but this is not necessarily true on a short-term basis where soils have very small reserves. Grass takes up N, P and K evenly from the whole surface soil, but the animal returns them to small areas; so areas of depleted soil alternate with smaller enriched patches. Overall production is liable to decrease because the extra growth of grass on the enriched patches may not balance the smaller yields on areas that have lost P and K. The greatest benefit from nutrients returned is with high stocking rates and long periods of grazing. Due to irregular distribution of excreta, some fertilizers may be needed to keep up production. The actual amounts needed may be quite small if soil reserves of P and K are large enough; all that is necessary then is to replace the P and K lost in the products that leave the field. On the other hand, larger dressings of P and K will be needed to maintain pasture on poor soils that have too little reserve to supply the herbage in the years between successive doses of excreta. On very poor soils the return through excreta must be ignored for a time and P and K dressings as large as those needed for cut herbage must be applied to maintain large yields.

Other things being equal, the P and K manuring needed to maintain the pasture will vary with the kind of stock and with grazing system. Least will be needed where sheep, or stores or beef cattle graze continuously and most where dairy cattle graze a field for only part of the day. Part of the plant food in herbage is lost under hedges, in gateways, and indoors with some systems of keeping dairy cattle. For swards grazed continuously by sheep or beef cattle annual dressings of 40–50 kg K_2O/ha should be enough on most land to maintain output and soluble K in the soil. Where grazing is by dairy cattle up to 80-100 kg K_2O per hectare *may* be needed each year, particularly on light land to maintain yield and amounts of soil K. But with intensive stocking and continuous grazing of good soils, none may be needed. The amount of phosphate that should be used to maintain output from soils that are 'satisfactory' in P status can only be guessed, 50 kg P_2O_5 per hectare given each year should be sufficient for intensive grazing by dairy cattle; half as much may be enough where only sheep, or stores or beef cattle use the herbage.

Soil analysis

Soil analysis is useful for detecting gross shortages of lime, phosphate and potassium under grazed grass. It is less reliable for assessing the need of better soils because it is difficult to obtain a representative sample and to interpret results, since excreta patches cause variations in soluble P and K. Nevertheless analyses can be useful on improved pastures which have been fertilized for a long time. After years of top-dressing, reserves accumulate in soil, and analyses done every three or four years show how amounts of soluble P and K in the soil are changing and give a basis for regulating top-dressings.

Potassium and phosphorus for cut grass

Potassium. When grass is cut continuously the full benefit from liberal nitrogen manuring may only be obtained when adequate potassium is also given. The amount of K needed by herbage is proportional to yield, and this is determined by the total supply of nitrogen from fertilizer plus clover; grass contains roughly equal weights of N and K. The need to use fertilizer-K will depend on the nature and value of soil K reserves and their rate of release. Some heavy soils are naturally rich in potassium, others have accumulated reserves from liberal dressings of FYM and fertilizers. It is possible to check whether a given manuring policy is maintaining soil K *levels* by periodical analyses, but no analytical method now used shows the size and value of K reserves in soil which can be used by grass.

Grass swards for cutting which receive large dressings of N (250 kg N/ha or more each year), will need 1 kg of K_2O to each kg of N on all soils that initially contain very little soluble K, and also on those where useful reserves cannot be built up (principally light sands and peats). On heavier soils with some reserves of useful K, 0.5 kg K_2O for each kg of fertilizer N should be enough to maintain yield of grass. On soils with larger reserves 0.3 kg K per kg N may be enough for a time at least.

With permanent grass used for continuous cutting there is, of course, no reason for using K-fertilizers unless the herbage responds to the dressings. But in ley-farming some K may have to be applied to grass so that soil K reserves are not depleted so seriously that following crops suffer loss of yield that cannot be made good with fertilizer-K. Potassium manuring may be less critical in continuous grass farming than in ley farming.

Phosphate. No general advice is possible; a sound policy appears to be to supply as much P as is removed in the crop, and 1 tonne of dry herbage may contain about 5 kg of P_2O_5. So if output is about $7\frac{1}{2}$ tonnes of dry matter per hectare (this will be obtained from a clover-grass sward given no fertilizer N in good grassland districts, or from 180 kg of N used on all-grass swards) about 40 kg P_2O_5 (300 kg of basic slag or 200 kg/ha of superphosphate) should be a satisfactory annual maintenance dressing. For grass receiving 375 kg of N per year, yielding 12.5 tonnes of dry matter per hectare and used intensively, 60 kg of P_2O_5 may be needed to maintain satisfactory amounts of soluble P in the soil.

Timing of dressings of P and K. A single winter or early spring dressing of phosphate has usually been as good as divided dressings. On soils with 'satisfactory' phosphate reserves it is not essential to dress pastures once a year. Where basic slag is used it is often convenient and quite efficient to apply large dressings less frequently.

Deficient soils which have little capacity to retain potassium need frequent small dressings rather than less frequent large ones; large dressings are easily lost by leaching from light land, they also lead to luxury uptake by the grass—the concentration of K in herbage rising without a commensurate increase in yield. This is wasteful, and may lead to hypomagnesaemia in grazing animals. The amounts of soluble potassium in soil are least at the end of the growing season; some of the potentially useful K in soil is liberated during the winter from soils with good reserves and soluble soil K is therefore at a maximum in spring. There is little point in applying K at this time to heavy soils when they are best able to supply the herbage. Certainly on farms that have a history of grass tetany outbreaks, little potassium should be applied in the first five months of the year; where possible, single dressings should be given in midsummer, and divided dressings should avoid the early spring months except on very poor soils, where herbage growth would be restricted by absence of K.

FERTILIZERS AND ANIMAL HEALTH

Fertilizers modify the composition of herbage, but simple precautions minimise any ill-effect that this may have. For a few weeks after heavy dressings of nitrate fertilizers have been given, herbage may have high concentrations of free nitrate-nitrogen that may be poisonous and upset grazing stock (later the nitrate vanishes as it is converted to protein by the plant, the smaller nitrate dressings that will normally be used in ammonium nitrate fertilizers have little effect of this kind). If large dressings of fertilizers containing *much* nitrate-nitrogen are given, herbage should not be grazed for a few weeks. Heavy potassium manuring applied in single doses leads to large concentrations of K in herbage, and Continental workers have associated these with outbreaks of grass tetany; such effects can be avoided by using K-fertilizers carefully and timing the dressings in ways described above.

Hypomagnesaemia or grass tetany is characterised by low magnesium in the blood serum of animals; it is unlikely when herbage contains more than 0.2% Mg in dry matter; with herbage having less magnesium than this, the trouble is more likely when grass is also high in crude protein and/or potassium. But many factors prevent any simple relationship between hypomagnesaemia and herbage composition, and different workers have come to different conclusions on the relative importance of different kinds of effects.

Hypomagnesaemia may be prevented by feeding magnesium oxide as a supplement. In Holland the trouble has been cured by dressings of magnesite to pastures, by using ammonium nitrate fertilizers diluted with dolomite instead of ordinary limestone and by manuring policies which prevent the accumulation of too much potassium in soils. Although many aspects of grass tetany remain obscure, British authorities consider that enough is known about methods of control to remove the problem from commercial dairy herds.

In intensive grassland farming nitrogen fertilizers *must* be used to produce extra grass for grazing *when it is needed* as well as to increase silage and hay. Potassium fertilizers *have* to be used to build up poor soils and to get maximum benefit from both clover and fertilizer nitrogen. These essential means to large production need not be disregarded because of fears for the health of stock; safe ways of using them are known.

FERTILIZING CROPS THAT FOLLOW GRASS

In planning fertilizer dressings for any cropping system the total nutrient balance must be calculated and allowance made for the effect of one crop on the nutrition of those that follow. The yields of crops that follow grass are affected by the type of herbage grown and the way it is used.

No simple and completely reliable advice can be given for fertilizing crops that follow grass. A short or long break under herbage crops does much good in controlling pests and diseases of arable crops and in improving soil structure, but it does make it more difficult to manure following crops correctly. Grasses grown alone and treated with much N for cutting often leave no residue of extra N in soils (sometimes they deplete soil N reserves for the following crop). These crops also deplete soil potassium, and it may not be desirable, or possible, to correct the potassium balance until the grass has been ploughed. Then it may be too late to build up soil K quickly enough for the first crop after the grass. Grazed pastures usually leave soil rich in K to benefit following arable crops. But grazing also builds up a large reserve of soil N which makes the correct dressing of fertilizer nitrogen for arable crops uncertain.

3. PRACTICAL RECOMMENDATIONS FOR FERTILIZING GRASSLAND

THE ESTABLISHMENT OF LEYS

Good establishment is vital for the success of a ley. The only opportunity for incorporating plant foods deeply in the soil that occurs in the life of a ley is when cultivating before sowing. The soil should be analysed before sowing. Any acidity should be corrected by working in liming materials and the analytical figures for soluble P and K should be used to plan manuring. If the soil contains less than 10 ppm of soluble P and less than 100 ppm of exchangeable K, *extra* dressings should be worked into the land. On soils deficient in phosphate 100 kg/ha of P_2O_5 should be given; 500 kg/ha of superphosphate (20% P_2O_5) or 700 kg/ha of basic slag (14% P_2O_5) will serve; if the soil is below pH 6.0, ground mineral rock phosphate may be used in wet areas. K-deficient soils should receive 100 kg of K_2O (170 kg/ha of muriate of potash) before sowing. *In addition* to these dressings, which are intended to build-up poor soils, 60 kg of P_2O_5 and 60 kg of K_2O should be worked into the seedbed before sowing to establish the crop; all soils except peats should receive 60 kg/ha of N. Water-soluble phosphate or basic slag should be used for these 'starter' dressings; insoluble phosphates may not act quickly enough.

Combine-drilling 100–200 kg/ha of superphosphate with the seeds sometimes gives better establishment and is worth a trial on soils poor in phosphate, but the N and K dressings *must not* be combine-drilled as they may damage germination.

GRASS FOR CUTTING

Nitrogen

Nitrogen manuring is essential to get a good yield of cut herbage. Dressings should be given for *all* hay or silage cuts to increase bulk cheaply. With swards that are cut continuously nitrogen should be used before each cut to achieve the desired yield. Clovers sown in the sward are worthwhile, they will increase output in the early life of the sward but are likely to be suppressed sooner or later.

Grass to be cut for hay should get 80–100 kg N per hectare when the field is closed; such dressings will increase yield from most swards by about 3 t/ha. Grass for silage should receive 100 kg of N per hectare. When the grass is harvested at the normal stage this amount of N will increase yield of a single cut from ordinary mixed herbage by about $1\frac{1}{2}$ tonnes to $2\frac{1}{2}$ tonnes of dry matter per hectare; if the grass is cut early there will be some residual effect from part of the nitrogen. Grass that is to be continuously cut should receive 80 kg of N per hectare before each cutting. With four or five cuts in the season at least 10–13 tonnes of dry matter per hectare will be produced and more may be obtained in the warmer and moister parts of the country. The crop should be treated like an arable crop, nitrogen fertilizer being used to obtain the yield, rate of growth and protein content of the herbage that is desired; the effect of the nitrogen on any legumes the sward may contain should be ignored.

Phosphorus and potassium

Before sowing a ley, or starting to use permanent grass intensively, gross shortages of lime and phosphate in the soil (detected by analysis) should be corrected with large fertilizer dressings. Later dressings should be planned to maintain yield and also a satisfactory amount of soluble P and K in the soil. A tonne of grass dry matter contains about 20 kg of K_2O and 5 kg of P_2O_5, it is wise to replace these amounts unless it is *certain* that soil reserves can supply the balance. The P and K removed in casual hay or silage cuts should also be replaced. In continuous cutting systems the results of periodical soil analyses should be used to balance manuring to the reserves of these nutrients in the soil and to the cutting systems.

Grass to be cut for hay should get about 80 kg/ha P_2O_5 (this will be more than enough to replace the P removed in the hay and to feed the grass for the rest of the year; annual dressings of this size will build up a reserve of P in the soil).

Except on soils poor in K, potassium is *not* applied in spring; this is to avoid it being taken up 'in luxury' by the hay crop, which grows at a time when the soil's natural reserves of soluble K are largest. It should be applied *after* hay or silage is removed as a straight K-fertilizer supplying 60–80 kg K_2O/ha. Alternatively a simple N fertilizer can be used in spring and 800 kg/ha of potassic basic slag applied after cutting (a 0–1–1 PK compound fertilizer would serve as well).

Manuring the aftermath

When a single cut of hay or silage is taken from grass fertilized with nitrogen and the sward is allowed to grow again, production from the aftermath is often less than where no nitrogen was applied to the grass before cutting; this is partly due to checking the growth of clovers. If the aftermath is required for grazing as soon as possible after cutting, the check can be avoided by giving a dressing of nitrogen. The potassium removed in the hay should be at least partially replaced at the same time. A convenient way is to use an NK compound fertilizer; 400 kg/ha of 25–0–16 would give nitrogen for regrowth of grass and replace the K in 3 tonnes of hay. Another possibility is to use a High-N NPK compound to supply the annual dose of phosphate as well and 500 kg/ha of a 20–10–10 fertilizer would be suitable.

Continuous cutting for silage, drying or feeding fresh

Each cut should receive 80–100 kg of N/ha. For 5 cuts 400–500 kg of N will be needed in a year. On land where soluble P and K are 10–15

TABLE 54

RECOMMENDATIONS FOR FERTILIZING GRASSLAND FOR CUTTING

	Amounts of plant nutrients									Type of fertilizer suitable (other materials supplying nutrients in the same proportions will also be suitable)	
	kg/ha			lb/acre			UNITS/acre			N % P2O5 % K2O %	
	N	P2O5	K2O	N	P2O5	K2O	N	P2O5	K2O		
Improvement and establishment dressings											
Extra dressings on deficient soils. Worked well in before sowing on a new ley, or broadcast in autumn or winter on established grass. (To be repeated as necessary until soil P and K are at 'satisfactory' levels)	—	125	125	—	110	110	—	100	100	0—20—20 / 0—12—10	PK compound / Potassic basic slag
Establishing leys of all kinds. Worked well into seedbed (additional to any dressings given to deficient soils; use no nitrogen on peats)	60	60	60	55	55	55	50	50	50	15—15—15 OR 0—12—10 +21—0—0	General purpose NPK compound / Potassic basic slag + straight N
Production and maintenance dressings											
Cutting clover-grass swards for hay or silage followed by grazing — In winter	0	80	60	0	70	60	0	60	60	0—15—0	Basic slag (or superphosphate)
In spring when field is closed	100	0	0	90	0	0	80	0	0	21—0—0	Straight N
After cutting	100	60	0	90	90	55	80	80	55	25—0—16	NK compound
Total in year	200	80	80	180	160	70	160	60	50		

Continuous cutting of swards of all kinds for silage, drying or feeding green

										Fertilizer
In winter	0	100	0	0	90	0	0	80	0	0—15—0 Basic slag (or superphosphate)
Before cut 1	80	0	0	70	0	0	60	0	0	21—0—0 ⎱ Straight N 34—0—0 ⎰
Before cuts 2, 3, 4 and 5	80	0	50	70	0	45	60	0	40	25—0—16 NK compound fertilizer
Total in year	400	100	200	350	90	180	300	80	160	

OR

										Fertilizer
Before cut 1	80	0	0	70	0	0	60	0	0	21—0—0 ⎱ 26—0—0 ⎬ Straight N 34—0—0 ⎰
Before cuts 2 and 4	80	50	50	70	45	45	60	40	40	20—14—14 High-N NPK compound
Before cuts 3 and 5	80	0	50	70	0	45	60	0	40	25—0—16 NK compound fertilizer
Total in year	400	100	200	350	90	180	300	80	160	

Adjustments for FYM or slurry (deduct these quantities)

25 t/ha (10 tons/acre) of FYM	40	50	90	35	45	80	30	40	70
10,000 litres/hectare (1,000 gallons/acre) of slurry	40	4	40	35	4	35	30	3	30

ppm of P and 100–150 ppm of K—('satisfactory', but not 'high') about 80 kg of P_2O_5 and 250 kg of K_2O will be needed in the year for maintenance.

In practice, this manuring with P and K may not match exactly the amounts lost, and reserves of these plant foods may rise (or fall). Soil analyses should be done every few years and the amounts of fertilizer adjusted as necessary. The N and K will be required regularly; some N alone should be applied before the first cut and NK compound fertilizer for later cuts. The phosphate can be applied as a single dose in winter to last the following season, 400 kg/ha of superphosphate ($20\% \ P_2O_5$) or equivalent basic slag will provide enough P. On *acid soils* in wet areas 400 kg/ha of ground rock phosphate would be suitable. A straight N-fertilizer should be given in February/March for the first cut. Before each subsequent cut 80 kg of N and 40 kg of K_2O should be given. An alternative method is to supply all the P and part of the K together with the nitrogen for the second cut by an NPK compound fertilizer applied after the first cut. A High-nitrogen NK compound should then be used before each subsequent cut. Suggested dressings are in Table 54.

GRASS FOR GRAZING

Nitrogen supply governs the level of output from grassland and before planning manuring, the overall level of production through the year and the amounts of grass needed for grazing at different times must be decided. The gross returns from the system also affect the amount that can profitably be spent on fertilizers. Heavy manuring, which may be profitable for intensive dairying and where silage is made, may be uneconomic where grass is used extensively for grazing by stores, beef cattle or sheep.

Large dressings of phosphorus and potassium fertilizers will be needed in the early years of a ley established on poor soil to maintain its productivity. Subsequently the management system affects the amount of phosphate and potassium that is justified. Whether these plant foods are returned to the soil in excreta to help to grow more grass, or whether they are partly or completely removed from the field, determines how much fertilizer is needed to maintain the system. The amounts will be greater where occasional cuts for hay or silage are taken. Under a long ley, or a permanent pasture, fertility will be built up due to the plant foods returned in dung and urine and to any extra nitrogen fixed by clover. After a number of years the increased quantities of plant foods in *circulation* will lessen the need for P and K. It is difficult to give advice on the P and K dressings needed by grazed grassland, but the suggestions made in Table 55 should be followed in the early years of an intensified grassland system. Periodical soil analyses should be made to see if the dressings used are maintaining soil P and K satisfactorily.

SUMMARY OF RECOMMENDATIONS

A summary of suggested recommendations for using fertilizers on grassland is in Table 54 (grass for cutting) and Table 55 (grazed grass). All the fertilizers suggested are simply illustrations of ways of carrying out the recommendations; other fertilizers that supply the same amounts of plant foods will be just as satisfactory.

However the grass is used, it will not yield well on soils that analysis shows to be deficient in phosphate or potassium, or both. Therefore both tables have at the top the large dressings of P and/or K needed to improve poor soils. These amounts are additional to maintenance manuring; when reclaiming poor land they may have to be applied (in the winter) for several seasons until analyses show that the soil is accumulating some reserve of these nutrients. The dressings suggested for establishing leys will serve both for one-year leys of ryegrass plus clover (or grass alone), and also for leys intended to stand for several years; they correspond to the first dressings of the year given to an established sward.

Dressings for production of grass and for maintenance of soil fertility are shown in the lower part of the tables. The nitrogen dressings can be used confidently, they are substantiated by the results of many recent experiments. The P and K dressings suggested are not so precise; if there is no local knowledge that offers a better basis they should be used for several years and then adjusted by comparing the results of soil analyses at the beginning and end of the period. The recommendations for grass grown for cutting (Table 54) allow for the large loss of potassium involved in removing herbage. No detailed recommendations can be made for using potassium on grazed grass because needs vary so much. If FYM or slurry are returned to grazed grass, less K-fertilizer will be needed.

Large dressings of nitrogen are recommended for all cut swards (Table 54) and where maximum yield is required, there is no reason for using less. Farmers who use little or no nitrogen on grazed grass will need to gain experience in using N-fertilizer and in managing and grazing the extra grass. Therefore three intensities of using N are shown in Table 55 to help farmers increase their yields gradually. As soon as experience and confidence have been gained a change should be made to the system using most N-fertilizer (at the bottom of the Table).

Fertilizer recommendations cannot be exact; for this reason the amounts of N, P_2O_5 and K_2O shown in Tables 54 and 55 in metric units are only approximately equivalent to amounts in Imperial units. Some recently-introduced fertilizers rich in nitrogen make it easier to fit recommendations to the compound fertilizers available. They are mentioned in the Preface (page xiv).

TABLE 55

SUMMARY OF RECOMMENDATIONS FOR GRAZED GRASSLAND

	Amounts of plant nutrients									Type of fertilizer suitable (other materials supplying nutrients in the same proportions will also be suitable)			
	kg/ha			lb/acre			UNITS/acre			N %	P₂O₅ %	K₂O %	
	N	P₂O₅	K₂O	N	P₂O₅	K₂O	N	P₂O₅	K₂O				
Improvement and establishment dressings Extra dressings on deficient soils Worked well in before sowing a new ley, or broadcast in autumn or winter on established grass. (To be repeated as necessary until soil P and K are at 'satisfactory' levels)	—	125	125	—	110	110	—	100	100	0	20	20	PK compound fertilizer
										0	12	10	Potassic basic slag
Establishing leys of all kinds Worked well into seedbed	60	60	60	55	55	55	50	50	50	15	15	15	General purpose NPK compound **OR**
										0	12	10	} Potassic basic slag + Straight N
										21	0	0	
Production and maintenance dressings Ordinary grazing of average clover-grass pastures Sheep, stores, beef cattle, dairying or mixed stock; stock on land all the time Late-autumn to mid-winter (maintenance dressing)	0	60	60	0	55	55	0	50	50	0	20	20	or } PK compound fertilizer
										0	12	10	Potassic basic slag
February, June or August. To extend grazing season or to get more summer grass. The same fields will not usually be dressed twice in a season	80	0	0	70	0	0	50	0	0	21	0	0	} Straight N
										26	0	0	
										34	0	0	
Total in year	80	60	60	70	55	55	50	50	50				
Moderately-intense grazing of good clover-grass pastures (Dairy cattle and followers, or intensive beef fattening) In February-March	50	0	45	40	0	0				21	0	0	Straight N

Late autumn-winter

Timing / operation	Recommended dressings									Fertilizer
Late autumn-winter *(top of block, partly cut off)*	0	60	60	0	55	55	0	50	50	0—12—10 Potassic basic slag
February-March	50	0	0	45	45	0	40	0	0	21—0—0 ⎫
Midsummer	80	0	0	70	70	0	60	0	0	26—0—0 ⎬ Straight N
August-September	50	0	0	45	45	0	40	0	0	34—0—0 ⎭
Total in year	180	60	60	160	55	55	140	50	50	

High-intensity grazing of all kinds of swards

Carefully managed rotational grazing and cutting for silage for dairy herds. Stock spending part of day elsewhere for milking. Clover sown but ignored afterwards.

Timing / operation										Fertilizer
Autumn-winter	0	50	50	0	45	45	0	40	40	0—20—20 PK compound
February-March	60	0	0	55	55	0	50	0	0	0—12—10 Potassic basic slag
After first grazing, for a silage cut	100	0	0	90	90	0	80	0	0	21—0—0 Straight N
After silage cut	80	0	50	70	0	45	60	0	40	26—0—0 Straight N
After each subsequent grazing except the last	60	0	0	55	55	0	50	0	0	34—0—0 ⎫ 25—0—16 ⎬ NK compound
Total in year (4 grazings, 1 cut)	360	50	100	325	45	90	290	40	80	

OR

Timing / operation										Fertilizer
February-March	60	0	0	55	0	50	0	0	0	21—14—14 High-N NPK compound
After first grazing, for a silage cut	100	0	0	90	0	80	0	0	0	21—0—0 Straight N
After silage cut	80	50	50	70	45	60	0	40		25—0—16 NK compound
After second grazing	60	0	0	55	0	50	0	0	0	21—0—0 Straight N
After third grazing	60	0	40	55	35	50	0	30		25—0—16 NK compound
Total in year (4 grazings, 1 cut)	360	50	90	325	45	80	290	40	70	

Adjustments for FYM or slurry used on grazed grassland

Deduct

for 25 t/ha of FYM (10 tons/acre)	40	50	90	35	45	80	30	40	70
for 10,000 litres/hectare of slurry (1,000 gallons/acre)	40	4	40	35	4	35	30	3	30

CHAPTER TWENTY-ONE

Vegetables, Glasshouse Crops and Fruit

Most of these crops are expensive to grow and their values per hectare are large, particularly the glasshouse crops. The costs of fertilizers are a very small part of total production costs in glasshouses, orchards and in many vegetable-growing enterprises. Farmers and growers should therefore plan to supply sufficient nutrients for the maximum yield of the desired quality. Care is often needed, particularly in glasshouses and for quickly-growing vegetables, to ensure that the large fertilizer dressings used cause no damage to growth; special fertilizers, slow-acting organic fertilizers, and potassium salts containing no chloride are often chosen for this reason. The soils used, and the conditions under which these crops are grown, vary so much that it is impossible to give general advice of much value to individual growers. This is not attempted in this Chapter which states some general principles and refers to sources of information. In addition to the papers listed here A.D.A.S. has published (ref. 43) recommendations that cover most practical conditions in Britain. The crops cover large areas and are very important financially. Nearly half a million acres (200,000 ha) of vegetables were grown in the open in U.K. in 1968/9; the area under protected cropping (glasshouses etc.) was practically 5,000 acres (2,000 ha). The total area under orchard fruit in 1968/9 was about 170,000 acres (70,000 ha), and a further 45,000 acres (18,000 ha) was under soft fruit. The total value of horticultural products in 1969/70 was about £271 million, vegetables accounting for well over half of this.

VEGETABLES

There are no results from large groups of field experiments on vegetables to provide general summaries of fertilizer responses such as are available for arable crops. F. Haworth wrote (ref. 86) that to determine optimum rates for each combination of crops, soil type and management would need too many field experiments; he considered that reasonable predictions of the potassium needed can be made from a knowledge of the amounts taken up by the crops and the amount in the soil. The maximum amounts of nutrients in the crops during the growing

246

season were more relevant than the amounts removed at harvest, since these quantities must be available in the soil from planting until maximum uptake occurs, and this may be well before harvesting. The amounts of nutrients in several common vegetables at the time of maximum nutrient content vary greatly as the following range of data show:

	kg/ha			lb/acre
Nitrogen	108 (lettuce)	–	314 (red beet)	96–280
Phosphorus	17 (lettuce)	–	56 (brussels sprouts)	15– 50
Potassium	162 (onions)	–	448 (carrots)	145–400
Calcium	25 (lettuce)	–	168 (brussels sprouts)	22–150
Magnesium	13 (onions)	–	50 (carrots)	12– 45

Some of these vegetables, like lettuce, have a very short growing season, often of only eight weeks; others that occupy the land longer grow as rapidly as lettuce for part of the season. The greatest need generally is for potassium and the soil must be able to supply these large amounts quickly enough to maintain growth rates. It is likely that the high reputation of farmyard manure for vegetable crops is due to its ability to supply large amounts of potassium in a readily 'available' form. Potassium sulphate is used to avoid the 'salt' damage to crops that sometimes occurs when the chloride is applied.

L. J. Hooper (ref. 46) has discussed the background for advice on vegetable manuring in England and Wales. Because the crops are valuable, policies for applying P and K should ensure that yields are never limited by shortages of these nutrients (and also avoid waste and damage by applying excesses). Magnesium deficiency is avoided by using Mg fertilizers; the deficiency is unlikely if exchangeable magnesium is above 50 ppm unless the ratio of exchangeable K:Mg is greater than 3. Leafy vegetable crops need much N-fertilizer (but using *too much* may damage their quality). Hooper gives a detailed account of how recommendations are made for brussels sprouts and for celery. For both crops the amounts of N, P_2O_5, K_2O and Mg recommended depend on soil type, and soil analyses.

A recent article (by Wellesbourne workers) on brussels sprouts shows that leafy vegetables grown on arable soils with less than 2 per cent of organic carbon need large dressings of nitrogen for satisfactory yields. Average yield without N-fertilizer was 8.0 t/ha (3.2 tons/acre). Yields were nearly directly proportional to N-fertilizer applied and were maximal at 15.5 t/ha (6.2 tons/acre) with 340 kg N/ha (270 UNITS N/acre). There was a sharp point of inflexion with this dressing and giving more N diminished yield. N-fertilizer had little effect on size or quality of sprouts, the dressings were efficient when all was applied at planting; top-dressing was not necessary. In other work on brussels sprouts aqueous ammonia used to supply the large nitrogen dressings needed was as efficient as solid fertilizer broadcast (ref. 121).

GLASSHOUSE CROPS

In a paper to the Fertiliser Society G. W. Winsor (ref. 87) stated that Britain, with nearly 6,000 acres (2,400 ha), has the second largest glasshouse industry in the world (Netherlands had 12,000 acres (4,800 ha), the U.S.A. fewer than 2,000 acres (800 ha)). Food crops grown in glasshouses are worth from £1,300–£8,000/acre (£3,200–£21,000/hectare), flower crops from £3,000–£24,000/acre (£7,500–£60,000/hectare). Capital costs are large: Winsor estimates about £25,000/acre (£60,000/hectare) for heated glasshouses. These facts are important in considering choice of soil and fertilizing policies; other costs are so great that soil may be improved economically by large dressings of organics, or it may be rep'aced altogether by another medium; ordinary fertilizer costs are trivial.

Because glasshouse crops are worth so much, their yield must not be limited by any factor that can be controlled. The method of deriving optimum fertilizer dressings for agricultural crops from response curves obtained in field experiments is not appropriate when fertilizer is such a very small part of total costs. Winsor also points out that it would be impossible to do enough field experiments to make satisfactory recommendations, and states 'The grower is thus prepared to build up soil nutrient levels to ensure that, even under the most favourable conditions he can create by manipulation of the environment, there is no risk that nutrient supply will ever become the limiting factor. The main danger here is the possibility of raising nutrient levels beyond the optimum and thus entering the region of excessive salinity'.

TABLE 56

SOIL P AND K LEVELS AT WHICH 'STANDARD RATES' OF FERTILIZERS WOULD
BE APPLIED TO VARIOUS TYPES OF CROP

Crop	ppm soil P	ppm soil K
Cereals	2–10	50–100
Root crops	5–20	50–100
Fruit	5–20	200–350
Vegetables	5–20	100–350
Glasshouse crops	40–70	500–700

Winsor shows the very large differences in the nutrient levels common in agricultural and horticultural soils. Table 56 (quoted from Winsor) gives the amounts of soil P and soil K determined by the modified 'Morgan' methods which he describes (not by the methods described in Chapter 14) at which standard rates of fertilizers would be applied when following the N.A.A.S. recommendations. Amounts of P judged standard are similiar for root crops, fruit and vegetables, but the two last groups need much more K than root crops. Glasshouse soils are very much richer in P and K than agricultural soils. Table

57 (also given by Winsor) comparing agricultural and horticultural crops, shows that the outdoor crops receive similar amounts of nitrogen and phosphorus but that potassium varies much more from crop to crop. Glasshouse crops receive very much more N, P and K. The average recommendation of K for tomatoes is about 50 times that for spring wheat. In early experiments at Cheshunt tomatoes removed in fruit and foliage about 750 kg of K_2O, 380 kg of N and 110 kg of P_2O_5 per hectare (670 lb of K_2O, 340 lb of N and about 100 lb of P_2O_5 per acre). Much of the potassium applied, perhaps 550 kg/ha of K_2O (500 lb K_2O/acre), was lost by leaching in drainage; much nitrate disappeared completely, and this could not be explained. Magnesium deficiency was recognised and cured by spraying with solutions of magnesium sulphate. Winsor showed how recommendations for managing

TABLE 57

RECOMMENDED FERTILIZER DRESSINGS FOR VARIOUS CROPS IN SOILS AT THE P AND K LEVELS SHOWN IN TABLE 56

Crop	Recommended fertilizer dressings kg per hectare			UNITS per acre		
	N	P_2O_5	K_2O	N	P_2O_5	K_2O
Spring wheat	0–125	38–75	38	0–100	30–60	30
Cabbage (winter or savoy)	75–125	75	150	60–100	60	120
Onions (bulb)	50–100	75	225	40– 80	60	180
Beet (main crop)	100–150	75	150	80–120	60	120
Lettuce (summer)	75–125	75	150	60–100	60	120
Apples (dessert)	50–150	20	63	40–120	15	50
Blackcurrants	75–150	38	63	60–120	30	50
Tomatoes						
Base fertilizer	0–120	150	650	0–95	120	520
Top-dressing { long season	900	–	1120	715	–	895
{ short season	560	–	780	445	–	625
Total } long season	1000	150	1775	810	120	1415
} short season	680	150	1440	540	120	1145
Carnations						
Base dressing	0–120	150	–	0–95	120	–
Top dressing	1120	–	1120	895	–	896

glasshouse crops, and particularly for liquid feeding, have developed from earlier work. Recently boron deficiency has affected some tomato crops in addition to carnations (where it is common).

The search for satisfactory slow-release N fertilizers continues, though Winsor says they may have no advantages over current methods and may even be worse. Slow-acting sources of N would probably be too slow for young plants, and would be superfluous where trickle irrigation and liquid feeds are used. Slow-release fertilizers decompose in soil at rates governed by biological, chemical and physical factors, most of which cannot be controlled; Winsor states 'once a heavy application of slow-release fertilizer has been given, the grower is no longer in full control of the nutritional situation'. American workers have warned that slow-release or coated fertilizers cannot be leached from soil and an overdose can injure roots and damage crops irreparably.

L. J. Hooper (ref. 46) discussed the A.D.A.S. basis for advising on manuring glasshouse crops and reviewed the British work on tomato nutrition. He gave the amounts of nutrients removed by the plants of a tomato crop yielding 125 tonnes/hectare (50 tons/acre) as 480 kg N, 78 kg P_2O_5, 980 kg K_2O per hectare (430 lb N, 70 lb P_2O_5, 875 lb K_2O per acre). Hooper showed how advice on fertilizer dressings for tomatoes is related to soil analyses, soil type, kind of crop and stage of growth.

FRUIT CROPS

L. J. Hooper (ref. 46) pointed out that it may take many years to assess the affects of fertilizers on tree crops. He discussed apple nutrition as an example.

Apples require relatively little phosphorus but recently responses have been obtained on P-deficient soils in the south of England. The crop needs much potassium, both for large yield and for good quality. Magnesium sulphate sprays are used for a quick control of Mg-deficiency but on soils with small exchangeable Mg, Mg fertilizers should be applied. For soils with 25 ppm of exchangeable Mg or less 63 kg of Mg/ha (56 lb/acre) is recommended, but none is given on soils with more than 100 ppm of exchangeable Mg.

Nitrogen dressings are influenced by management, summer rainfall, variety, rootstock, irrigation and the cover crop. Grass covers compete with the trees for N and at East Malling it was found that 134 kg N/ha (120 lb N/acre) was not enough to compensate for N removed by a perennial ryegrass sward. Dessert apples need less nitrogen than cooking apple trees in cultivated orchards, these need less N than trees in orchards with strips of grass, and these in turn need less than trees in orchards which are grassed down. In areas with more than 350 mm (14 in.) of summer rain, apples need less fertilizer-N than in drier areas. The figures in Table 58 were given by Hooper to show the factors that alter fertilizer recommendations.

Leaf analysis

C. Bould (refs. 59 and 60) has published good accounts of fruit tree nutrition and pointed out the value of plant tissue analyses in determining whether trees have sufficient nutrients. Because trees occupy the soil permanently but explore only a small fraction with their roots, and because nutrient uptake is altered by very many factors of weather and management, and by interactions with other nutrients, the nutritional status of trees cannot be predicted with any certainty even when the amounts of soluble nutrients in soils are known exactly. The value of leaf analysis in planning the manuring of fruit trees is that it can diagnose incipient or 'sub-clinical' deficiencies in trees which appear normal and show no deficiency symptoms, but in which growth and yield will not be maximal because the tree has insufficient nutrients. By using leaf analysis to diagnose the *current* nutritional status of a tree it is also possible to detect excessive fertilizer applications and to find whether applied fertilizers are actually being taken up by the tree. Bould stresses the value of nutrition control by leaf analysis 'With the possi-

TABLE 58

FERTILIZER RECOMMENDED FOR DESSERT APPLES IN AREA WITH LESS THAN 350 MM OF SUMMER RAIN

Nitrogen needed	kg N/ha	UNITS N/acre
(i) Cultivated orchard	50	40
(ii) Orchard with strips of grass, and cultivated strips treated with herbicide	75	60
(iii) Grassed orchard	125	100

Phosphorus needed			
Index	Soluble P in soil (ppm)	kg P_2O_5/ha	UNITS P_2O_5/acre
0	0 – 2.0	75	60
1	2.1 – 5	38	30
2	5.1 – 10	19	15
>2	>10	0	0

Potassium needed			
Index	Soluble K in soil (ppm)	kg K_2O/ha	UNITS K_2O/acre
0	0 – 50	250	200
1	51 – 100	188	150
2	101 – 200	94	75
3	205 – 350	31	25
>3	>350	0	0

Magnesium needed			
Index	Soluble Mg in soil (ppm)	kg Mg/ha	UNITS Mg/acre
0	0 – 25	63	50
1	26 – 50	44	35
2	51 – 100	31	25
>2	>100	0	0

bility in future of having to produce higher yields on lower, or the same, acreage, it is envisaged that nutritional control will be used more frequently as an integral part of fruit orchard management. This practice is accepted as routine with crops such as sugar cane'.

Examples given by C. Bould show the effect of cover crops and soil potassium reserves on leaf analyses and the relationship between leaf analyses, and effects of fertilizers on yield. The figures in Table 59 illustrate the concentrations accepted for diagnosing nutritional status in apples (ref. 59).

In Netherlands advice on fruit-tree nutrition is also based on leaf analysis; soil analysis is not considered reliable for showing how much fertilizer is needed. Some critical concentrations used to guide advisors in Netherlands are also shown in Table 59. The limits chosen to define 'deficient' trees, and those trees that have 'sufficient' nutrients are similar in the two countries for N, P, K and Mg; critical concentrations of calcium differ. The following dressings are recommended in Netherlands for high-yielding apple trees in full production:

	kg/ha
N	200–300
P_2O_5	40–60
K_2O	200–400
Mg	25–50

The amounts of N and K used are much larger than are advised in England (Table 58).

TABLE 59

RANGES OF NUTRIENT CONCENTRATIONS IN APPLE LEAVES IN JULY-AUGUST

Nutrient	English data		Dutch data	
	Deficient (showing symptoms)	Sufficient	Low	Satisfactory
	as % in dry matter			
N	1.7 – 2.0	2.4 – 2.8	2.2	2.7
P	0.07 – 0.10	0.20 – 0.25	0.13	0.17
K	0.4 – 0.7	1.3 – 1.6	0.8	1.2
Ca	0.50 – 0.75	1.0 – 1.6	1.6	2.0
Mg	0.06 – 0.15	0.25 – 0.30	0.12	0.24
	as ppm in dry matter			
Cu	1 – 3	5 – 10	–	–
Zn	1 – 5	15 – 25	–	–
B	5 – 15	25 – 30	–	–
Mn	5 – 20	30 –100	–	–

Soil analysis and leaf analysis compared. Few investigations have compared soil and leaf analysis for advising on fertilizing fruit trees, but it is commonly believed that leaf analysis is better for assessing the fertility of orchard soils. C. R. Frink (ref. 87a) prefaced a critical comparison of the two methods on apple orchards in Connecticut (U.S.A.) by stating 'we have no substantial reason for choosing one over the other if our goal is distinguishing well nourished prosperous trees from poorly nourished and unproductive trees. The interpretation of a leaf analysis is at least as difficult as the interpretation of the corresponding soil analysis'. Often no adverse effects on trees occurred when nutrient concentrations were only a half or a third of concentrations published as standards. In Frink's work 'available' nutrients in the orchard soils were not well correlated with nutrient concentrations in the leaves. The (Morgan method) soil tests used in Connecticut distinguished good from poor apple trees better than leaf analyses did for potassium, calcium and magnesium. Leaf P analyses were slightly better than soil P analyses for distinguishing the two groups. Leaf analyses for nitrogen divided the trees into three groups, but this could have been done visually. As Frink could not show that the more elaborate tissue tests were superior, he concluded the logical choice was a soil test to distinguish between the pH, P, K, Ca and Mg contents of good and poor orchard soils and a visual observation to estimate the concentration of N in the foliage. It is clear that more work is needed to discover the relationships between the vigour of a tree, the concentrations of nutrients in its leaves and the supplies of nutrients in the soil in which the tree grows.

CHAPTER TWENTY-TWO

Fertilizing Crops grown in Warm Climates

The suggestions for fertilizing British crops in Chapters 19, 20 and 21 are relevant to manuring these crops wherever they are grown in cool temperate climates. Most of the World's important agricultural areas are warmer than Britain and many have larger rainfall. These conditions suit crops such as rice which cannot be grown here and maize, which can be grown satisfactorily in only a small part of Southern England. They also suit the many crops grown in tropical and sub-tropical regions for industrial processing and the food crops that serve local populations and often leave surpluses for export. This Chapter does not attempt to advise on fertilizing all these crops as the many texts available cover both tropical crops as a whole (e.g. refs. 88, 89, and 90) and the specialised crops grown on a large scale on estates (e.g. oil palm, ref. 91). The Chapter deals with a few crops to show the principles that have been established for profitable manuring; it also emphasises the new developments that are making larger yields possible and lead to closer control of crop nutrition.

Most of the important crops of the world are listed in Table 60 which gives F.A.O.'s estimates (ref. 92) of the areas grown and yields in 1968. Cereal-growing dominates the world's arable agriculture, the area used is about twice that used for all other crops in Table 60. The important cereals are maize, wheat and paddy rice, roughly equal total quantities of each are produced. New developments in cereal growing in warm countries are discussed here; brief accounts are given of fertilizing some of the important annual food and industrial crops. The Chapter also deals with the nutritional control of tree crops grown on industrial scale on estates—oil palm, rubber and sugar cane.

CEREALS

In most countries where cereals are grown larger yields have been made possible in the last two decades by the introduction of new varieties that respond better than varieties they replaced to increased fertilizer dressings. Better ways of growing the crops have also been developed; irrigation has been used where needed and modern chemicals to control weeds.

254

MAIZE

Maize grown for grain yields well in countries which have a long-enough frost-free growing season, high temperatures in summer, and sufficient water from rain or irrigation. Parts of south-east England are just on the fringe of the areas where present varieties of maize produce grain satisfactorily. Further north summer temperatures are not con-

TABLE 60

CROP AREAS AND YIELDS IN THE WORLD IN 1968

(from FAO Production Yearbook)

	hectares millions	yield 100 kg/ha
Wheat	227.5	14.6
Rye	22.4	15.0
Barley	74.9	17.4
Oats	32.3	16.8
Maize	106.0	23.7
Millet + sorghum	111.2	7.7
Paddy rice	132.2	21.5
(Total cereals	711.3	–)
Potatoes	22.8	138.0
Sweet potatoes + yams	16.0	84.0
Cotton (lint)	31.6	3.6
Total pulses	62.7	–
Total oilseed crops	112.6	–
Sugar cane	10.3	522
Sugar beet	7.9	319
Cassava	9.8	87
Bananas	1.7	150
Tea	1.3	–
Tobacco	4.1	11.6
Soybeans	33.6	13.0
Groundnuts	17.6	8.5

sistently high enough for satisfactory grain production, but the crop produces large green yields for silage. Nitrogen fertilizer, water supply, spacing of plants and summer temperatures are the most important factors in obtaining maximum yields. Sometimes two crops a year are taken in the humid tropics but published experiments suggest that total yields from them are often no more than from a single well-grown crop in warm temperate or sub-tropical regions.

Improved cultural practices

In developing countries yields can be much increased by comparatively simple modifications to practice, some of which cost very little. Allan (ref. 93) gave a good example of gains from good husbandry practices based on experiments in Kenya; Table 61 is adapted from his figures.

Simple improvements in cultural methods, and a better type of maize had much larger effects than fertilizers. In such conditions there would be no benefit from fertilizer unless a good type of crop was properly grown.

TABLE 61

GAINS FROM IMPROVED PRACTICES IN MAIZE GROWING IN EAST AFRICA

	Yield kg/ha	Gain kg/ha	Value of* gain shillings/ hectare	Cost of* treatment shillings/ hectare
Correct planting time				
Start of rains	5830	2430	746	very little
4 weeks later	3400	–	–	–
Larger plant population				
39,600 } /hectare	5130	1030	319	20
19,800 }	4100	–	–	–
Type				
Hybrid maize	5450	1660	511	30
Local variety	3790	–	–	–
Weeds				
3 weedings	5200	1160	361	49
1 weeding	4040	–	–	–
Fertilizers				
56 kg/ha P_2O_5	4660	90	27	79
No P_2O_5	4570	–	–	–
78 kg/ha of N	4900	570	180	178
No N	4330	–	–	–

*Kenya shillings. The value of maize was 31 shillings/100 kg.

Improved plant material

Large improvements in potential yields of maize have occurred through breeding programmes. Two types of maize are available: (1) Hybrid maize is bred each year with controlled pollination from selected stocks. The advantage of the hybrid vigour of the first (F1) generation is lost in the second generation, so fresh hybrid maize seed must be bought each year; the farmer cannot get the advantage of hybrid vigour by saving his own seed. (2) New varieties which are open pollinated are bred normally for yield, disease resistance etc.; they retain their advantages for more than one season and a farmer may save his seed. The new varieties and hybrids now produced in many maize growing countries have often given two or three times as much yield as local varieties; they also respond much better to N. For example in India 90–125 kg N/ha are recommended for new varieties and hybrids but only 40–60 kg N/ha for old local varieties. In a series of experiments in

India the local maize normally grown by farmers, and adapted to the district, yielded 2807 kg/ha whereas the yields of nine hybrids newly developed ranged from 3978 to 4732 kg/ha. With the best of the new hybrids the financial return was doubled. Responses to nitrogen by hybrid maize were usually at least 20 kg of grain/kg of N, and ranged up to nearly 50 kg.

Plant density

The returns from nitrogen are only maximal when there are sufficient plants. Experiments in Illinois (U.S.A.) gave these results:

Plants/hectare	N applied kg/ha			
	0	90	180	270
	yields, tonnes/hectare			
33,040	7.46	7.80	8.27	8.34
41,500	7.73	8.67	8.87	9.01
54,000	7.53	8.67	9.61	10.29

Maximum yield was only obtained with the largest rate of N and largest plant population. Very large fertilizer dressings were only justified when plant population was also large. Such large yields can only be obtained when other soil conditions are favourable; in dry areas of U.S.A. the available water may only be enough for 15,000 plants/ha. But with adequate moisture for well-grown crops 60,000 plants/ha are recommended in the 'Corn Belt' of U.S.A.

Fertilizer dressings used for maize in developed agriculture

The main nutrient needed for high yields from maize in developed agricultural systems is *nitrogen*. Total amounts of nutrients are given below in crops yielding 9,400 kg/ha of grain (these large yields are often produced in U.S.A. on field scale).

	In grain	In stover
	kg/ha	
N	129	61
P	31	8
K	39	157
Ca	1.5	37.5
Mg	11	34
S	12	10

Of the 190 kg N/ha the total crop needs, two-thirds is removed in grain, but of the equally large quantity of *potassium* only a fifth is in grain. So if nutrients in stover are returned to the land fertilizers need only supply moderate amounts of other nutrients along with the large dressings of N.

In U.S.A. increases in yield of 30 kg of maize grain are expected per kg of N (much larger increases have been obtained in some experiments). Because yield potentials are so large 220 kg N/ha is often recommended in southern U.S.A. Where such large dressings are used, part of the N is given when planting together with the P and K needed; the rest of the N is given as later side-dressings when the crop is about 50 cm high. This avoids both injury to young plants and losses by early leaching.

Phosphate is particularly important where soils contain little soluble P, which are common where modern agriculture is only now being developed, and also to promote root development where the length of the season is limited by frost or drought. In U.S.A. the amount of P applied is usually settled by soil analysis and experience on the land. On very poor soils band placement of phosphate produces larger yields and makes more efficient use of the fertilizer. Potassium and magnesium dressings are also settled by soil analysis. If maize is grown for silage the green crop removes very much K (200 kg/ha of K or more) and this must be replaced.

Leaf analysis has been used to determine fertilizer dressings; one authority suggested optimum leaf contents were 3.1 per cent N and 0.29 per cent P.

Recommendations for fertilizing maize in *three countries with very contrasted climates* are (per hectare):

Indonesia: 90 kg N, 60 kg P_2O_5, 20 kg K_2O.

Israel: 240 kg N + 120 kg P_2O_5 on alluvial clay.

South Africa: from 40 kg N in dry areas to 100 kg N/ha in the wetter climate of Natal.

RICE

Rice is the most important cereal in many tropical and sub-tropical areas. Modern varieties, good cultural methods, and liberal fertilizing lead to yields of 3–6 t/ha in Japan, Taiwan, Mediterranean countries and U.S.A. But in much of Asia and most of Africa yields are usually only a third as much. While better management and fertilizing would undoubtedly raise these yields, potential yields have remained small in many traditional rice growing areas because the *Indica* varieties used have tall weak stems and much N fertilizer cannot be used because of risks of lodging. The *Japonica* varieties grown in Japan, U.S.A., and the Mediterranean area are shorter, respond well to nitrogen and yield well. But they are not suited to day-lengths and temperatures of the tropics.

Plant material

Very successful breeding work at the International Rice Research Institute (IRRI) in the Philippines has changed the outlook for rice-eating countries in the tropics (ref. 94). The new dwarf varieties of Indica rice developed have very much larger potential yields, they

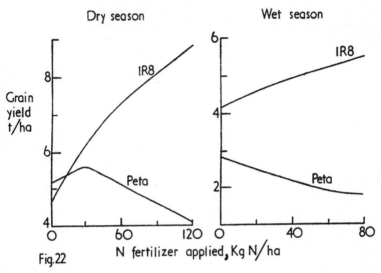

Fig. 22   The effects of N-fertilizer on the yields of an old variety of rice
(Peta) and a new dwarf variety (IR8) grown in experiments in
both wet and dry seasons. (Ref. 94)

respond well to nitrogen fertilizer and are not easily lodged. Fig 22
summarises the basis of the success of this work by comparing the
response to nitrogen of the variety IR8 (recently developed by the IRRI)
with that of one of its parents, Peta, a typical tall variety of Indica rice
developed in Indonesia and widely grown there and in Philippines.
Peta yielded 5.1 t/ha without N in dry season experiments and 5.5 t/ha
with 30 kg N/ha; giving more N lessened yields. Variety IR8, which
replaced it, yielded 4.6 t/ha without N-fertilizer but 8.5 t/ha with 120
kg N/ha. In the wet season, Peta yielded 2.9 t/ha and its yield was
depressed by giving N-fertilizer; IR8 gave 4.1 tonnes without N-
fertilizer and 5½ tonnes with 80 kg N/ha. Other varieties developed by
IRRI seem even more promising, being more resistant to disease and
more acceptable to the consumer. Trials in many African countries, in
the Middle East and in Asia, have commonly shown that the new dwarf
varieties yield twice as much as the older local varieties.

Using fertilizers on flooded land

Paddy rice grows in soil saturated with water; for convenience in water
management and to control weeds the land is usually flooded. This has
several implications for fertilizer use:

Forms of nitrogen: As the land is waterlogged it is anaerobic and
conditions are ideal for N to be lost from nitrates by denitrification
(Chapter 1). Ammonium salts must therefore be used; urea is a very
suitable fertilizer for rice, it dissolves in the water immediately and

none is lost as ammonia as may happen when urea is used for top-dressing ordinary cereals. (If during the growing season the land is allowed to dry, nitrates will be formed from ammonium salts and this nitrate may be lost when the soil is wetted again.) When denitrification of nitrate occurs IRRI results show that from 20 to 700 kg N/ha may be lost. Neither nitrates, nitrophosphates nor ammonium nitrate are as suitable as ammonium salts. Ammonium sulphate is still one of the best rice fertilizers, but liming is essential to correct the acidity it causes. Urea has been equally good.

Time of application of N. To minimise risks of denitrification ammonium salts should be applied immediately before flooding so that there is no time for nitrification to occur. Dressings drilled or disced well below the soil surface are even more effective as they are in reducing layers of the soil. Top-dressings of part of the N are often given; these should be applied after fields have been drained to avoid loss by leaching, particularly if urea is used. On average the best time for top-dressing seems to be when the panicles are forming.

Optimum dressings of nitrogen. The optimum dressing for new stiff-strawed varieties is generally about 120–160 kg N/ha; the minimum needed seems to be about 90 kg N/ha. By contrast the tall varieties traditionally grown cannot use more than 30–40 kg N/ha without severe lodging occurring. All rice of course gets some nitrogen from soil, decomposition of crop residues, and organic manures, and from irrigation water; further quantities come from non-symbiotic fixation by blue-green algae (common estimates for such fixation are 15–20 kg N/ha annually, but some workers have suggested that much more than this may be fixed). Average increases in yields of dwarf Indica varieties have been 20–30 kg of grain per kg of N applied.

Phosphate and other nutrients are more mobile in waterlogged soil than they often are in 'dry land' agriculture; nutrient ions have maximum opportunities to diffuse to roots.

Responses to phosphate are less common and smaller than those to N; however many soils in Asia, Africa and Latin America contain too little soluble P. Generally 20–30 kg P_2O_5/ha are sufficient and rarely is more than 40 kg P_2O_5/ha needed.

Potassium: The nutrients in the total crop which produces 1 tonne of rice grain have been given by IRRI workers:

19 kg N	7.2 kg Ca
4.3 kg P	5.3 kg Mg
47 kg K	

Most of the potassium is in the straw, but less than half the N and P. If straw remains in the field only 5 per cent of the total K in the crop is removed. Where *soil analysis* is used about 80 ppm of exchangeable K is considered enough. Irrigation water may supply much K and this may be enough for the rice. (Water flowing from volcanic soils may supply up to 50 kg K_2O/ha.) In many rice-growing areas potassium fertilizer is not needed except on sandy soils.

Indian experiments have shown that the new high-yielding varieties of rice (and the new dwarf wheats discussed in the following section) often need some K to secure the maximum response to N. Average results from large series of fertilizer experiments are given below. The crops all had basal dressings of 120 kg N/ha and 60 kg P_2O_5/ha.

	Rice	Wheat
	Yield of grain, kg/ha	
Yield without K	4431	2936
With 30 kg K_2O/ha	4718	3031
With 60 kg K_2O/ha	4857	3173

The responses to K by both crops were reported to be profitable under Indian conditions.

Dryland rice

This crop is grown in many countries which also produce rice on flooded land; much is grown in Brazil. The same varieties are often suitable; cultivations are the same as for other cereals. Yields are usually more erratic and average considerably less than from flooded rice. Productive varieties (particularly the new dwarfs) justify from 60–100 kg N/ha. The soils used are often deficient in P, particularly in Africa and in South America and adequate P-fertilizer is essential.

DWARF WHEATS

The wheats suitable for cool temperate climates, (advice on fertilizing them is discussed in Chapter 19), do not grow well in the tropics. But possibilities of growing wheat in warm countries have been revolutionised by the dwarf wheats produced by the plant breeding programme in Mexico that is now well known. Possibilities of cereal improvement were well discussed by J. Hutchinson (ref. 95).

Seed of new dwarf wheats that do not lodge was taken to the United States from Japan in 1946 and used in breeding programmes. Gaines wheat, introduced in Washington State in 1961, was one result; it is now much grown in Western States where farmers often harvest '100 bushel crops' (about $7\frac{1}{2}$ t/ha) and more from it. Parallel developments in breeding in Mexico so increased yields that the country changed in a few years from importing to exporting wheat although the population increased considerably. Mexican seed of the dwarf wheats was introduced in 1963 to India where national yields had been about 800 kg/ha for several decades. After testing in 1963–1965, two million hectares were sown in 1967/68. The older tall varieties had maximum yields of about 3000–4000 kg/ha; the new dwarf wheats have doubled this limit, yielding up to 7000–8000 kg/ha. Other changes in farming in India have had to be made to achieve these larger maximum yields. The timing of irrigation was changed, the seed had to be sown more shallowly and more nitrogen fertilizer used.

Elsewhere the dwarf wheats have been equally successful and large areas were being grown in India, Pakistan and the Middle East in 1968/69. The varieties respond to much nitrogen and between 20 and 40 kg of extra grain are commonly produced per kg of N applied. The gains from the new wheats are shown by these figures for trials in India of the Mexican Sonora 64 against an Indian variety NP876 which was a product of much breeding work in the 1930s:

kg N/ha	NP876	Sonora 64
	tonnes of grain/hectare	
40	2.3	3.8
80	2.8	4.8
120	2.5	5.3
160	1.6	5.2

120–150 kg N are recommended for the Mexican wheats, but only 60 kg N/ha for the improved Indian varieties.

Wherever these dwarf wheats are grown the optimum dressing seems to be about 120 kg N/ha. If the wheat follows a green manure or other crop that leaves residues of N, only 80 kg N is needed. Unless the soil is known to have enough soluble P, 40 kg P_2O_5/ha is usually recommended.

SORGHUM

A number of different types of cereals with small grains are covered by the names sorghum and millet. They need warm climates and usually thrive where maize does but they yield better where water is insufficient for maize. Most is used for human food in Asia and Africa but in other countries (e.g. U.S.A.) sorghum is used mainly for animal feed. In Africa most sorghum yields are small, often because the types grown have low potentials. Farmers' yields are about 700 kg/ha; yields in trials with fertilizers have been from 2000 to 3000 kg/ha.

About 5 million hectares of sorghum are grown in U.S.A. Breeders have developed hybrids, as with maize, that have greatly increased potential yields. The best hybrids when well-grown, have yielded on farm scale 11,000 kg/ha of grain and in field experiments plots have yielded twice as much as this.

The fertilizer needed by sorghum depends greatly on the yields that are possible and these depend on the reserves of soil moisture and the rain the crop will receive. A crop of 9000 kg/ha of sorghum will contain:

	Grain	'Straw'
	kg/ha	
N	130	90
P_2O_5	45	30
K_2O	35	180
S	16	12

200 kg N and 45 kg P_2O_5/ha are usually needed. Potassium is needed on deficient soils and sulphur has to be applied in deficient areas of U.S.A. The large amount of K taken up by the crop (sorghum extracts K from soil more easily than maize does) is mainly in the straw and, unless this is removed, the potassium is returned to the soil. Half of the large dressings of N needed is usually applied when planting, and half as a later side-dressing.

Some types of sorghum and millet are used for silage and for direct feeding. When crops are removed for such purposes they take very large amounts of K and this must be replaced to maintain fertility.

FOOD CROPS GROWN IN TROPICAL COUNTRIES

The amounts of nutrients removed by common yields of food crops in the tropics are shown in Table 62 (the data are from ref. 96 and other sources). In all examples of food crops grown on small farms yields are small, much smaller than can be obtained when these crops are grown better on the land of experimental stations or on large well-mechanised farms. Often the nutrients involved in producing such crops for local food are only a tenth of the amounts in modern yields of temperate crops or in the yields of some tropical crops grown on large estates for processing or export also shown in Table 62.

The results of the Freedom from Hunger Campaign programme of experiments have shown fertilizers can do much to increase the yields that farmers obtain. Some examples of responses have already been given in Table 30. In the series in West Africa on maize, rice, millet and yams the soils were so poor that responses to each of the three nutrients tended to be of equal size, whereas in more developed agriculture on richer soils responses to N fertilizer are usually larger than to P or K fertilizers.

Cassava

Cassava or tapioca (*Manihot utilissima* Pohl) is a native of South America but is used as food in tropical Africa; some is exported. The roots often contain 33 per cent starch but have only 1.7 per cent protein. Cassava needs 1000–2000 mm of rain/year.

Cassava takes large amounts of nutrients from soil and has a greater capacity to extract them from poor soils than most other crops. For this reason cassava is often the last crop in a shifting cultivation system; it then produces 5–10 t/ha, but on better land it can produce 40–50 t/ha. Analyses of large crops (40 t/ha) showed they removed 85 kg N, 60 kg P_2O_5, 280 kg K_2O and 60 kg Ca/ha. The large amounts of potassium removed can have a serious effect on soil fertility (cassava is able to extract the quantities shown from very poor soils); much phosphate is also removed. Unless these nutrients are replaced the result of growing and selling several crops of cassava on permanently cultivated land can be a disastrous fall in soil fertility that will greatly diminish the yields of following crops. Few experiments have been done to determine how

TABLE 62

NUTRIENTS IN HARVESTS OF SOME COMMON TROPICAL CROPS

Crop	Yield per hectare	N	P	K	Ca	Mg
			kilogrammes per hectare			
Food crops grown on small farms						
Maize	1100 kg of grain	17	3	3	0.2	1
Rice	1100 kg of paddy	13	4	4	0.9	2
Groundnuts	600 kg of kernels	28	2	3	0.3	1
	200 kg of shells	2	0.2	2	0.7	–
Cassava	11 tonnes of tubers (30% dry matter)	25	3	65	6	–
Yams	11 tonnes of tubers (30% dry matter)	38	3	39	0.7	–
Cacao	600 kg of beans	13	3	11	–	–
	600 kg of husks	11	1	25	–	–
Cash crops grown on large farms						
	Yield per hectare equivalent to					
Oil palm	2.5 tonnes of oil	162	30	217	36	38
Sugar cane	88 tonnes of cane	45	25	121	–	–
Coconuts	1.4 tonnes of dry copra	62	17	56	6	12
Bananas	45 tonnes of fruit	78	22	224	–	–
Rubber	1.1 tonnes of dry rubber	7	1	4	–	–
Soybeans	3.4 tonnes of grain	210	22	60	–	–
Coffee	1 tonne of made coffee	38	8	50	–	–
Tea	1300 kg of dried leaves	60	5	30	6	3

much fertilizer should be applied to cassava. Recommendations for crops in Kerala (India) are 60 kg N, 60 kg P_2O_5 and 120 kg K_2O per hectare—no more than a moderate crop removes.

Sweet potatoes (*Ipomoea batatas*)

This crop is grown in many parts of the tropics and sub-tropics. Too few fertilizer experiments have been done for precise advice on fertilizing to be given. Recommendations and the rates used by farmers vary much. Taiwan workers for example recommend these ranges of fertilizer dressings (per hectare): 30–90 kg N, 25–50 kg P_2O_5 and 80–200 kg K_2O/ha. They report, however, that little fertilizer is used by farmers.

In North Carolina 60–70 kg N, 50–60 kg P_2O_5, and 140 kg K_2O are recommended per hectare. Different sets of recommendations agree in suggesting that the crop needs much potassium, particularly on the lighter soils that suit it.

Yam (*Dioscorea* spp.)

Yams are often the first crop after fallow in shifting cultivation and need more plant nutrients for satisfactory yields than cassava does. In F.F.H.C. experiments yield seemed to range from 9000 to 15,000 kg/ha.

Most yam crops are still manured only with FYM and composts and experiments suggest that fertilizers can increase yields in all regions where they are grown. In experiments in the Savannah zone of Nigeria 15,000 kg/ha could be produced with 50–55 kg N/ha; in the F.F.H.C. fertilizer demonstrations, NK fertilizer raised yields by 42 per cent in this part of the country, and by 25 per cent in the forest zone.

Groundnuts *(Arachis hypogea)*

These are legumes and need good supplies of phosphorus; sulphur is also needed as a fertilizer in West Africa and in some other countries. Most fertilizer recommendations are modest; for example, a general recommendation in U.S.A. is 300–500 kg/ha of 4–12–12 or 5–10–15 compound (mixed) fertilizer.

Foliar analysis has been used in West Africa to guide fertilizing of groundnuts. Critical values above which no response is expected are 3.5 per cent for N, 0.225 per cent for P, 0.8–1.0 per cent for K, 1.2 per cent for Ca and 0.5 per cent for Mg, all expressed on leaf dry matter.

Soybeans *(Glycine max. L.)*

This crop has been grown for 5000 years in China, but U.S.A. now has more than half of the world's total area. Soybeans need nearly neutral soil (pH 6–6.5). As with other legumes grown for their seeds, in most experiments the responses to fertilizers have not been large. The crop responds a little to nitrogen and mostly receives 20 kg N/ha. A general recommendation in U.S.A. is to give about 400–500 kg of 5–10–15 per hectare.

A crop of soybeans yielding 3400 kg/ha contains:

	N	P_2O_5	K_2O
		kg/ha	
Grain	210	50	70
Straw	70	17	45
Stubble and roots	33	11	22
Total	313	78	137

The nitrogen is mostly fixed by symbiotic bacteria in nodules on the roots. The crop takes up much P and K but it seems that it normally secures the amounts shown from residues of fertilizers left in the soil after growing other crops that normally receive large fertilizer dressings.

TROPICAL CROPS GROWN FOR SALE OR MANUFACTURE

The tropical crops that are grown for export are often better grown and receive much more fertilizer than most food crops that are consumed locally. Export crops are often grown on large estates which have supported scientific work on plant breeding, nutrition and cultural methods.

The best varieties are planted; usually the crops are very well grow and adequately fertilized. The results of field experiments and so analyses are used for making fertilizer recommendations, nutrition o many estate crops is now controlled by regular plant analyses. Som large estates have their own laboratories and growers on small area have benefited from the experience of larger organisations. Larg amounts of nutrients are removed in the good yields that are obtaine and some data are in Table 62. Three crops are discussed below a examples.

OIL PALM (*Elaeis guinensis*)

One and a half million tonnes of palm oil were produced in the worl in 1969; about a fifth of this in West Malaysia where much experi mental work has been done on the crop (ref. 98). The oil is in th mesocarp of harvested fruits (which generally contain about 20 per cen of oil). The bunches of fruit contain much nutrients (particularly potas sium) which must be replaced to maintain soil fertility and yield Nutrients in trees, roots and fruit of oil palm grown in Nigeria have been given in Table 42.

About 150 palms/hectare are planted 12-14 months after the seec has germinated. For the following year the plants grow vegetatively They begin to flower in their second year and harvesting of fruit begin: about $2\frac{1}{2}$ years after planting. Maximum yields are usually achievec after 4 to 6 years of harvesting. Fertilizer programmes have to be designed to grow the young palms *and* to provide for fruit production in the early years; later they can be designed to replace nutrients removed in fruit. Total dry matter produced annually has been estimated as about 30 t/ha for 6–17 year old palms in Malaysia and 45 t/ha from 17–20 year old palms in Nigeria. Annual nutrient uptakes estimated in Malaysia are:

Year	N	P	K	Mg
	Total annual uptake (in grammes/palm)			
1	68	6	95	17
2–3	547	63	1174	140
4	1063	137	1967	322
5	1139	162	1956	365
6	1309	177	1940	391
7	1296	176	1788	408

Some of the *total* uptake is returned to the soil in prunings of the fronds and in male flowers. In addition to the nutrients *removed* in bunches, some are *immobilised* because they are stored in the trunk. (Nutrients are withdrawn from the trunk as the palm becomes older.) The partition of nutrients taken up in one year by 148 palms on one hectare of land in Malaysia was found to be:

	N	P	K	Mg	Ca
	Annual uptake (kg/hectare)				
Cumulative vegetative growth	41	3	69	12	14
Pruned fronds	67	9	86	22	62
Fresh fruit (25 tonnes)	73	12	93	21	20
Male flowers	11	2	16	7	4

A fertilizer programme proposed to obtain maximum yields from young plants has been published (ref. 98). The figures below are an example of the nutrients needed by palms grown on soils ranging from sandy loam to sandy clay in West Malaysia. The dressings are given in terms of amounts of the *fertilizers*, ammonium sulphate (N), Christmas Island rock phosphate (P), muriate of potash (K), kieserite (Mg), and not as weights of the elements whose symbols are used to identify them. (The rock phosphate contains about 38% P_2O_5; analyses of the other fertilizers are in Chapters 4 and 6.)

Years in field	kg of *fertilizer* per palm/year			
	(N)	(P)	(K)	(Mg)
1	0.68	0.68	0.60	0.80
2	1.10	0.92	1.90	0.34
3 and 4	1.20	1.20	3.10	1.15
5 and 6	1.48	1.50	3.30	1.45
7 and 8	2.10	1.80	3.30	1.60

Recommendations for other soils in Malaysia are based on results of field experiments; for example on coastal alluvial soils in Malaysia there was no response to P or Mg and so none is advised. Most of the responses found in experiments would be economically justified unless the price of palm oil became very small.

Variations in recommendations to allow for local conditions in one area are made by using both leaf analysis and soil analysis; neither method leads to *certain* fertilizer recommendations. *Soil analysis* differentiates between soils that are particularly rich or poor in each nutrient, and helps in deciding between large and small dressings to individual sites. *Leaf analysis* shows whether the existing fertilizer regime is roughly correct and in balance; the method is well established as an aid to manuring wherever oil palm is grown. Establishing critical *ranges* of nutrient concentrations in leaves is more useful than are attempts to settle critical *single values* that define 'deficiency' as weather and nutrient interactions alter the actual amounts found.

Nitrogen manuring needed in the early years varies with the soil and the cover crop used when planting the palms. The amounts shown in the recommendations above are for areas where leguminous cover crops are planted and maintained until the canopy closes; if a legume is not planted, twice as much N may be needed. Adequate fertilizing in the first years is particularly important so that the young palms grow as quickly as possible and give their maximum yields early, so minimising the effects on profits of overhead costs of establishment.

RUBBER

In one respect rubber (*Hevea brasiliensis*) is a complete contrast to oil palm; the harvest is latex which removes little plant nutrients. In other ways the crops are similar; both grow through an immature period when there is no yield and then are regularly harvested for many years. Both are grown on estates with close scientific control; much has been done to improve planting material, control disease and to use fertilizers both to lessen the period of immaturity, and to secure maximum yield afterwards.

Good accounts of nutritional work on rubber have been published (refs. 62 and 69) and E. Bellis pays 'tribute to those who, over the past 100 years, have transformed a botanical curiosity of the South American jungle to the pioneer plantation crop and the economic mainstay of much of tropical Asia'. Early yields were about 400 kg/ha. Replanting in Malaysia has raised average estate yields to 900 kg/ha but some commercial crops are now yielding 2 to 4 times as much. Annual yields from intensive exploitation of 7200 kg/ha have been recorded and calculations of potential photosynthesis, tree growth and rubber production suggest that yields of 10,000 kg/ha are possible.

Immature trees. About 400–600 trees/hectare are planted but these are thinned to about half the number. Leguminous covers usually planted between the trees supply much nitrogen, often sufficient for the crops in the later stages of immaturity. The trees are tapped when the trunk girth is 50 cm at 150 cm above the graft. Liberal fertilizer dressings are needed by immature rubber on most soils in West Malaysia except the alluvial soils near the coast. There have been good responses to K on inland sandy soils and magnesium deficiency symptoms occur often enough to justify Mg-fertilizers. There is no evidence of calcium supplies being deficient, although rubber-growing soils are very acid. Manganese deficiency is not uncommon and is corrected by dressings of manganese sulphate supplying 1.5–15 kg Mn/ha.

The full manuring programme outlined by E. Bellis (ref. 100) is:
(1) 18 grammes of phosphorus in each planting hole.
(2) Either 2 applications of compound fertilizer supplying NPKMg during the 3 months before budding on green bark, or 4 applications over 8 months before budding on brown bark.
(3) 6 applications during 18 months to seedlings, transplants and buddings.
(4) 6 applications during 4 years as the canopy develops.
These dressings supply (in grammes/tree of the *elements* shown):

Stage	N with legume cover	N without legume cover	K rich soil	K poor soil	P	Mg
1	–	–	–	–	18	–
2 (4 dressings)	48	48	14	35	37	9
3	100	150	30	75	80	15
4	440	1200	100	250	270	50

In 6½ years the immature trees will have received:

	Heavy soils	Light soils
	kg/ha	
N	214	188
P	109	108
K	54	118
Mg	25	25

Mature rubber in tapping. Nitrogen fertilizers increase growth in the first one or two years, but effects on yield are not obtained until the third or fourth year or later; then nitrogen often increases yield only when potassium is applied as well. Average annual fertilizer recommended for mature rubber in West Malaysia is about 34 kg N, 9 kg P, 25 kg K and 8 kg Mg/hectare. It is however necessary to discriminate between plantings which need much fertilizer and those that justify little or none (and where fertilizer-N may lessen yield or make trees liable to wind damage). Differential recommendations suited to soil, climate and management of mature rubber are based on plant nutrient cycles, soil analysis, leaf analysis and soil surveys of estates; all of these are linked to measurements on the crops.

Nutrient cycles. Average amounts of nutrients in an annual yield of rubber latex, and the amounts needed each year for growing the trees, have been given in Chapter 16 (p. 175). Annual growth needs about as much nutrients as some temperate crops, and these amounts are stored in trunk and branches; the latex removes very little. Components of the nutrient cycle in a rubber plantation in Malaysia have been discussed by G. A. Watson (ref. 69). A crop of 33-year old trees contained about 1600 kg N, 230 kg P, 1050 kg K, 1800 kg Ca and 320 kg Mg per hectare. The trees contained nearly as much magnesium and potassium, and nearly three times as much calcium as the top 100 cm of the soil contained on which they were grown. The crop also accumulated much nitrogen (about 55 kg N/ha), and about 7 kg P/ha annually. Weights of other nutrients accumulated by 33-year old trees were:

	kg/ha		kg/ha
Sulphur	239	Boron	1.7
Manganese	16	Zinc	5.1
Iron	14	Copper	1.1

Amounts of molybdenum were much less (41 grammes/ha) than amounts of the other micronutrients.

When rubber is replanted the usefulness of the large store of nutrients in trees to the next planting depends on the fate of the old trees. If the trees are removed all the nutrients are lost; if burned, most of the nitrogen and sulphur in the trees will be lost and other nutrients will be distributed irregularly in the ash left at the burning sites. Even if the trees are left to decompose, mineralisation of the organic matter, erosion and leaching will remove much of the store of nutrients originally accumulated by the jungle and used to produce the first crop of rubber.

Both oil palm and rubber need much fertilizer in the earlier years for establishing and growing the trees. After tapping begins, rubber is manured so that is continues to grow well; most of the nutrients supplied are retained on the site and in the later years of the life of a mature plantation the yield removes little and fertilizers will be needed only to replace losses in the nutrient cycle. In contrast fertilizers for oil palm must provide both for the continued growth of the trees until they are mature, and also must replace the large amounts of nutrients removed in fruit throughout the life of the crop if soil fertility is not to be depleted.

Leaf analysis is widely used on about 0.6 million ha in Malaysia for estimating the nutrient status of mature rubber (*soil analysis* is of much less value). The amounts of nutrients in mature leaves that are considered to indicate satisfactory nutrition have been given in Chapter 15 (p. 165).

Soil series. On smaller areas where management does not permit close control of manuring by leaf analysis, fertilizer dressings can be adjusted by taking account of soil series where surveys have been made which discriminated between rich and poor soils growing mature rubber. Most immature plantings benefit from as much or more fertilizer as is now used, but more efficient use of fertilizer on mature crops will only be obtained by using information on soil and leaf composition to assess local fertility.

SUGAR CANE

Sugar cane occupies a larger area of the World's farmland, and produces more sugar, than beet. Much of the cane crop is grown on large estates under very careful scientific control; chemical and physical measurements, and observations on the crops are used to guide fertilizing and irrigation. The interpretation of these data is based partly on field experiment results and partly on experience with successful crops.

Crop logging in Hawaii

The intensive control practised in Hawaii, called 'crop logging', has been described by H. F. Clements (ref. 101). Crop logging aims to show 'all the important factors of a plant's physiology which the crop experiences from the time it is started until it is brought in for harvest and in such a precise and simple manner that plantation management can use it as a guide for the profitable production of cane sugar'. Fields are sampled every 35 days. In 1961 one group of control laboratories did 100,000 analyses/year costing $50,000. The plantations produced 400,000 (short) tons of sugar/year and the cost of scientific control was 30 cents/ton of sugar for laboratory work plus collecting and drying samples. Sugar was then worth $135/ton so the cost of scientific control of crop growth was less than 0.25 per cent of the value of the crop. Leaf analyses were much more useful than soil analyses in planning the manuring of sugar cane in Hawaii.

Nitrogen. The nitrogen concentration in the leaves and their moisture content are related; equations have been developed from which the expected nitrogen concentration in an excellent crop can be calculated if per cent moisture in the leaf sheath and the age of the leaf is known (this is called the Normal Nitrogen Index). Control is planned so that soil moisture and nitrogen supply are never factors that limit growth. By comparing the actual N concentrations in the leaf with the Normal Nitrogen Index expected in leaves with the moisture contents found, a grower knows whether his crop is receiving too much or too little N or is being fertilized correctly. The example below given by H. F. Clements in 1964 (ref. 102) shows how N-fertilizer affected %N in leaf dry matter; the Normal Nitrogen Index expected when the crop had optimal moisture supply was 1.60 per cent N, so it seems that 450 kg N/ha was adequate for this crop at the stage when it was sampled.

kg N/ha	Leaf N % in dry matter
225	1.48
450	1.63
675	1.66
900	1.65
1125	1.70

Potassium. Similar experimental work done with potassium showed the crop had too little K when the index fell below 2.25 per cent K in sugar-free dry matter. An example is below:

K_2O applied as fertilizer kg/ha	K-index (% K in sugar-free dry matter)
60	1.87
112	2.92
225	3.02
340	3.28

Later, it was found that the potassium index depended on leaf moisture (and therefore on weather). By making allowance for leaf sheath moisture, correlations between K-index and growth were improved.

Phosphorus. P concentrations in leaves were found to be related to sheath moisture and to total sugars in the leaves; standard phosphorus indices were based on these measurements. Later a correction was made for that part of the P present which was not translocated to new tissue.

Other nutrients. 'Crop logs' have been used for other nutrients, including calcium, magnesium, boron, zinc and copper. Aluminium is potentially toxic, largest concentrations of Al are in the roots of sugar cane and these have been used to establish 'aluminium indices'.

Fertilizer practice in Hawaii

The use made of measurements on sugar cane is best illustrated by summarising Clements' (ref. 101) description of the pattern of fertilizing cane grown on a field where logs of previous crops were used to determine the initial dressings:

Pieces of cane stalk 55 cm long are planted which carry about 3 buds, the first N dressing of 20–40 kg N/ha is placed *under* them. The first regular dressing, given after a month, is 90 kg N/ha (with 100 kg K_2O/ha if the last log showed K was needed). The second dressing of N (and K if needed) is applied at 3 months. The third and last routine application, at 6 months, supplies 140 kg N and 100 kg of K_2O/ha. Tractors cannot be used as the cane is then too big; dressings are applied in irrigation water or by airplane (using urea pellets and pellets of muriate of potash). Irrigation is controlled by the results of soil moisture measurements. Crop logs begin at 3 months and continue at 35-day intervals to harvest-time; measurements of moisture, total sugar, nitrogen and potassium are recorded; when the crop is growing fastest, P, Ca, Mg and trace elements are determined at 3 consecutive logs.

Between the third application 6 months after planting and the last at 10–12 months, no more N and K are given unless the log shows they are needed. The final dressings of N and K (from 80–160 kg N/ha and from none to 100 kg K_2O/ha) given to the crop 10 to 12 months after planting also depend on the log of the current crop. No more N is applied unless the index falls seriously below normal before the crop is 15 months old. After 15 months nothing is given irrespective of the indices measured.

Ripening. On unirrigated plantations the only control of ripening is by careful fertilizing. In irrigated plantings samples are taken every week for sheath moisture and irrigation is controlled so that sheath moisture will be 73 per cent at the intended harvest date.

Dressings of liming materials and phosphate are based on logs of previous crops. Where the calcium index is low and soil pH is 4.6–5.2 3300 kg/ha of coral stone are broadcast when the soil is being prepared. Phosphate is placed under the 'seed' at planting at rates settled by phosphorus indices of the previous crop. From none to 450 kg P_2O_5/ha are given. If the soil pH is below 5.5 rock phosphate plus superphosphate are applied; when soil has pH about 5.5 triple superphosphate is used. Ammonium phosphate is suitable where soil calcium is adequate or excessive. If the phosphorus index of the current crop falls seriously ammonium phosphate pellets are applied from the air.

This detailed account of the control of growth of cane by regular measurements on the crop has been given because it shows how much can be done for a special crop grown continuously in large areas under unified management. Clements stated that crop physiology and ecology must be used to understand fertilizer responses; he also wrote that the performance expected from the crop under known local conditions must be standardised. The principles established for cane need to be applied to other crops. Similar methods can be developed for other crops in sampling of smaller less homogeneous areas is organised and if farmers of smaller areas (who are usually more independent than estate managers) can be persuaded to work together and to accept advice.

Sugar cane in other countries

In experiments done on sugar cane in other countries, nitrogen has generally increased yields and lessened sugar percentages, and the largest responses to N have been when irrigation has been applied. Phosphate has increased yields of cane grown on certain types of soil; P fertilizer has had no regular effect on sugar percentage but K fertilizer has generally increased it. Sugar cane removes large amounts of plant nutrients. The nutrients removed in an average crop of about 90 t/ha of millable cane (Table 62) can be greatly exceeded; for example a record crop of 125 tonnes of dry matter produced in Hawaii in under two years removed 330 kg N, 90 kg P and 900 kg K per hectare. The nitrogen dressings recommended as 'optimum' vary from about 100 kg N/ha for unirrigated crops in some countries to over 300 kg N/ha in other countries where irrigation is possible. In Queensland and South Africa much phosphate has been needed. The amount of potassium fertilizer needed by sugar cane varies from none to large amounts, depending on the soils used. If supplies of potassium are too small yields will decline slowly and it is said that too much K as fertilizer is more economical than risking using too little. Leaf analyses are used as a basis for fertilizer advice in most countries and one example has already been given in Chapter 15 based on data obtained in Guyana.

COTTON (*Gossypium* spp.)

Cotton is the best known example of a number of crops which are grown in warm climates to produce fibres. Cotton (and kapok) fibres are derived from the fruits of these plants; jute and ramie are examples where the fibres are in the stems, while sisal contains the fibres in its leaves. Although the yield of cotton is very sensitive to weather, it is widely grown in tropical and sub-tropical countries. The crop is particularly useful where agriculture is being developed, because it can be grown satisfactorily on small farms. Local farmers who are attempting to diversify their cropping often find that cotton is a very good cash crop to be grown in rotation with food crops for local use. The harvest is of *seed cotton* from which the *lint* is separated; the lint yield is usually slightly more than one-third of the yield of 'seed cotton'. Yields may be reported either as seed cotton, or lint; they commonly range from a few hundred kilogrammes/hectare of seed cotton, where the crop is grown without fertilizer on poor soils, up to 3000 kg/ha where well-fertilized and irrigated crops are grown on good soils. Very large yields from irrigated cotton have been reported from Israel (up to 1600 kg/ha of *lint*); even larger yields have been measured in some experiments in U.S.A.

Nitrogen is the most important fertilizer nutrient for cotton and in

U.S.A. it is stated that 1 kg of N increases yield of seed cotton by 12 kg; in the Sudan the gain is said to be 8 kg of seed cotton/kg of N. Adequate phosphorus is essential, but on good soils cotton often gets enough from reserves in the soil. Potassium is very important, as for other fibre crops, but in many countries the soils used for cotton contain enough soluble K. The plants often contain rather more sulphur than phosphorus and sulphur deficiency is common in South America and Central Africa. The crop is sensitive to soil acidity.

Nutrients removed

Cotton takes up much nutrients. Good crops (2000 kg/ha of seed cotton) take up roughly 180 kg N, 30 kg P, 150 kg K, 180 kg Ca, 50 kg Mg and 40 kg of S. The large crops grown in Israel have taken up 250–300 kg N/ha and corresponding amounts of other nutrients. Half or more of this large stock of nutrients remains in stems, leaves and roots and is left in the field after harvest. One authority gives the nutrients removed in the amount of seed cotton providing 1 bale of lint (= 480 pounds, or 215 kg) as about 15 kg of N, 6 kg of P_2O_5 and 6 kg of K_2O; these amounts of P and K must be replaced by fertilizers if soil fertility is not to be depleted.

Practical recommendations

The most stress is always on using enough nitrogen fertilizer, but amounts recommended vary greatly with the potential of the crop on different sites. In India 25 kg N/ha is said to be a common dressing, with twice as much used on irrigated crops. In African countries recommendations range from 20 to 200 kg N/ha. In U.S.A. it seems that 80–120 kg N/ha are commonly used; up to 240 kg N/ha has been profitable on crops receiving full irrigation, but giving much nitrogen may delay maturity and increase disease. In areas where cotton is successful it appears that well-grown crops kept free of pests and diseases, and irrigated if necessary, should receive at least 120 kg N/ha.

Leaf analyses have been used in many areas to advise on fertilizing but the published figures that separate plants with too little nutrients from those that have enough vary greatly. It seems that local conditions vary too much for figures that are useful in one area to be used confidently in another district. Interpreting leaf analyses depends greatly on the skill and experience of the adviser.

There is no general agreement on the other nutrients needed by cotton; amounts of fertilizers required depend on reserves in the soils, and these must be assessed by field experiment results and by soil analyses. From 30 to 80 kg P_2O_5/ha is commonly used and K-fertilizers are used on deficient soils. It is essential that all acid soils should be limed, otherwise fertilizers may not give satisfactory increases in yield.

TROPICAL GRASSLANDS

Much of the world's grassland consists of indigenous grasses, herbs and shrubs which receive little or no fertilizer and produce little feed for stock. Yield is limited both by natural poverty of soil and by the unproductive species in the pastures; it is usually limited too by the long seasonal droughts which make 'grass' rather than forest the natural vegetation for these areas. In some countries spectacular improvements have resulted from replacing indigenous grasses by more productive species and by using fertilizer. Elsewhere high-yielding modern strains of grasses and legumes have been grown in hot climates where rainfall is large or irrigation is possible. These plantings, given adequate fertilizer, have produced very large yields of forage and show that the humid tropics may have a great potential in future for producing large amounts of animal food per hectare very cheaply.

The problems of growing large yields of grass in the tropics are not very different from the problems of intensifying grassland in Britain which were discussed in Chapter 20.

Tropical grasses in Puerto Rico

One example of large production from tropical grassland has already been shown in Fig. 9. Much experimental work on high-yielding tropical grasses done in Puerto Rico has shown that very large yields are possible. For example Napier grass (*Pennisetum purpureum*) has responded well up to total amounts of about 900 kg N/ha used in the year in split applications applied to successive cuts. Average yields for Napier grass cut every 60 days were:

N applied in year kg/ha	Crude protein in herbage %	Yield of dry matter t/ha
0	6.5	17.0
448	7.9	41.1
896	9.7	50.0
1345	11.9	52.2

The amount of N that could profitably be used (about 900 kg/ha) was twice the maximum that is generally useful in Britain and dry matter yield with this amount of N was nearly four times as much as can be grown by using N on grass in Britain. Temperate grasses however normally contain much more protein. When the Napier grass was cut every 60 days, using a total of 448 kg N/ha in the year gave the large return of 60 kg of extra dry matter per kg of N used. With twice as much N as this the large yield (50 t/ha of dry matter) removed large amounts of nutrients (750 kg of N, 625 kg of K and 135 kg/ha each of Ca, Mg and P). Unless these nutrients are returned in animal manures, very large fertilizer dressings are necessary.

Coastal bermuda grass in U.S.A.

A very successful recent introduction in U.S.A. is a grass (propagated vegetatively by planting the underground stems) called Coastal bermuda grass (*Cynodon dactylon*) which now occupies over 3 million acres in the Southern States. It is rapidly becoming the principal hay crop where lucerne is not dependable. It does not take up as much phosphorus and potassium as most cool-season grasses (usually it contains 1–2 per cent K and 0.18 to 0.23 per cent P in dry matter). As the grass is a 'strong feeder' on P and K reserves in soil it does not respond to these nutrients as fertilizers until supplies of soluble P and K in soil are very small. Coastal bermuda grass makes good use of nitrogen fertilizer and responds to very large dressings; it often gives increases of 30–40 kg of extra dry matter per kg of N applied as fertilizer. Fertilizer dressings with ratios of about $\%N : \%P_2O_5 : \%K_2O$ of 10 : 2 : 5 maintain soluble P and K in the surface soil.

The advance in potential yields from grassland which were made possible by introducing this species in warm areas is shown by these results of an experiment in Virginia (U.S.A.) which compared Coastal bermuda grass with a fescue. Mean annual yields of dry matter under comparable conditions were:

Nitrogen applied kg/ha	Coastal bermuda grass t/ha	Ky. 31 fescue t/ha
112	9.0	3.8
224	12.8	6.5
448	18.1	8.5
896	19.8	8.3

The optimum annual rates were judged to be 225 kg N/ha for fescue and 450 kg/ha for Coastal bermuda grass; the fescue recovered only 65 per cent of the applied nitrogen, the bermuda grass 89 per cent. Coastal bermuda grass also recovered applied potassium more efficiently. Although Coastal bermuda grass removed more N, P and K from soil than fescue did (because it yielded twice as much) the *percentages of* N, P and K in dry herbage were considerably less in the bermuda grass, showing, as other evidence has done, that this grass can produce a unit of yield with less fertilizer than is needed by the grasses grown in cooler climates.

The problems involved in increasing production from tropical grasslands are less difficult than those involved in producing much larger yields of crops grown for human food. Planting productive strains of grasses and giving adequate fertilizer will increase yields almost everywhere. Problems of using the extra herbage to best advantage are much more difficult, particularly in developing countries. Capital is needed to buy livestock, buildings and equipment; the skill and experience needed to manage and feed cattle efficiently cannot be gained quickly.

Bibliography

Reference
1. MITCHELL, R. L. (1963) 'Soil aspects of trace element problems in plants and animals.' *Journal of the Royal Agricultural Society of England*, Vol. 124, pp. 75-86.
2. WILLIAMS, R. J. B. (1971) 'The chemical composition of water from land drains at Saxmundham and Woburn, and the influence of rainfall upon nutrient losses.' *Report of Rothamsted Experimental Station for 1970*, Part 2, pp. 36-67.
3. BARBER, S. A. (1964) 'Water essential to nutrient uptake.' *Plant Food Review*, Vol. 10, (no. 2), pp. 5-7.
    also
    BARBER, S. A., WALKER, J. M. and VASEY, E. H. (1963) 'Mechanisms for the movement of plant nutrients from the soil and fertilizer to the plant root.' *Journal of Agricultural and Food Chemistry*, Vol. 11, pp. 204-207.
4. COOKE, G. W. (1967) 'The Control of Soil Fertility'. London: Crosby Lockwood Ltd.
5. HEMINGWAY, R. G. (1961) 'The mineral composition of farm-yard manure.' *Empire Journal of Experimental Agriculture*, Vol. 29, pp. 14-18.
6. BERRYMAN, C. (1965) 'Composition of organic manures and waste products used in agriculture.' *N.A.A.S. Advisory Papers* No. 2.
7. BENNE, E. J., HOGLUND, C. R., LONGNECKER, E. D. and COOK, R. L. (1961) 'Animal manures, what are they worth today?' *Michigan Agricultural Experiment Station Circular Bulletin* No. 231, pp. 1-15.
8. TINSLEY, J. and NOWAKOWSKI, T. Z. (1959) 'The composition and manurial value of poultry excreta, straw-droppings composts and deep litter.' *Journal of the Science of Food and Agriculture*, Vol. 10, pp. 145-150, 150-167, 224-232 and 232-241.
8a. WEBBER, J. and BASTIMAN, B. (1967) 'Experiments testing poultry manure as a source of nitrogen for grass." *Experimental Husbandry*, No. 15, pp. 11-18.

8b. GARNER, H. V. (1970) 'Experiments with kiln-dried poultry manure on agricultural crops and vegetables at Rothamsted, Woburn and other centres in 1933-39.' *Experimental Husbandry,* No. 19, pp. 13-28.

9. MINISTRY OF AGRICULTURE, FISHERIES AND FOOD. (1968) 'Slurry Handling and Disposal on the Dairy Unit.' *Bridgets Experimental Husbandry Farm Annual Report,* No. 9, pp. 10-13.

10. DAVIES, H. T. (1970) 'Experiments on the fertilizing value of animal slurries.' *Experimental Husbandry,* No. 19, pp. 49-68.

11. RILEY, C. T. (1970) 'Slurry.' *Journal of the Farmers' Club,* April 1970, pp. 27-37.

12. AMERICAN CHEMICAL SOCIETY. (1969) 'Cleaning our environment, the chemical basis for action.' Report by the Sub-Committee on Environmental Improvement.

13. MATTINGLY, G. E. G. (1957) 'The agricultural value of sewage sludge and town refuse.' *Proceedings of the Institution of Civil Engineers,* Vol. 8, pp. 414-420.

14. BUNTING, A. H. (1963) 'Experiments on organic manures, 1942-49.' *Journal of Agricultural Science,* Vol. 60, pp. 121–140.

15. WEBBER, J. (1961 and 1963) 'An experiment to compare bulky organic manures.' 1. Efford and Stockbridge House, *Experimental Horticulture,* No. 5, pp. 53-65.
2. Luddington and Rosewarne, *Experimental Horticulture,* No. 9, pp. 39-56.

16. DJOKOTO, R. K. and STEPHENS, D. (1961) 'Thirty long-term fertilizer experiments under continuous cropping in Ghana. 1. Crop yields and responses to fertilizers and manures.' *Empire Journal of Experimental Agriculture,* Vol. 29, pp. 181-195.

17. MINISTRY OF AGRICULTURE, FISHERIES AND FOOD. (1973) 'The Fertilisers and Feeding Stuffs Regulations 1973.' Statutory Instruments 1973, No. 1521. London: H.M.S.O.

18. FOOD AND AGRICULTURE ORGANISATION OF THE UNITED NATIONS (1970) 'Fertilizers.' An annual review of world production, consumption, trade and prices, 1969. F.A.O. Rome. Italy.

19. MINISTRY OF AGRICULTURE, FISHERIES AND FOOD. (1969) 'Landlord and Tenant. The Agriculture (Calculation of Value for Compensation) Regulations 1969.' Statutory Instruments, 1969, No. 1704, London: H.M.S.O.
also
'The Agricultural (Calculation of values for Compensation) (Amendment) Regulations 1972'. Statutory Instruments 1972. No. 864.

20. WIDDOWSON, F. V. and PENNY, A. (1969) 'Effects on barley and kale of NPK fertilizers containing different proportions of urea and ammonium nitrate and either triple superphosphate or mono-urea phosphate.' *Journal of Agricultural Science,* Vol. 73, pp. 125-132.

21. TOMLINSON, T. E. (1970) 'Urea—agronomic applications.' *Proceedings of the Fertiliser Society,* No. 113, pp. 3-76.

22. VAN BURG, P. F. J. (1969) 'The agronomic value of anhydrous ammonia in Western Europe.' *Outlook on Agriculture*, Vol. 6, No. 2, pp. 55-59.

23. WIDDOWSON, F. V. (1968) 'Why starve grass?' *Dairy Farmer*, February, pp. 34-36, 55.

24. HIGNETT, T. P. (1969) 'Trends in technology.' *Proceedings of the Fertiliser Society*, No. 108, pp. 4-43.

25. FARRAR, K. (1969) 'Trace elements and magnesium in basic slag and their value to plants.' *N.A.A.S. Advisory Papers* No. 6. London: Ministry of Agriculture, Fisheries and Food.

26. GARDNER, H. W. and GARNER, H. V. (1953) *The use of Lime in British Agriculture*. London: E. & F. N. Spon Ltd.

27. BOLTON, J. (1971) 'Long-term liming experiments at Rothamsted and Woburn.' *Report of Rothamsted Experimental Station for 1970*, Part 2, pp. 98-112.

28. MEHLICH, A. (1969-70) 'Crop response to sulphur in Kenya.' *Sulphur Institute Journal*, Vol. 5, No. 4, pp. 10-13.

29. BIXBY, D. W. and BEATON, J. D. (1970) 'Sulphur-containing fertilizers, properties and applications.' Sulphur Institute, Technical Bulletin No. 17.

30. WILLIAMS, R. J. B., STOJKOVSKA, A., COOKE, G. W. and WIDDOWSON, F. V. (1960) 'Effects of fertilizers and farmyard manure on the copper, manganese, molybdenum and zinc removed by arable crops at Rothamsted.' *Journal of the Science of Food and Agriculture*, Vol. 11, pp. 570-575.

31. ANDERSON, A. J. (1970) 'Trace elements for sheep pastures and fodder crops in Australia.' *Journal of the Australian Institute of Agricultural Science*, Vol. 36, No. 1, pp. 15-29.

32. SAUCHELLI, V. (1969) *Trace Elements in Agriculture*. New York: Van Nostrand Reinhold Company.

33. SCHÜTTE, K. H. (1964) *The biology of the trace elements. Their role in nutrition*. London: Crosby Lockwood Ltd.

34. INTERNATIONAL POTASH INSTITUTE (1970) Proceedings of Ninth Congress 'Role of fertilizing in the intensification of agriculture,' Antibes, September 1970.
    CHAMINADE, R. 'Nutrition des plantes en relation avec le sol.'
    BOYD, D. A. 'Some recent ideas on fertilizer response curves.'

35. BOYD, D. A. and DERMOTT, W. (1964) 'Fertilizer experiments on maincrop potatoes 1955-61.' *Journal of Agricultural Science*, Vol. 63, pp. 249-263.

36. CROWTHER, E. M. and YATES, F. (1941) 'Fertilizer policy in war-time.' *Empire Journal of Experimental Agriculture*, Vol. 9, pp. 77-97.

37. F.A.O. (1966) *Statistics of crop responses to fertilizers*. Rome: Food and Agriculture Organisation of the United Nations, pp. 1-112.

38. BIRCH, J. A., DEVINE, J. R., HOLMES, M. R. J. and WHITEAR, J. D. (1967) 'Field experiments on the fertilizer requirements of

maincrop potatoes.' *Journal of Agricultural Science,* Vol. 69, pp. 13-24.

39. BOYD, D. A., TINKER, P. B. H., DRAYCOTT, A. P. and LAST, P. J. (1970) 'Nitrogen requirement of sugar beet grown on mineral soils.' *Journal of Agricultural Science,* Vol. 74, pp. 37-46.

40. COUSTON, J. W. (1969) 'F.F.H.C. Fertilizer Programme: Crop responses and economic returns.' *Phosphorus in Agriculture,* No. 52, pp. 31-35.

40a. MEREDITH, R. M. (1965) 'A review of the responses to fertilizer of the crops of Northern Nigeria.' Samaru Miscellaneous Paper No. 4, Institute for Agricultural Research, Ahmadu Bello University, Samaru, Northern Nigeria.

41. ENGELSTAD, O. P. and TERMAN, G. L. (1966) 'Fertilizer nitrogen: Its role in determining crop yield levels.' *Agronomy Journal,* Vol. 58, pp. 536-539.

42. INKSON, R. H. E. and REITH, J. W. S. (1966) 'Estimating optimal nutrient rates for potatoes.' Transactions of Commissions 2 and 4, International Society of Soil Science, Aberdeen 1966, pp. 377-384.

43. MINISTRY OF AGRICULTURE, FISHERIES AND FOOD. (1973) 'Fertilizer recommendations for agricultural and horticultural crops.' Bulletin No. 209. London: H.M.S.O.

44. BOYD, D. A. (1968) 'Experiments with ley and arable farming systems.' *Report of Rothamsted Experimental Station for 1967,* pp. 316-331.

45. SCHRADER, W. D., FULLER, W. A. and CADY, F. B. (1966) 'Estimation of a common nitrogen response function for corn (*Zea mays*) in different crop rotations.' *Agronomy Journal,* Vol. 58, pp. 397-401.

46. HOOPER, L. J. (1970) 'The basis of current fertilizer recommendations.' *Proceedings of the Fertiliser Society,* No. 118.

47. AN FORAS TALUNTAIS (1970) *Fertilizer Manual.* (Prepared by Soils Division, Johnstown Castle, Wexford.)

48. JOHNSTON, A. E. and others (1970) 'The value of residues from long period manuring at Rothamsted and Woburn.' *Report of Rothamsted Experimental Station for 1969,* Part 2, pp. 5-90.

49. COOKE, G. W. (1967) 'The value and valuation of fertilizer residues.' *Journal of the Royal Agricultural Society of England,* Vol. 128, pp. 7-25.

50. STEPHENS, D. (1969) 'Changes in yields and fertilizer responses with continuous cropping in Uganda.' *Experimental Agriculture,* Vol. 5, pp. 263-269.

51. WARREN, R. G. and COOKE, G. W. (1962) 'Comparisons between methods of measuring soluble phosphorus and potassium in soils used for fertilizer experiments on sugar beet.' *Journal of Agricultural Science,* Vol. 59, pp. 269-274.

52. MINISTRY OF AGRICULTURE, FISHERIES AND FOOD. (1965) 'Soil phosphorus'. *Technical Bulletin* No. 13, London: H.M.S.O.

53. OLSEN, S. R., COLE, C. V., WATANABE, F. S. and DEAN, L. A. (1954) 'Estimation of available phosphorus in soils by extraction with sodium bicarbonate.' *U.S. Department of Agriculture, Circular* No. 939.

54. DEWIS, J. and FREITAS, F. (1970) 'Physical and chemical methods of soil and water analysis.' Soils Bulletin No. 10. Food and Agriculture Organisation of the United Nations. Rome.

55. WILLIAMS, R. J. B. and COOKE, G. W. (1962) 'Measuring soluble phosphorus in soils, comparisons of methods, and interpretation of results.' *Journal of Agricultural Science*, Vol. 59, pp. 275-280.

56. DRAYCOTT, A. P., DURRANT, M. J. and BOYD, D. A. (1971) 'The relationship between soil phosphorus and response by sugar beet to phosphate fertilizer on mineral soils.' *Journal of Agricultural Science*, Vol. 77, pp. 117-121.

57. BINGHAM, F. T. (1962) 'Chemical tests for available phosphorus.' *Soil Science*, Vol. 94, pp. 87-95.

58. MINISTRY OF AGRICULTURE, FISHERIES AND FOOD. (1967) 'Soil potassium and magnesium.' *Technical Bulletin*, No. 14, London: H.M.S.O.

59. BOULD, C. (1966) In: 'Fruit Nutrition' (N. F. Childers, Editor) pp. 651-684. Horticultural Publications, Rutgers State University, New Jersey.

60. BOULD, C. (1965) 'The nutrient requirements of fruit crops.' *Span*, Vol. 8, pp. 81-83.

61. CLEMENT, C. R. and HOPPER, M. J. (1968) 'The supply of potassium to high-yielding cut grass.' *N.A.A.S. Quarterly Review*, No. 79, pp. 101-109.

62. BELLIS, E. (1968) 'Rubber, an example of progress in fertilizer use in tropical agriculture.' *Transactions of the Ninth International Congress of Soil Science*, Vol. 4, pp. 77-84.

63. LE POIDEVIN, N. and ROBINSON, L. A. (1964) 'Foliar analysis procedures as employed on the Booker Group of sugar estates in British Guiana. Part I Sampling and analytical techniques; Part II The interpretation of results.' *Fertilité*, No. 21, pp. 3-11 and 12-17.

64. COOKE, G. W. (1958) 'The Nation's plant food larder.' *Journal of the Science of Food and Agriculture*, Vol. 9, pp. 761-772.

65. RICHARDSON, H. L. (1938) 'The nitrogen cycle in grassland soils: with especial reference to the Rothamsted Park Grass Experiment.' *Journal of Agricultural Science*, Vol. 28, pp. 73-121.

66. COLE, D. W. and GESSEL, S. P. (1968) 'Cedar River Research. A program for studying the pathways, rates and processes of elemental cycling in a forest ecosystem.' Forest Resources Monograph. Seattle: University of Washington, College of Forest Resources.

67. WOLTON, K. M. (1965) 'Techniques in grassland experimenta-

tion.' Bulletin of documentation of the International Superphosphate Manufacturers' Association No. 41, pp. 1-13.

68. WEBBER, J., HERBERT, R. F., ROTHWELL, J. B. and JONES, D. A. G. (1963) 'Losses of nutrients from glasshouse soils.' *Experimental Horticulture*, No. 8, pp. 19-26.

69. WATSON, G. A. (1964) 'Maintenance of soil fertility in the permanent cultivation of *Hevea brasiliensis* in Malaya.' *Outlook on Agriculture*, Vol. 4, pp. 103-109.

70. FRANCIS, A. L. (1969) 'Fertilizing Bridgets in the 70's'. *Bridgets Experimental Husbandry Farm Annual Report*. No. 10, 1969, pp. 25-30. Ministry of Agriculture, Fisheries and Food.

71. DEVINE, J. R. and HOLMES, M. R. J. (1964) 'Field experiments comparing autumn and spring applications of ammonium sulphate, ammonium nitrate and calcium nitrate for winter wheat.' *Journal of Agricultural Science*, Vol. 63, pp. 69-74.

72. BATEY, T. and BOYD, D. A. (1967) 'Placement of fertilizers for potatoes.' *N.A.A.S. Quarterly Review*, No. 78, Winter 1967, pp. 47-56.

73. BOYD, D. A., HILL, J. M. and BATEY, T. (1968) 'The effect on yield of maincrop potatoes of different methods of fertilizer application.' *Experimental Husbandry*, No. 16, pp. 13-20.

74. COOKE, G. W., JACKSON, M. V., WIDDOWSON, F. V. and WILCOX, J. C. (1956) 'Fertilizer placement for horticultural crops.' *Journal of Agricultural Science*, Vol. 47, pp. 249-256.

74a. LUGG, G. W. (1970) 'Economic considerations.' Proceedings of Anhydrous Ammonia Symposium, Silsoe, December 1970, pp. 122-123.

75. HOLMES, M. R. J. (1970) 'Field experiments on the nitrogen requirements of sugar beet.' *Fisons Agricultural Technical Information*, Spring 1970, pp. 7-9.

76. DRAYCOTT, A. P. and DURRANT, M. J. (1969 and 1970) 'Magnesium fertilizers for sugar beet.' *British Sugar Beet Review*, Part I, Vol. 37, pp. 175-179; Part II, Vol. 38, pp. 175-180.

77. COOKE, G. W. (1960) *Fertilizers and Profitable Farming*, London: Crosby Lockwood Ltd.

78. WHITEHEAD, D. C. (1970) 'The role of nitrogen in grassland productivity.' Bulletin No. 48, Commonwealth Bureau of Pastures and Field Crops, Hurley, Berkshire.

79. REID, D. (1970) 'The effects of a wide range of nitrogen application rates on the yields from a perennial ryegrass sward with and without white clover.' *Journal of Agricultural Science*, Vol. 74, pp. 227-240.

80. HOLMES, W. (1968) 'The use of nitrogen in the management of pastures for cattle.' *Herbage Abstracts*, Vol. 38, (no. 4), pp. 265-277.

81. RAYMOND, W. F. and SPEDDING, C. R. W. (1965) 'The effect of fertilizers on the nutritive value and production potential of forages.' *Proceedings of the Fertiliser Society*, No. 88.

82. BLAXTER, K. L. (1968) 'Fertilizers and animal production.' *Proceedings of the Fertiliser Society*, No. 105.

83. RAYMOND, W. F. (1968) 'Grassland research and practice.' *Journal of the Royal Agricultural Society of England*, Vol. 129, pp. 85-105.

84. COOKE, G. W. (1970) 'Fertilizers for grassland.' *Journal of the Royal Agricultural Society of England*, Vol. 131, pp. 120-125.

85. HOOPER, L. J. (1970) 'Potassium level in herbage as a guide to fertilizer requirement.' Proceedings of Potassium Institute Limited, Symposium, Hurley, June 1970, pp. 47-56.

86. HAWORTH, F. (1964) 'Potassium nutrition of vegetables.' *N.A.A.S. Quarterly Review*, No. 64, pp. 151-156.

87. WINSOR, G. W. (1968) 'The nutrition of glasshouse and other horticultural crops.' *Proceedings of the Fertiliser Society*, No. 103.

87a. FRINK, C. R. (1965) 'Apple orchard soil and leaf analysis.' *Connecticut Agricultural Experiment Station Bulletin*, No. 670.

88. (i) DE GEUS, J. G. (1967) *Fertilizer guide for tropical and sub-tropical farming*.
(ii) DE GEUS, J. G. (1970) *Fertilizer guide for food grains in the tropics and sub-tropics*. Published by Centre d'Etude de l'Azote, Zurich.

89. RUSSELL, E. W. (1968) 'The place of fertilizers in food crop economy of tropical Africa.' *Proceedings of the Fertiliser Society*, No. 101.

90. WEBSTER, C. C. and WILSON, P. N. (1966) *Agriculture in the tropics*. London: Longmans.

91. HARTLEY, C. W. S. (1967) *The oil palm*. London: Longmans.

92. F.A.O. (1970) *Production Yearbook*, Vol. 23, F.A.O., Rome, Italy.

93. ALLAN, A. Y. (1968) 'Maize research in Kenya: 3 Agronomy Research.' *Span*, Vol. 11 (no. 3), pp. 147-149.

94. INTERNATIONAL POTASH INSTITUTE (1970) Ninth Congress at Antibes, 1970. Proceedings.
(i) VON UEXKÜLL, H. R. 'Role of fertilizer in the intensification of rice cultivation.'
(ii) CHANDLER, R. F. 'Overcoming physiological barriers to higher yields through plant breeding.'

95. HUTCHINSON, J. (Sir) (1971) 'High cereal yields.' *Journal of the Royal Society of Arts*, Vol. 119, pp. 104-114.

96. NYE, P. H. and GREENLAND, D. J. (1960) 'The soil under shifting cultivation.' Commonwealth Bureau of Soils, Harpenden, Technical Communication No. 51.

97. BOLTON, J. (1964) 'The manuring and cultivation of *Hevea brasiliensis*.' *Journal of the Science of Food and Agriculture*, Vol. 15, pp. 1-8.

98. NG, SIEW KEE (1970) 'Greater productivity of the oil palm with efficient fertilizer practices.' International Potash Institute Ninth

Congress, Antibes, 1970, Proceedings.

99. NG, SIEW KEE and THAMBOO, S. (1967) 'Nutrient contents of oil palms in Malaya.' *Malaysian Agricultural Journal* Vol. 46, (no. 1), pp. 3-45.

100. BELLIS, E. (1970) 'Intensification of plantation rubber production by *Hevea brasiliensis* through manuring.' International Potash Institute Ninth Congress, Antibes, 1970, Proceedings.

101. CLEMENTS, H. F. (1961) 'Crop logging of sugar cane in Hawaii.' Plant Analysis and Fertilizer Problems. American Institute of Biological Sciences, pp. 131-147.

102. CLEMENTS, H. F. (1964) 'Foundations for objectivity in tissue diagnosis as a guide to crop control.' Plant Analysis and Fertilizer Problems No. IV, pp. 90-110.

103. COATES, W. H. (1974) 'A survey on contemporary production and related problems of the British fertiliser industry.' *Proceedings of the Fertiliser Society* No. 144.

104. CHURCH, B. M. and CALDWELL, T. H. (1973) 'Lime requirements and the lime status of soils in England and Wales 1963–68.' *Experimental Husbandry* No. 24, pp. 48–53.

105. CROMACK, H. T. H. and CLARE, R. W. (1974) 'Nitrogen manuring of spring wheat'. *Experimental Husbandry* No. 26, pp. 48–59.

106. DAVIES, D. B., EAGLE, D. J. and FINNEY, J. B. (1972) *Soil management.* Ipswich: Farming Press Ltd.

107. DRAYCOTT, A. P. (1972) *Sugar-beet nutrition.* London: Applied Science Publishers Ltd.

108. DRAYCOTT, A. P. (1974) 'Sugar-beet nutrition'. *Proceedings of the Fertiliser Society* No. 143.

109. DRAYCOTT, A. P. and DURRANT, M. J. (1974) 'The influence of previous cropping and soil texture on the nitrogen requirement of sugar beet.' *Experimental Husbandry* No. 25, pp. 41–51.

110. DYKE, G. V. (1974) *Comparative experiments with field crops.* London: Butterworth.

111. GASSER, J. K. R. (1973) 'An assessment of the importance of some factors causing losses of lime from agricultural soils.' *Experimental Husbandry* No. 25, pp. 86–95.

112. HARRIS, P. B. (1973) 'Effects of copper on cereal yields.' *Experimental Husbandry* No. 24, pp. 7–11.

113. HARROD, M. F., JOHNSON, E. W. and WILKINSON, B. (1974) 'The manuring of maincrop carrots: I. Sandy soils.' *Experimental Horticulture* No. 26, pp. 60–72.

114. HARROD, M. F. (1974) 'The manuring of main crop carrots: II. Peat soils.' *Experimental Horticulture* No. 26, pp. 73–81.

115. HORNE, B. (1973) 'Leys and soil fertility. Part I. Crop production.' *Experimental Husbandry* No. 23, pp. 86–103.

116. LANG, R. W. and HOLMES, J. C. (1973) 'Effect of nitrogen application at different growth stages on the yield of winter wheat.' *Experimental Husbandry* No. 23, pp. 31–36.

117. LEHR, J. R. and MCCLELLAN, G. H. (1972) 'A revised laboratory

reactivity scale for evaluating phosphate rocks for direct application.' *Bulletin* Y-43 Tennessee Valley Authority, Muscle Shoals, Alabama.

118. MORRISON, J., JACKSON, M. V. and WILLIAMS, T. E. (1975) 'The role of nitrogen in grassland productivity. Variation in the response of grass to fertiliser-N in relation to environment.' *Proceedings of the Fertiliser Society* No. 141.

119. MUNDY, E. J. and SELMAN, M. (1974). 'Intensification of cereals: cereal monoculture duration and nitrogen rate. Part I. Winter wheat.' *Experimental Husbandry* No. 25, pp. 22–40.

120. MYLONAS, D. M. (1973) 'Fertilizer legislation'. *Soils Bulletin* No. 20. Rome: FAO.

121. PAGE, E. R., TATHAM, P. B. and WOOD, M. B. (1974) 'Aqueous ammonia as a nitrogen fertilizer for single harvested Brussels sprouts.' *Experimental Horticulture* No. 26, pp. 82–90.

122. RUSSELL, E. W. (1973) *Soil conditions and plant growth*. 10th Edn. London: Longman.

123. SHORT, J. L. (1974) 'Straw disposal trials at the Experimental Husbandry Farms.' *Experimental Husbandry* No. 25, pp. 103–136.

Index